AF682675

Birkhäuser

Static & Dynamic Game Theory: Foundations & Applications

Series Editor
Tamer Başar, University of Illinois, Urbana-Champaign, IL, USA

Editorial Advisory Board
Daron Acemoglu, MIT, Cambridge, MA, USA
Pierre Bernhard, INRIA, Sophia-Antipolis, France
Maurizio Falcone, Università degli Studi di Roma "La Sapienza," Italy
Alexander Kurzhanski, University of California, Berkeley, CA, USA
Ariel Rubinstein, Tel Aviv University, Ramat Aviv, Israel; New York University, NY, USA
William H. Sandholm, University of Wisconsin, Madison,WI, USA
Yoav Shoham, Stanford University, CA, USA
Georges Zaccour, GERAD, HEC Montréal, Canada

More information about this series at http://www.springer.com/series/10200

Mindia E. Salukvadze • Vladislav I. Zhukovskiy

The Berge Equilibrium: A Game-Theoretic Framework for the Golden Rule of Ethics

Birkhäuser

Mindia E. Salukvadze
Department of Engineering Sciences and Information Technology
Georgian National Academy of Sciences
Tbilisi, Georgia

Vladislav I. Zhukovskiy
Department of Optimal Control
Faculty of Computational Mathematics and Cybernetics
Moscow State University
Moscow, Russia

ISSN 2363-8516 ISSN 2363-8524 (electronic)
Static & Dynamic Game Theory: Foundations & Applications
ISBN 978-3-030-25545-9 ISBN 978-3-030-25546-6 (eBook)
https://doi.org/10.1007/978-3-030-25546-6

Mathematics Subject Classification (2010): 49-XX, 49N70, 49N90, 90-XX, 90C47, 90B50, 91-XX, 91A10, 91A23, 91A80, 91Bxx, 91B06, 91B30

© Springer Nature Switzerland AG 2016, 2020
Original Russian edition published by Georgian National Academy of Sciences, Tbilisi, Georgia, 2016
This work is subject to copyright. All rights are reserved by the Publisher, whether the whole or part of the material is concerned, specifically the rights of translation, reprinting, reuse of illustrations, recitation, broadcasting, reproduction on microfilms or in any other physical way, and transmission or information storage and retrieval, electronic adaptation, computer software, or by similar or dissimilar methodology now known or hereafter developed.
The use of general descriptive names, registered names, trademarks, service marks, etc. in this publication does not imply, even in the absence of a specific statement, that such names are exempt from the relevant protective laws and regulations and therefore free for general use.
The publisher, the authors, and the editors are safe to assume that the advice and information in this book are believed to be true and accurate at the date of publication. Neither the publisher nor the authors or the editors give a warranty, expressed or implied, with respect to the material contained herein or for any errors or omissions that may have been made. The publisher remains neutral with regard to jurisdictional claims in published maps and institutional affiliations.

This book is published under the imprint Birkhäuser, www.birkhauser-science.com by the registered company Springer Nature Switzerland AG.
The registered company address is: Gewerbestrasse 11, 6330 Cham, Switzerland

Mindia E. Salukvadze (May 3, 1933–December 27, 2018)

Dedicated to the memory of Mindia E. Salukvadze.

Biography of Mindia E. Salukvadze

A man is measured by his deeds and contribution to the global pool of values. This is an axiom. Just listing all the academic degrees, posts and titles of Mindia E. Salukvadze forms the picture of an extraordinary personality.

Mindia E. Salukvadze was born in 1933 in Tbilisi, and grew up as an orphan. His parents were subjected to repression in 1937 (subsequently both were fully exonerated). In 1955, he received Diploma with honors from the Physics Faculty of Tbilisi State University. His postgraduate studies were continued at the Institute of Electronics, Automation and Remote Control of the USSR Academy of Sciences (presently, Trapeznikov Institute of Control Sciences of the Russian Academy of Sciences), where he defended first the Candidate's Dissertation (1963) and then the Doctor's Dissertation (1974) in Engineering.

In 1983, Mindia E. Salukvadze was elected a Corresponding Member of the Academy of Sciences of the Georgian SSR; in 1993, a Full Member of the Georgian National Academy of Sciences. Later on, he became the Academician-secretary of the Department for Applied Mechanics, Machine Building, Energy and Control Processes of the Academy and also a Member of the Academy's Presidium. In 1996, he was awarded the Nikoladze Prize of the Academy, which was established as back as 1973 for the best scientific works in engineering. In 1996 and again in 2004, he was awarded the State Prize of Georgia in the field of science and technology. In 2014, he and Vladislav I. Zhukovskiy became the winners of the International Contest for the Best Scientific Book, held in Russia.

For many years, Mindia E. Salukvadze was the Head of the Georgian Section of the International Federation of Automatic Control (IFAC) as well as a member of the editorial boards of several scientific journals such as *Moambe* (*Bulletin of the Georgian National Academy of Sciences*), *International Journal of Information Technology and Decision Making*, and *Automation and Remote Control* (Trapeznikov Institute of Control Sciences of the Russian Academy of Sciences). In addition, he was a member of several scientific councils and organizations. Mindia E. Salukvadze had close cooperation with a series of research centers all over the world and participated in leading international conferences and symposia. He was a member of different international academies of sciences, including the New York Academy of Sciences (since 1994).

For 25 years, Academician Mindia E. Salukvadze was Director of the Eliashvili Institute of Control Systems (1981–2006), and then Chairman of the Institute's Scientific Council.

The research interests of Mindia E. Salukvadze covered the stability of control systems and the theory of optimal control. He authored over 140 scientific papers, 13 monographs and 6 textbooks, known in Georgia and also abroad. Salukvadze's method, Salukvadze's solution, Salukvadze's principle—these terms were introduced by American and Russian researchers.

In 1975, Metsniereba Press (Tbilisi) published Mindia E. Salukvadze's well known monograph *Zadachi vektornoi optimizatsii v teorii upravleniya*, which was translated into English under the title *Vector-Valued Optimization Problems in Control Theory* and published by Academic Press in 1979. Another prominent monograph by Mindia E. Salukvadze, *Vector-Valued Maximin* (in co-authorship with Vladislav I. Zhukovskiy), was published by Academic Press in 1994.

Mindia E. Salukvadze was a fruitful educator, holding the position of Professor at Tbilisi State University. He supervised a series of Doctor's and Candidate's Dissertations.

The major books by Mindia E. Salukvadze are as follows.

6 Textbooks

1. Gugushvili, A., Salukvadze, M., and Chichinadze, V., *Optimal and Adaptive Systems. Book I. Static Optimization*, Tbilisi: Teknikuri Universiteti, 1997, 290 p. (in Georgian).
2. Gugushvili, A., Salukvadze, M., Chichinadze, V., *Optimal and Adaptive Systems. Book II. Optimal Control of Dynamic Systems*, Tbilisi: Teknikuri Universiteti, 1997, 437 p. (in Georgian).
3. Gugushvili, A., Salukvadze, M., Chichinadze, V., *Optimal and Adaptive Systems. Book III. Optimal Control of Stochastic Systems. Adaptive Control of Systems*, Tbilisi: Teknikuri Universiteti, 1997, 325 p. (in Georgian).
4. Gugushvili, A., Topchishvili, A., Salukvadze, M., Chichinadze, V., and Jabladze, N., *Optimization Methods*, Tbilisi: Teknikuri Universiteti, 2002, 634 p. (in Georgian).
5. Zhukovskiy, V.I. and Salukvadze, M.E., *Otsenka riskov i garantii v konfliktakh* (Estimation of Risks and Guarantees in Conflicts), Moscow: Yurait, 2018, 302 p. (in Russian).
6. Zhukovskiy, V.I. and Salukvadze, M.E., *Otsenka riskov i mnogoshagovye pozitsionnye konflikty* (Estimation of Risks and Multistage Positional Conflicts), Moscow: Yurait, 2018, 306 p. (in Russian).

13 Monographs

1. Salukvadze, M.E., *Zadachi vektornoi optimizatsii v teorii upravleniya* (Vector-Valued Optimization Problems in Control Theory), Tbilisi: Metsniereba, 1975, 201 p. (in Russian).

2. Salukvadze, M.E., *Vector-Valued Optimization Problems in Control Theory*, New York: Academic Press, 1979, 219 p.
3. Ioseliani, A.N., Mikhalevich, A.A., Nesterenko V.V., and Salukvadze, M.E., *Metody optimizatsii parametrov teploobmennykh apparatov AES* (Parameter Optimization Methods for the Heat-Exchange Systems of Nuclear Power Plants), Minsk: Nauka i Tekhnika, 1981, 144 p. (in Russian).
4. Salukvadze, M.E., *Zadacha A.M. Letova o sinteze optimal'nykh sistem avtomaticheskogo upravleniya* (A.M. Letov's Problem on Optimal Automatic Control Systems Design), Tbilisi: Metsniereba, 1988. 381 p. (in Russian).
5. Zhukovskiy, V.I. and Salukvadze, M.E., *Mnogokriterial'nye zadachi upravleniya v usloviyakh neopredelennosti* (Multicriteria Control Problems under Uncertainty), Tbilisi: Metsniereba, 1991. 128 p. (in Russian).
6. Zhukovskiy, V.I. and Salukvadze, M.E., *The Vector-Valued Maximin*, New York: Academic Press, 1994, 404 p.
7. Zhukovskiy, V.I. and Salukvadze, M.E., *Optimizatsiya garantii v mnogokriterial'nykh zadachakh upravleniya* (Optimization of Guarantees in Multicriteria Control Problems), Tbilisi: Metsniereba, 1996. 475 p. (in Russian).
8. Zhukovskiy, V.I. and Salukvadze, M.E., *Nekotorye igrovye zadachi upravleniya i ikh prilozheniya* (Some Game-Theoretic Control Problems and Their Applications), Tbilisi: Metsniereba, 1998. 462 p. (in Russian).
9. Salukvadze, M., Topchishvili, A., and Maisuradze, V., *Duality in Nonscalar Optimization Problems*, Tbilisi: Modesta, 2000, 168 p. (in Georgian).
10. Zhukovskiy, V.I. and Salukvadze, M.E., *Riski i iskhody v mnogokriterial'nykh zadachakh upravleniya* (Risks and Outcomes in Multicriteria Control Problems), Tbilisi: Intelekti, 2004, 356 p. (in Russian).
11. Zhukovskiy, V.I. and Salukvadze, M.E., *Riski v konfliktnykh sistemakh upravleniya* (Risks in Conflict Control Systems), Tbilisi: Intelekti, 2008, 456 p. (in Russian)
12. Zhukovskiy, V.I., Salukvadze, M.E., and Beltadze, G.N., *Matematicheskie osnovy Zolotogo pravila nravstvennosti* (Mathematical Foundations of the Golden Rule of Ethics), Tbilisi: the Georgian National Academy of Sciences, 2017, 343 p. (in Russian).
13. Zhukovskiy, V.I., and Salukvadze, M.E., *Dinamika Zolotogo pravila nravstvennosti* (Dynamics of the Golden Rule of Ethics), Tbilisi: the Georgian National Academy of Sciences, 2018, 400 p. (in Russian).

Mindia E. Salukvadze organized a series of international scientific conferences and meetings.

Mindia E. Salukvadze was a well-known public figure. At different times, he was elected a Deputy to the Supreme Soviet of the Georgian SSR, Tbilisi Soviet, and district Soviets.

The aforesaid characterizes Mindia E. Salukvadze as a talented researcher, manager, and public figure. His true portrait includes greatheartedness, rarely encountered honesty, and unselfishness. For all of us—his relatives, friends, and colleagues—his decease will always be an irreparable loss.

Preface

Imagine that there are three sellers in a market, namely, a man (husband), his wife, and their son. At their disposal they have resources $\mathrm{X_h}$, $\mathrm{X_w}$, and $\mathrm{X_s}$, respectively, and to gain some profit they allocate parts of their resources $x_i \in \mathrm{X}_i$ $(i = \mathrm{h, w, s})$ to each family member. Which of these values will yield the greatest (possible) profit for each member? Really, the profit (revenues) $P_i(x_\mathrm{h}, x_\mathrm{w}, x_\mathrm{s})$ $(i = \mathrm{h, w, s})$ directly depends on the chosen values x_h, x_w and x_s.

The concept of Nash equilibrium has been "reigning" in such decision problems or game situations so far. A Nash equilibrium is a strategy profile $x^\mathrm{e} = (x_\mathrm{h}^\mathrm{e}, x_\mathrm{w}^\mathrm{e}, x_\mathrm{s}^\mathrm{e})$ that satisfies the three equalities

$$\begin{cases} \max\limits_{x_\mathrm{h}} P_\mathrm{h}(x_\mathrm{h}, x_\mathrm{w}^\mathrm{e}, x_\mathrm{s}^\mathrm{e}) = P_\mathrm{h}(x_\mathrm{h}^\mathrm{e}, x_\mathrm{w}^\mathrm{e}, x_\mathrm{s}^\mathrm{e}), \\ \max\limits_{x_\mathrm{w}} P_\mathrm{w}(x_\mathrm{h}^\mathrm{e}, x_\mathrm{w}, x_\mathrm{s}^\mathrm{e}) = P_\mathrm{w}(x_\mathrm{h}^\mathrm{e}, x_\mathrm{w}^\mathrm{e}, x_\mathrm{s}^\mathrm{e}), \\ \max\limits_{x_\mathrm{s}} P_\mathrm{s}(x_\mathrm{h}^\mathrm{e}, x_\mathrm{w}^\mathrm{e}, x_\mathrm{s}) = P_\mathrm{s}(x_\mathrm{h}^\mathrm{e}, x_\mathrm{w}^\mathrm{e}, x_\mathrm{s}^\mathrm{e}). \end{cases}$$

Here, a selfish nature clearly appears because everyone seeks to increase (maximize) his/her *own* profit only, ignoring the interests of the others.

The concept of Berge equilibrium put forward in this book is the exact opposite of Nash equilibrium. A Berge equilibrium is a strategy profile $x^\mathrm{B} = (x_\mathrm{h}^\mathrm{B}, x_\mathrm{w}^\mathrm{B}, x_\mathrm{s}^\mathrm{B})$ defined by the three equalities

$$\begin{cases} \max\limits_{x_\mathrm{w}, x_\mathrm{s}} P_\mathrm{h}(x_\mathrm{h}^\mathrm{B}, x_\mathrm{w}, x_\mathrm{s}) = P_\mathrm{h}(x_\mathrm{h}^\mathrm{B}, x_\mathrm{w}^\mathrm{B}, x_\mathrm{s}^\mathrm{B}), \\ \max\limits_{x_\mathrm{h}, x_\mathrm{s}} P_\mathrm{w}(x_\mathrm{h}, x_\mathrm{w}^\mathrm{B}, x_\mathrm{s}) = P_\mathrm{w}(x_\mathrm{h}^\mathrm{B}, x_\mathrm{w}^\mathrm{B}, x_\mathrm{s}^\mathrm{B}), \\ \max\limits_{x_\mathrm{h}, x_\mathrm{w}} P_\mathrm{s}(x_\mathrm{h}, x_\mathrm{w}, x_\mathrm{s}^\mathrm{B}) = P_\mathrm{s}(x_\mathrm{h}^\mathrm{B}, x_\mathrm{w}^\mathrm{B}, x_\mathrm{s}^\mathrm{B}). \end{cases}$$

It is precisely these conditions that implement the Golden Rule of ethics, which states, "do to others as you would like them to do to you." According to these conditions, each family member has to maximize the profit of the other members so that they act in the same way, maximizing his/her profit. Thus, choosing $x_\mathrm{h} = x_\mathrm{h}^\mathrm{B}$

the husband does his best to maximize the profit of his wife and son, as dictated by the equalities

$$\max_{x_h, x_s} P_w(x_h, x_w^B, x_s) = P_w(x^B), \quad \max_{x_h, x_w} P_s(x_h, x_w, x_s^B) = P_s(x^B).$$

The wife and son reciprocate with $x_w = x_w^B$ and $x_s = x_s^B$, respectively, maximizing the husband's profit, that is,

$$\max_{x_w, x_s} P_h(x_h^B, x_w, x_s) = P_h(x^B).$$

The wife has the same behavior: choosing $x_w = x_w^B$, she maximizes the profits of her husband and son. Following the ethical lead of the parents who maximize his payoff

$$\max_{x_h, x_w} P_s(x_h, x_w, x_s^B) = P_s(x^B),$$

the son also strives to maximize their profits using $x_s = x_s^B$, i.e.,

$$\max_{x_w, x_s} P_h(x_h^B, x_w, x_s) = P_h(x^B), \quad \max_{x_h, x_s} P_w(x_h, x_w^B, x_s) = P_s(x^B).$$

Thus, in a Berge equilibrium each of the members maximizes the profit of the other two members and receives the same response from them. In other words, the concept of Berge equilibrium matches well the Golden Rule.

As it will be evident from Sects. 1.1–1.5 of the book, the Golden Rule features in many fields of human activity. Moreover, in a series of cases the decision-making procedures based on the Golden Rule yield more profitable solutions in competitive economic models than the generally accepted Nash equilibrium; see Chap. 4. In this respect we fully agree with B. Russell,[1] who admitted that "in all affairs it's a healthy thing now and then to hang a question mark on the things you have long taken for granted."

Tbilisi, Georgia — Mindia E. Salukvadze
Moscow, Russia — Vladislav I. Zhukovskiy

[1] Bertrand Arthur William Russell, (1872–1970), was a British philosopher, logician, social reformer, and Nobel laureate in Literature.

Basic Notations

$\mathbb{R}^l$—the l-dimensional Euclidean space with the Euclidean norm $|\cdot|$;
$\mathbb{N} = \{1, \dots, N\}$—the set of players;
$\mathbb{N}\backslash\{i\} = \{1, \dots, i-1, i+1, \dots, N\}$;
$\mathbb{K} = \{i_1, \dots, i_K \mid i_l \in \mathbb{N}\ (l = 1, ..., K)\}$—a coalition of players from $\mathbb{N}$;
$\mathbb{K}(i)$—a coalition that includes player i;

$$\mathcal{P} = \{\mathbb{K}_1, \dots, \mathbb{K}_r \mid \mathbb{K}_i \bigcap \mathbb{K}_j = \varnothing\ (i, j = 1, \dots, r; i \neq j) \wedge \bigcup_{i=1}^{r} \mathbb{K}_i = \mathbb{N}\};$$

—a coalitional structure;
X_i—the set of strategies x_i of player i;
Y—the set of uncertain factors y;
X^{e}—the set of Nash equilibria x^{e};

$$\overline{\mathcal{A}}=\{\gamma=(\gamma_1, \dots, \gamma_N) \in \mathbb{R}^N \mid \gamma_j=\mathrm{const} \geqslant 0 \ \ (j\in 1, \dots, N) \wedge \sum_{j=1}^{N} \gamma_j > 0\};$$

$\mathbb{R}^N_{\geqslant} = \{f = (f_1, \dots, f_N) \in \mathbb{R}^N \mid f_j \geqslant 0 \ \ (j = 1, \dots, N)\}$;
$\mathbb{R}^N_{\leqslant} = -\mathbb{R}^N_{\geqslant}$;
L = Slater, Pareto;
$\mathrm{Y_L}$—the set of L-minimal uncertainties y_L;
$f_i(x, y)$—the payoff function of player i;
$f(x, y) = (f_1(x, y), \dots, f_N(x, y))$—the column vector whose components are the payoff functions of players $1, \dots, N$;
$f^{\mathrm{L}} = f(x^{\mathrm{e}}, y_{\mathrm{L}}), \quad \Phi^{\mathrm{L}} = \Phi(x^{\mathrm{e}}, y_{\mathrm{L}})$;
$\mathrm{X} = \prod_{i\in\mathbb{N}} \mathrm{X}_i$—the set of strategy profiles $x = (x_1, \dots, x_N)$;
2^{X}—the set of all subsets of the set X;
Y^{X}—the set of functions $y(x)$ with a domain of definition X and a codomain Y;
$(x||z_i) = (x_1, \dots, x_{i-1}, z_i, x_{i+1}, \dots, x_N)$;
$\varnothing$—empty set;
comp $\mathbb{R}^k$—the set of compact sets in $\mathbb{R}^k$;
$\{f \in \mathcal{F} \mid \pi(f)\}$—the collection of elements f from a set $\mathcal{F}$ that satisfy a condition $\pi(f)$;

$\Leftrightarrow$—logical equivalence ("if and only if");
$\Rightarrow$—logical implication ("implies");
$\wedge$—logical conjunction ("and");
$\vee$—logical disjunction ("or");
$\urcorner$—logical negation ("not");
$\forall$—universal quantifier ("all");
$\exists$—existential quantifier ("exist(s)");
$\in$—inclusion relation ("belong(s) to");
$\notin$—non-inclusion relation ("do(es) not belong to");
$\bigcup$—union;
$\bigcap$—intersection;
For two N-dimensional vectors $f^{(k)} = (f_1^{(k)}, \ldots, f_N^{(k)})$ $(k = 1, 2)$,

$$f^{(1)} = f^{(2)} \Longleftrightarrow [f_i^{(1)} = f_i^{(2)} \quad (i \in \mathbb{N})];$$
$$f^{(1)} \neq f^{(2)} \Longleftrightarrow \urcorner(f^{(1)} = f^{(2)});$$
$$f^{(1)} \geqslant f^{(2)} \Longleftrightarrow [f_i^{(1)} \geqslant f_i^{(2)} \quad (i \in \mathbb{N})];$$
$$f^{(1)} \geq f^{(2)} \Longleftrightarrow (f^{(1)} \geqslant f^{(2)}) \wedge (f^{(1)} \neq f^{(2)});$$
$$f^{(1)} \ngeq f^{(2)} \Longleftrightarrow \urcorner(f^{(1)} \geq f^{(2)});$$
$$f^{(1)} > f^{(2)} \Longleftrightarrow [f_i^{(1)} > f_i^{(2)} \quad (i \in \mathbb{N})];$$
$$f^{(1)} \not> f^{(2)} \Longleftrightarrow \urcorner(f^{(1)} > f^{(2)});$$

$A_{ij},\ B_{ij},\ C_i$—constant matrices;
$a_i,\ b_i,\ c_i$—constant vectors;
$'$—transposition;
$A > 0$ $(<,\ \geqslant,\ \leqslant)$—a quadratic form $z'Az$ that is positive definite (negative definite, positive semidefinite, negative semidefinite, respectively);
$\det A$—the determinant of a matrix A;
$\dfrac{\partial \varphi(x, y)}{\partial x}$—the gradient of a scalar function φ with respect to the elements of a vector x;
$\dfrac{\partial^2 \varphi(x, y)}{\partial x^2}$—the Hessian matrix (matrix of all second-order partial derivatives of a scalar function φ with respect to the components of a vector x);
0_n—a zero column vector of dimension n;
$0_{l \times k}$—a zero matrix of dimensions $l \times k$;
E_n—an identity matrix of dimensions $n \times n$;
$f_i^{\mathrm{o}} = \max\limits_{x_i \in \mathrm{X}_i} \min\limits_{x_{\mathbf{N}\setminus\{i\}} \in \mathrm{X}_{\mathbf{N}\setminus\{i\}}} f_i(x)$;
$\mathrm{Idem}\{y \to y^*\}$—the result of replacing y with y^* in a bracketed expression;
$\mathbb{R}^{n \times m}$—the set of constant matrices of dimensions $n \times m$;
$\lambda \odot f$—the vector $(\lambda_1 f_1, \lambda_2 f_2, \ldots, \lambda_N f_N)$;
For sets A and B from the space $\mathbf{R}^N$,

$A \cup B$—the union of A and B;
$A \cap B$—the intersection of A and B;
$A + B = \{z \in \mathbb{R}^N \mid z = a + b,\ a \in A,\ b \in B\}$;
$A - B = \{z \in \mathbb{R}^N \mid z = a - b,\ a \in A,\ b \in B\}$;
$A \times B$—the Cartesian (direct) product of A and B;

$\min\limits_{y\in \mathrm{Y}}{}^{(\mathrm{L})} f(x, y) = f(x, y(x))$—the value of a vector criterion $f(x, y)$ for a given x and some uncertainty $y(x) \in \mathrm{Y}$ that is L-minimal in a multicriteria choice problem

$$\langle\, x, \mathrm{Y}, f(x, y) \,\rangle\,; \qquad (*)$$

$\mathrm{Argmin}\limits_{y\in \mathrm{Y}}{}^{(\mathrm{L})} f(x, y) = y(x)$—an uncertainty $y \in \mathrm{Y}$ that is L-minimal in problem $(*)$;
NE—Nash equilibrium;
BE—Berge equilibrium;
NGU—a noncooperative game under uncertainty;
MCPU—a multicriteria choice problem under uncertainty;
DM—a decision maker;
■—end of proof;
∗—reference to the short biographies at the end of this book.

Introduction

This book is written at the junction of two sciences that have little in common with each other (at least, at first glance), namely, philosophy (the Golden Rule of ethics) and cybernetics (mathematical theory of noncooperative games). The authors were motivated by the IX Moscow Festival of Science held on October 10, 2014, at Moscow State University. The program of that event in the fundamental library of MSU included lectures by Nobel laureates chemists Kurt Wüthrich (USA) and Jean-Marie Lehn (France) and biochemist Sir Richard Roberts (USA) as well as by RAS Academicians Mikhail Ya. Marov ("The Chelyabinsk meteor") and Lev M. Zelenyi ("Exoplanets: Searching for a second Earth"), Doctors of Sciences Alexander V. Markov ("Why does a human need such a big brain") and Yury I. Aleksandrov ("Neurons, humans, and cultures"). Among the other lecturers, RAS Academician Abdusalam A. Guseinov, Director of the RAS Institute of Philosophy, delivered his talk "The Golden Rule of ethics." Being inspired by the perfectly organized and delivered lecture, one of the authors of this book addressed to the speaker the following somewhat "impudent" question, "Are You interested in a mathematical theory of the Golden Rule?" The answer was affirmative. The fact is that the asker carried in his pocket the Candidate of Sciences Dissertation of Konstantin S. Vaisman, his former postgraduate, who defended it back in 1995. The dissertation was devoted to our early attempts to study a new solution concept for noncooperative games, called Berge equilibrium. The term "Berge equilibrium" arose as the result of reviewing Claude Berge's book *Théorie générale des jeux á n personnes* [202], which was originally published in 1957 and translated into Russian in 1961. We are deeply convinced that the notion of Berge equilibrium matches well the main requirements of the Golden Rule. Unfortunately, Vaisman's sudden death at the age of 35 suspended further development of Berge equilibrium in Russia. At that time, however, the concept of Berge equilibrium was "exported from Russia" by two Algerian postgraduates of V. Zhukovskiy, M. Radjef and M. Larbani. Later on, it was actively used by Western researchers. As shown by their publications, most of investigations are focused on the properties of Berge equilibria, the specific features and modifications of this concept, and relations with Nash equilibria. It seems that the nascent theory of Berge equilibrium is getting close to becoming a consistent and rigorous mathematical theory. Probably, an intensive accumulation of facts will lead to an evolutionary inner development. At this stage, tradition requires us to answer two fundamental questions:

1°. Does a Berge equilibrium exist?
2°. How can it be calculated?

As a matter of fact, precisely these questions are treated in the book. Note that the book was awarded the first prize in the control and economics section at the 2016 All-Russia Best Scientific Book Contest organized by the Foundation for National Education Development. This book reveals the internal instability of the set of Berge equilibria. To eliminate this negative feature, we suggest a method to construct a Berge equilibrium that is Pareto-maximal with respect to all other Berge equilibria. The method reduces to the computation of a saddle point for an auxiliary zero-sum two-player game that is effectively designed using the original noncooperative game. We establish the existence of such a (Pareto-refined) Berge equilibrium in mixed strategies under standard assumptions of mathematical game theory, i.e., compact strategy sets and continuous payoff functions of players. This provides the answer to both questions!

Much attention in the book is also paid to Berge equilibria in the games under uncertainty as a brand-new research direction.

Finally, at the end of this book we consider applications to competitive economy models (the Cournot and Bertrand oligopolies) as well as three new approaches to important problems of mathematical game theory and multicriteria choice, namely, payoff increase with simultaneous risk reduction, stability of coalitional structures in cooperative games without side payments under uncertainty, and integration of the "selfish" Nash equilibrium with the "altruistic" Berge equilibrium.

Thus, the readers are offered five independent parts of the book as follows.

Chapter 1 discusses general philosophical issues related to the Golden Rule of ethics.

Next, Chap. 2 introduces a practical design method for the Berge–Pareto equilibrium and proves its existence in mixed strategies.

Then Chap. 3 presents results of a pioneering research of guaranteed Berge equilibria in conflicts under interval uncertainty.

Chapter 4 studies the explicit forms of Berge equilibria in the mathematical models of Cournot and Bertrand oligopolies, including their setups under uncertainty.

Chapter 5 considers three new approaches to important problems of mathematical game theory and multicriteria choice, which are described in four sections (Sects. 5.1–5.4). The first approach ensures payoff increase with simultaneous risk reduction in the Savage–Niehans sense in multicriteria choice problems (Sect. 5.1) and noncooperative games (Sect. 5.2). The second approach allows us to stabilize coalitional structures in cooperative games without side payments under uncertainty (Sect. 5.3). The third approach serves to combine the "selfish" Nash equilibrium with the "altruistic" Berge equilibrium. Note that the investigations in Sects. 5.2–5.4 involve a special Germeier convolution of payoff functions and calculation of its saddle point in mixed strategies.

However, we are still far from exclaiming "*Acta est fabula!*"[1] Our research efforts were based on the *Non multa sed multum* principle[2] and the *Nune aut nonquam* slogan.[3] The results presented in this book form a part of the mathematical theory of the Golden Rule that describes the static case. Some relevant issues remain untouched, such as risk consideration, the dynamic case of the Golden Rule (particularly, for the multistage games) and a gamut of other problems arising in the modern theory of differential positional games. Our intention is to cover these issues in a separate book.

We are grateful to Alexander Yu. Mazurov, Candidate of Sciences (Physics and Mathematics) for his careful translation of the Russian text, editorial changes and valuable contribution to the English version of the book.

At the end of the Introduction, let us quote Sir Richard Stone, who believed that with a mathematical description of processes "our decisions may eventually come to rest a little more on knowledge and a little less on guesswork than they do at present."[4]

[1] Latin "The play is over!".

[2] Latin "Not many, but much," meaning not quantity, but quality.

[3] Latin "Now or never."

[4] Sir John Richard Nicholas Stone, (1913–1991), was a British economist and the father of national income accounting, who in 1984 received the Nobel Prize in Economic Sciences. A quote from *Scientific American*, 1964, vol. 211, no. 3, pp. 168–182.

Contents

Chapter 1
What Is the Golden Rule of Ethics?

Mais comme tout est compencé dans
le meilleur des mondes possibles.[1]

Quod tibi fieri non vis, alteri ne feceris.[2]

First of all, the essence of the Golden Rule is elucidated. Then its connections with philosophy, morality, duty, ethics, and politics are considered.

1.1 Scribitur ad narrandum, non ad probandum[3]

Do as you would be done by.
—English proverb[4]

The negative and positive statements of the Golden Rule are identified and its history is traced back [36–40].

[1]French "There are doubts whether everything is really compensated in the best of all possible worlds"; from A.I. Herzen's letter to N.A. Herzen, June 7, 1851. An ironic combination of two famous quotes from *Des compensations dans les destinées humaines* by French philosopher P.H. Azaïs (1766–1845) and *Candide* by French Enlightenment writer, historian, and philosopher Voltaire (1694–1778).

[2]Latin "Do not do unto others what you don't want others to do unto you." A favourite phrase of Roman emperor Marcus Aurelius Severus Alexandrus (222–235 A.D.).

[3]Latin "Is written to narrate, not to prove." A quote from *Institute of Oratory* X: 1, 13, by Roman rhetorian Marcus Fabius Quintilianus (appr. 35–95). He used this phrase to discriminate between the tasks of history and eloquence.

[4]Considered by many ethicists and moralists, not only Christians, as a foundation of proper behavior. The world would be almost ideal if everybody obeyed this proverb.

© Springer Nature Switzerland AG 2020
M. E. Salukvadze, V. I. Zhukovskiy, *The Berge Equilibrium: A Game-Theoretic Framework for the Golden Rule of Ethics*, Static & Dynamic Game Theory: Foundations & Applications, https://doi.org/10.1007/978-3-030-25546-6_1

In fact, the idea of this book occurred spontaneously, like those of many books dedicated to cybernetics. Our motivation emanated from the plenary lecture "The Golden Rule of ethics," delivered by RAS Academician A. Guseinov[5] on October 10, 2014, during the opening session of the IX Moscow Festival of Science in the fundamental library of Moscow State University. In the early 1970s, Guseinov was the first Russian philosopher to pioneer research on the Golden Rule [34, 35]. Epigraph no. 2 represents the quintessence of this rule in its negative statement. However, there exists a positive statement, "Behave unto others as you would like them to behave unto you."

The Golden Rule of ethics is not only a topic of academic studies, but also a subject of contemplation for any thinking person (even if he forgets this rule or simply does not realize its role in everyday life). The Golden Rule was suggested by prominent sages in ancient times. It still remains topical in our days. This rule dates back to the middle of the first millennium B.C., a period of humanistic revolution. The Golden status was assigned to it in the eighteenth century.

As is well known, in tribal communities people followed the custom of blood vengeance called talion (the law of retribution in kind). This severe law restricted the wars of tribes through an equivalent punishment for any crime. When tribal relations started to disappear, it became difficult to discriminate between "friends" and "foes." The economic relations beyond a community had gradually become more significant than family ties. As a result, communities strived to bear no responsibility for the actions of individual members. That processes made the use of talion inefficient, and communities needed a fundamentally new principle to regulate interpersonal relations regardless of tribal membership. The solution was provided by the Golden Rule.

1.2 World Religions About the Golden Rule

Those who cannot remember the
past are condemned to repeat it.
—Santayana[6]

[5]Abdusalam A. Guseinov, RAS Academician and Director of RAS Institute of Philosophy. Formulated the hypothesis about the phased origin of ethics based on the isolation of an individual from a tribal community as an active person. Has been developing the concept of non-violence ethics since the late 1980s. Associated with a series of original ideas, namely, an interpretation of the classical European ethics as different experiences in the spiritual overcoming of contradictions between happiness (bliss) and goodness (virtue); a justified consideration of ethics and moral reasoning as a single spiritual complex that lies outside the framework of science and its subject; a description of moralizing as a fetishistic form of cognition. In the recent years, has been working on an ethical concept that substantiates a particular role of bans and negative actions in morality.

[6]Jorge Agustín Nicolás Ruiz de Santayana y Borrás, well-known in the English speaking world as George Santayana, (1863–1952), was a Spanish-American philosopher, poet, and humanist who made important contributions to aesthetics, speculative philosophy, and literary criticism.

The original statements of the Golden Rule from leading world religions are presented.

Let us discuss the statements of the Golden Rule that can be found in ancient religions.

The New Testament, see *the Gospel of Matthew*, Chapter 7:12: "Therefore **all things whatsoever ye would that men should do to you, do ye even so to them**: for this is the law and the prophets."

The New Testament, see *the Gospel of Luke*, Chapter 6:31: "And **as ye would that men should do to you, do ye also to them likewise**."

The Babylonian Talmud, Shabbat 31a: "Once there was a gentile who came before Shammai, and said to him: "Convert me on the condition that you teach me the whole Torah while I stand on one foot." Shammai pushed him aside with the measuring stick he was holding. The same fellow came before Hillel, and Hillel converted him, saying: "**That which is despicable to you, do not do to your fellow**, this is the whole Torah, and the rest is commentary, go and learn it."

An earliest mention of the Golden Rule can be found in the Old Testament, see *The Book of Tobit*, Chapter 4:14–15. Tobit exhorts his son Tobias, "Be careful, my child, in all you do, well-disciplined in all your behaviour. **Do to no one what you would not want done to you**." Most of modern biblical scholars date *The Book of Tobit* to a period between the fifth and third centuries B.C.

The same (or even earlier) period is assigned to the teachings of Confucius, see *The Analects* (*Lun Yu*), Chapter XV, 24: "Zi Gong [a disciple] asked: "Is there any one word that could guide a person throughout life?" The Master replied: "How about 'reciprocity'! **Never impose on others what you would not choose for yourself**."

Similar statements also appeared in old Indian and Muslim texts. A saying of the Buddha reads, **"As one teaches others so should one do oneself"** (see *Dhammapada* XII: 159). A hadith of the Prophet Muhammad states, "None of you has faith until he **loves for his brother what he loves for himself**" (see Hadith 13 in *Forty Hadith An-Nawawi*).

Of course, we should also mention numerous modern statements, from "you scratch my back and I'll scratch yours" to "reciprocal altruism." Modern ethologists believe that reciprocal altruism is the result of human evolution from natural egoism.

Without going deep into the history of the Golden Rule (time, place and origin), let us emphasize **the fundamental difference between the statements in the New and Old Testaments**. Many regard these statements as identical and even think that the Golden Rule appeared in the New Testament from the Old. Despite superficial resemblance, they have different, one might say, opposite sense. In the New Testament, the statement of the Golden Rule is positive: **do** to others what is good for you. But good for us does not always mean good for others. On the other hand, the Old Testament suggests the negative statement: **do not** do to others what is bad for you. Following this principle, one never does evil to anybody, even to an unknown person. This principle is more universal and well-grounded in relations with akins and friends as well as with strangers.

1.3 The Golden Rule and Philosophy

Philosophy is the science
which considers truth.
—Aristotle[7]

The connection between the Golden Rule and philosophy is considered.

Interestingly, the Golden Rule of ethics can be also found in philosophy. **Thales of Miletus**, the first among the seven famous Greek sages and philosophers, answered the question "What method must we take to lead a good life?" in the following way: "To do nothing we would condemn in others." Aurelius Augustinus (St. Augustine), a philosopher and theologian of the fourth to fifth centuries A.D., wrote, "The rule of love is that one should wish his friend to have all the good things he wants to have himself, and should not wish the evils to befall his friend which he wishes to avoid himself" (see *Of True Religions*, Chapter XLVI).

Thomas Hobbes, an outstanding philosopher of the New Time, noted, "...yet to leave all men unexcusable, they [natural laws] have been contracted into one easie sum, intelligible even to the meanest capacity; and that is, "Do not that to another, which thou wouldest not have done to thy selfe" (see *Leviathan*, Chapter XV).

Finally, **Lev Tolstoy** quoted the Golden Rule in his *What Is Religion, of What is Its Essence?* in the following way: "The truths of the religion common to everyone today are so very simple, intelligible and close to the hearts of all men; the practical law of which is that **man must behave towards others as he would wish others to behave towards him**."

Many other thinkers also mentioned the Golden Rule in certain form. The greatest sages on the Earth that are generally recognized as the teachers of mankind underlined the crucial role of this rule in human life. Possibly, to a large extent this was the core of their wisdom.

Let us summarize the Golden Rule in its conventional statements. These statements reflect the common basis of the rule as well as its nuances.

1. **Sympathy rule**: "Never impose on others what you would not choose for yourself" (this statement goes back to Confucius).
2. **Autonomy rule**: "Do nothing you would condemn in others" (this statement goes back to Thales of Miletus).
3. **Reciprocity rule**: "As you would that men should do to you, do also to them likewise" (this statement goes back to the Gospels).

Essentially, all these rules are suggesting the same. A common feature is that, while making a decision in complicated or ambiguous situations, a man should be guided by his beliefs, assessments and desires regarding the best relations among the people.

[7] Aristotle, Greek Aristoteles, (384–322 B.C.), was an ancient Greek philosopher and scientist. One of the greatest intellectual figures of Western history.

Summarizing this historical review, we also note that the Rule figured as an aphorism, fundamental principle, commandment, etc. *In the seventeenth century it was called Golden in the European culture and still exists under this name*. True, the Golden Rule of ethics is also associated with other statements. Some philosophers tried to suggest a "metallic family" of ethical rules. For instance, American theologian and church historian Leonard I. Sweet introduced the following system of rules.

1. "Do unto others before they do unto you." (**The Iron Rule**).
2. "Do unto others as they do unto you." (**The Silver Rule**).
3. "Do unto others as you would have them do unto you." (**The Golden Rule**).

In addition, he formulated **the Titanium Rule**: "do unto others as Jesus has done to us." Here the key principles are selflessness and self-sacrifice.

1.4 What Does the Golden Rule Suggest?

We know the truth, not only by
the reason, but also by the heart.
—Pascal[8]

The essence of the Golden Rule is emphasized and its intended use is described.

What does the Golden Rule suggest? From what comprehension of human nature does it stem? This model relies on the following hypotheses.

1. Every man is the cause of all his deeds: before doing anything, every man makes a corresponding decision. This, of course, does not imply the absence of exogenous determinative factors, as they do exist; but the behavior of every man is conscious and reasonable and all his deeds are the result of his own decisions.
2. Every man strives for good, that is, best deeds according to his beliefs (particularly, best deeds for himself).
3. Best deeds are the deeds of intrinsic value, i.e., they will never turn into evil for the man performing them. Best deeds yield internal rewards and no man will regret them.
4. The deeds of intrinsic value remain such for any man striving for good. This hypothesis is of crucial importance. In other words, whenever a man finds a best decision within his reasoning-based ethical aspiration, this decision will be acknowledged by every man who follows the principles of good and rational argumentation.

[8]Blaise Pascal, (1623–1662), was a French mathematician, physician, religious figure, and writer.

The Golden Rule is a mechanism that allows every man to answer the following question in any ambiguous situations. Are seemingly valuable deeds of real value? Am I mistaken?

What are the intended use and capabilities of the Golden Rule?

The Golden Rule cannot change an immoral man into moral one. In fact, this task seems impossible for any rule. The intended use of the Golden Rule is to assist every man who strives to act ethically in keeping self-respect, i.e., choosing a correct decision. The Golden Rule is not a requirement applied to others. In the first place, it is imposed on oneself. Not coincidentally the linguistic form of the Golden Rule has two moods, namely,

1. imperative ("do"–"do not")
   and
2. subjunctive ("would"–"would not").

Imperativeness concerns one's own deeds while subjunctiveness the deeds of others. That is, we should be judgemental about our own deeds. For the deeds of the others, we may only hope and lead them by our own example.

In an ambiguous situation, the Golden Rule calls to mobilize imagination and carry out a mental experiment, exchanging roles in order to assess the relative significance (ergo, ethical purity) of a prospective deed. This approach allows one to remove all doubts and make a responsible and judgemental decision.

The Golden Rule of ethics is not an abstract norm. On the contrary, it is very specific and applicable to real situations, doubts, temptations, or enticements. People do not need special training or skills to use this rule, as it is not a logical formula but a **working scheme of behavior**. Everybody knows and recognizes this rule because it is present in our experience.

We resort to the Golden Rule while trying to deter another man from a bad deed. In short, the Golden Rule of ethics is a fundamental principle of our everyday life based on morality.

In conclusion, we note that some researchers endeavor to overcome the Golden Rule, belittling its importance as the quintessence of ethics, and suggest alternative regulation rules for moral behavior. Here a widespread approach is the Platinum Rule introduced by American culturologist Milton J. Bennett, which states, "Do unto others as they would have you do unto them." Russian culturologist Mikhail N. Epstein proposed the Diamond Rule in the following form: "Act in such a way that you yourself would like to become an object of your actions but no one else could be their subject." In other words, "Do what others need and no one else can do in your place." Both rules emphasize some autonomy for the ethical aspect of any action, i.e., it is assumed that each man has to simulate ambiguous situations in his mind, like a game played with himself. No doubt, these statements are important and reflect crucial points of our moral life, but still do not overcome the Golden Rule of ethics: the Platinum and Diamond Rules lose reciprocity. Indeed, according to the Golden Rule, each man should behave taking into account the expected effect on other people. Following the Platinum or Diamond rule, each man uses the others just to form his own autonomous behavior, though they do not define the canon of ethics; so this reciprocity is naturally lost.

Thus, it is too early to write off the Golden Rule. Particularly for the reason that the existence and application of this rule have no concern with academic studies: the Golden Rule accompanies our real life, relations and everyday experience.

The remainder of Chap. 1 consists mostly of translated fragments from the book *The Golden Rule of Behavior* written by Russian philosopher, Professor Lev E. Balashov; see [2] for the original version in Russian.

1.5 The Golden Rule as the Key Principle of Social Life

What is not good to ye, do not make ye to a friend.
—Clerk (deacon) Joannes[9]

Nam tua res agitur, paries cum proximus ardet.[10]

Connections between the Golden Rule and moral philosophy, ethics, the sense of duty, law, and a healthy way of life are considered.

Let us summarize the outcomes. Recall that, in the positive form, the Golden Rule precepts, "Behave to others as you would like them to behave to you." While the negative form is, "Do not behave to others as you would not like them to behave to you."

The Golden Rule gives an integral and concentrated view of ethics by capturing its major aspect—*the relation to others as to oneself.* This rule establishes, fixes and defines a *measure* of human nature in everybody as well as morally *equalizes* all people and *likens* them to each other. In Guseinov's opinion, whenever one speaks about moral equality one is concerned with only one thing—each individual is worthy of the right to happiness and "the mutual acknowledgement of this right is a prerequisite for moral communication." The Golden Rule demands "from an individual to put himself/herself in place of other individuals and behave unto them as if he/she would be in their place." "The mechanism of the Golden Rule can be defined as assimilation, as a requirement to mentally take the place of another individual." [34, p. 134]. Moral equalization is a *quantitative* procedure while moral assimilation a *qualitative* procedure. Their combination yields a *measuring* process: the Golden Rule suggests each man to *harmonize* his deeds with the deeds of the others, using his "yardstick" for their deeds and, conversely, their "yardstick" for his own deeds. Following this rule, every man should find a *common measure* for his own deeds and the deeds of the others, always acting in accordance with this common measure.

---

[9]A quote from *Izbornik of Sviatoslav*, 1073.

[10]Latin "It is your concern when your neighbor's wall is on fire." A quote from *Epistles* I: 18, 84, by Quintus Horatius Flaccus (65–8 B.C.), an outstanding Roman lyric poet and satirist, well-known in the English speaking world as Horace.

The negative statement of the Golden Rule establishes *the lowest admissible hurdle* or bound for the moral attitude of every man to the others, *prohibits doing evil*, thereby specifying the *minimum ethical requirements* to individual behavior.

In turn, its positive statement establishes the *highest admissible hurdle* for the moral attitude of every man to the others, *encourages doing good*, thereby providing *the maximum ethical requirements* to individual behavior.

Therefore, the Golden Rule covers the whole range of moral deeds and is a basis for discriminating between the ethical categories of *good* and *evil*. (J. Korczak wrote, "Many times I thought what "being good" means. To my mind, a good man is a man who has imagination and understands others, who can feel like others do."[11] A quote from the book [84].)

The same function is performed by the Golden Rule subject to *the sense of duty*. To explain this, just consider it from another viewpoint—how does this rule *commensurate* the deeds of every man with the deeds of the others? Such a commensuration *proceeds* from the following line of reasoning adopted by every man. "I was born and set up in life by parents, people and society (fed, dressed, shod, educated, etc.), i.e., they all did good unto me, just as I *would like* the others do. So, I am going or *must* do unto them (parents, people, society) at least in the same way, i.e., my behavior *must not* deteriorate or reduce the quality and amount of life given to me and the others. Moreover, as much as possible, I *must* apply every effort to improve or increase the quality and amount of life (mine and of the others, of the whole society)." In this context, we also translate into English a good quote by P. Lavrov: "In the course of his development, an intellectually mature man must pay a considerably higher price than the cost of this development for the mankind."[12] See the original in [91, p. 417].

This is a general understanding for the sense of duty. Of course, there exist different duties, depending on the meaning of "others." If "others" are our parents, then the matter concerns our duty to them; if our nation or country, our duty to the Motherland; if all people in the world, our duty to the mankind.

A duty is a "normal deviation" from an optimal norm, like a need. In turn, a need is a deviation from an optimal norm subject to a healthy way of life of an individual. Likewise, a duty is a deviation from an optimal norm subject to a healthy way of life of a society. Duties fulfilment by specific people has the same value for a healthy society as satisfaction of needs for a healthy individual. In his youth, every man accumulates duty, as he mostly takes from the others and gives almost nothing in return. At mature age, every man repays by doing his duty.

While *moral philosophy* (*ethics*) regulates the relations among people as well as maintains a healthy society in a small neighborhood of an optimal norm (realization

[11] Janusz Korczak, the pen name of Henryk Goldszmit, (1878–1942), was a Polish–Jewish doctor, writer, and child advocate [106] who, in order to maintain his orphanage, refused to escape Nazi-occupied Poland during World War II.

[12] Pyotr Lavrov, original name Pyotr Lavrovich Mirtov, (1823–1900), was a Russian Socialist philosopher, theorist of narodism, and publicist.

and fulfilment of duty), *law* does the same in a wider sense—bans, prevents, and cures *pathological* deviations from the optimal norm, often called offences and (or) crimes. Actually, *offences* and *crimes* have the same effect on a healthy society as *diseases* on a healthy individual. If many offences and crimes occur in a society, it is *de jure* sick. Such a society would hardly be healthy in the ethical sense.

The Golden Rule establishes a correlation between a healthy individual and a healthy society. It declares that the life and health of a society are formed by the people that compose it; that *morality* is valuable not by itself but as the result of a healthy way of life of a specific individual, as **a natural continuation** of this life and health. On the one hand, moral health is a part of social health (a group of people, a nation, etc.); on the other, a constituent of the individual health of every man belonging to a given society. *Law* is also not valuable by itself. It represents *a natural continuation* of morality and, like the latter, relies on the Golden Rule. T. Hobbes wrote that a man should "be contented with so much liberty against other men, as he would allow other men against himselfe." (see *Leviathan*, Chapter XIV). Nearly the same was claimed by an ancient political and juridical rule: "Everybody must obey only the law he/she has agreed with." This rule may perhaps seem somewhat dogmatic yet it is correct in substance, being based on the Golden Rule. Compare it with another rule: "Observing the rights of the others, we protect our own rights" (from a movie by Jacques–Yves Cousteau, 1984). This rule is used by thousands of diggers in the Amazon goldfields, and thefts are a rarity there. A detailed analysis of its meaning shows that this rule is a particular case of the Golden Rule in the negative statement. Consequently, in the deep sense, ***law is a mutual admission and restriction of freedom.*** A mutual admission of freedom yields various *human rights*, whereas a mutual restriction of freedom results in various *human duties*.

The Golden Rule is also remarkable for **self-sufficiency, self-connectedness and self-groundedness**. In particular, it combines an accidental "I want to..." with a necessary "I have to...." This combination finally gives what we call *freedom*. The Golden Rule is ***the formula of freedom***. Being combined in the Golden Rule, "I want to..." and "I have to..." complement and restrict each other as well as establish a measure and *moderate* each other.

With this combination of "I want to..." and "I have to...," the Golden Rule also eliminates the ethical dilemma of *happiness versus duty*. It *demands* from every man only what he *wants* to be done unto himself. Not without reason it is called Golden.

A negative ectype of the Golden Rule is found in popular expressions, such as "an eye for an eye, a tooth for a tooth," "Vengeance is mine; I will repay" and proverbs "as you sow so shall you reap", etc. [5]. Their essence is that **if you were done evil, you have the right to or should repay in kind**. *Despite a superficial similarity with the Golden Rule, such an approach is actually its antipode*. This "rule" works when the Golden Rule is violated. Its destructive power for human relations can be illustrated by *vengeance* (if you do evil unto me, my response will be the same). In this sense, the most dangerous phenomenon is *blood* vengeance, which may cause annihilation of entire families.

One wonders: if the Golden Rule is so good, why do people infringe it on a regular basis, doing evil and not fulfilling their duty? Here we may draw an analogy with a healthy way of life and diseases. The latter do not make our health less valuable; on the contrary, a sick man tries to recover from his disease as soon as possible. Similarly, a breach of the Golden Rule does not reduce its value. In the total balance of human deeds, the deeds based on the Golden Rule outweigh the deeds that violate it. Otherwise, our society would be far gone and dying.

The Golden Rule is not so trivial as it may seem at a first glance. For this rule to work efficiently, at least two conditions are required:

1. *Man must be normal and healthy; if not, he must take into account any abnormality and lack of moral health while choosing his attitude to the others.* The attitude to the others is the attitude to oneself.
2. *Man must be able to mentally put himself in place of the others, thereby making appropriate corrections in his behavior.* This procedure is not easy. Frequently people do harm to others not maliciously, but due to thoughtlessness, in particular, because they are unable to put themselves in place of the others.

Finally, it should be emphasize that the Golden Rule prohibits killing *in any form*. Indeed, no normal man wants to die, much less to be killed. If you do not want to be killed, you should not wish or do it unto others. Therefore, malicious or reckless killing, as well as enemy annihilation in war or execution of death penalty—all these contradict the Golden Rule.

1.6 Moral Decline of Modern Society

Do not treat others like you would not have them treat you.
—Russian proverb [8]

The moral level of modern society is discussed.

Nowadays, people often say that modern society suffers from a moral decline and even from a continuous destruction of ethical norms [116].

According to the Merriam-Webster's definition, "ethics is the discipline dealing with what is good and bad and with moral duty and obligation; the principles of conduct governing an individual or a group; a set of moral issues or aspects (such as rightness)." At present times, almost anybody speaking about ethics will be blamed for hypocrisy and dissimulation. Obeying moral norms is no longer fashionable or prestigious. The elderly note that just several decades ago people were different—not hesitating to be gentle and admonitory to each other. Today we often feel awkward to offer our arm to a woman, to assist a blind person cross the road, etc., against the typical attitude of every man, his true nature.

The dynamics of these destructive processes of the human nature are well described by a Chinese poem:

In the 1950s people helped each other.
In the 1960s people competed with each other.
In the 1970s people betrayed each other.
In the 1980s people cared only about themselves.
In the 1990s people exploited for their benefit everybody they met.

Since the early 2000s, the moral sphere of modern society has been considerably devalued in the whole world. This is a direct consequence of the prevailing economic problems and related ideological and political issues: almost all actions of people are aimed at accumulation of material goods.

In a continuous pursuit of wealth, man has neglected spirituality and stopped thinking about inner self-development, ignoring the ethical aspect of his deeds. This trend dates back even to the end of the nineteenth century. Famous Russian writer and philosopher F.M. Dostoyevsky wrote about an uncontrollable itch for money that seized the people of that period up to stupefaction; see *The Idiot*.

Most people forgot (many had never been aware of!) the essence of the Golden Rule. The destructive processes in modern society may cause a serious stagnation for our civilization; what is more dangerous, further evolution may even reach an impasse.

An essential role in the fadeaway of society's morality, e.g., in Russia and Germany, was played by corresponding ideologies adopted by bolsheviks and nazi, respectively. A low ethical level of people often manifests itself at critical periods of history (revolutions, civil wars and external military conflicts, instable political regimes, etc.). For example, we mention the crying violation of state norms in Russia during the Civil War (1918–1921), World War II (1939–1945), Stalin's industrialization (1920s–1930s) and also nowadays, in the form of an epidemic of terrorist acts. All these events led to a deplorable result—the mass mortality of innocent people.

The ethical aspects are often disregarded in the management of state affairs, i.e., in the course of economic, social, agricultural and industrial reforms. As a rule, this has a negative impact on the environment.

In some countries, a currently unfavorable condition in many spheres of human life is a direct consequence of governmental miscalculations (incorrect decisions) given the current ethical level of the society. We are observing a deterioration in the criminal situation: a growing number of killings (including contract and brutal murders), tortures, thefts, rapes, corrupt practices, acts of vandalism, etc. In many cases, these actions go unpunished, as the crime detection and punishment rate went down. As a somewhat funny example of disorder and chaos, consider a much-talked-of story that occurred in the middle of the 1990s in one country. Two men were caught in the government house for stealing a cardboard box with \$500,000. After an official announcement that the owner of that money did not show up, the criminal case was closed and further investigation terminated. As a result, the two criminals became "the benefactors of the state" because they found "a buried treasure"; and the money were redirected to government's coffers. Clearly, the owner of the money acquired it in a most underhand manner (otherwise, he/she would immediately claim the right for it). The public prosecutor's office had to identify the source of that cardboard box with a large sum of money. In fact, no investigation was conducted and the officials maintained a discreet silence why. To all evidence, police, courts and public prosecutor's office were unable to control the criminal situation in the country, the reason apparently being the high level of corruption of many public officials.

1.7 The Golden Rule and Policy

> Power will intoxicate the best hearts, as wine the strongest heads. No man is wise enough, nor good enough to be trusted with unlimited power.
>
> —Colton[13]

> Nobody should go into politics unless he has a hide like a rhinoceros.
>
> —Roosevelt[14]

This section was written under the impression of the lecture "The Golden Rule of Ethics and Its Interpretation in Policy" by Academician A. Guseinov, which was delivered on March 31, 2015, live on *Vmeste–RF*, the official channel of the Council of the Federation, the Upper Chamber of the Parliament of the Russian Federation.

Social Policy There exists a moral judgement of human activity in accordance with the behavioral rules accepted in a given society. The deeds of every man can be moral (worthy, noble, proper) and immoral. The criteria used to discriminate between them are called moral norms. In fact, morality is multiform, it can be treated as wordly wisdom, the divine commandments, a tool for maintaining social order, honesty in human relations, the supreme sense of human life, the inner voice of conscience or even obsolete requirements preventing us from being ourselves.

Morality is based on conscience (the ethical sense that allows every man to assess his deeds in terms of good and evil) and duty (the ethical will to act following one's own idea of correct behavior). Most of the world peoples have features of ethical behavior such as selflessness, courage, truthfulness, modesty, humanism, wisdom,

[13] Charles Henry Colton, (1848–1915), was an American clergyman of the Roman Catholic Church and writer.

[14] Franklin Delano Roosevelt, (1882–1945), was an American statesman and political leader who served as the 32nd President of the United States from 1933 until his death in 1945; he won a record four presidential elections.

to name a few. The qualities that are disapproved by many peoples (vices) include foolishness, self-interest, vanity, flattery, and so on.

The fundamental categories of morality are the representations about good and evil. These are most general concepts for assessing the actions and deeds of people. Good is the major value of every man, his moral sacred thing. Good stands against evil.

Each of us chooses between the paths of virtue or vice independently, but we still bear responsibility for this choice.

The generally accepted ethical requirements and guidelines of moral deeds constitute the universal component of moral consciousness that is common to all mankind. They express the demands of an ethical ideal as the supreme moral aim (the Golden Rule of ethics). Since the ancient times until present, the Golden Rule of ethics had underwent many changes, but today it still keeps the ideas of freedom and equality of all people, the self-esteem and dignity of each individual. As repeatedly mentioned, in its most general form it states, "Behave to others as you would like them to behave to you."

A special feature of morality is that it involves values, i.e., the preferences of people in accordance with their goals and ideals. The ethical values proceed from a comprehension of welfare (the supreme form of good, the state of complete harmony between a man and reality). This yields kindness, generosity, compassion, concern for one's neighbor, honesty, calmness, hope, and so on. All these values can be called virtues. They stand against such vices as hate, envy, pride, surfeit, egoism, greediness, and others.

For every man, moral perfection consists in shifting his internal proportions of good and evil towards the former. However, to do this every man should make his personal moral choice.

1.8 Is Ethical Policy Possible?

Nous dansons sur un volcan.[15]

The character of contradictions between policy and morality depends on the implementation processes of state power as well as on the types of ethical and political consciousness. At the same time, these conditions do not fully determine the matching of moral criteria with the fundamental principles of state authorities.

[15]French "We are dancing on a volcano." This famous phrase was addressed by French diplomat Narcisse-Achille de Salvandy to Louis Philippe, then the Duke of Orleans, at a grand ball given to the King of Naples. The words turned out to be a prophecy, as 2 months later French King Charles X was disthroned by the July Revolution of 1830. A similar expression is associated with Maximilien de Robespierre, (1758–1794), one of the leaders of the French Revolution: "Nous marchons sur des volcans."

Indeed, every social group is guided by its own ethical standards that justify or direct the activities of its members. This results in several centers of ethical energetics in politics. First of all, we may discuss the political ethics of different social groups—intellectuals, youth, working class and others—which characterizes the degree of assimilation of collective values by an individual. Moreover, in every country there exist public moral norms that are acknowledged by most of population as the main goals of life and activity. In turn, they can match to some degree the universal ethical rules that embody the supreme principles of humanism and unite people, despite their social, national, religious and other differences. These principles are "thou shalt not kill," "thou shalt not steal," and others.

From a political standpoint, the problem is to correlate these types of ethical reflection that prioritize human behavior in the field of state power. Perhaps the most acute problem concerns the role of different collective moral norms, as the supreme ethical ideals of a group pretend to replace public moral norms. In addition, separate groups may acknowledge the right of other groups for their own ideals, or may not. In the latter case, the representatives of such groups often believe that it is possible to compel people "for their own good" (due to their ignorance, blindness and misunderstanding of true goals) or may consider any contacts and compromises with political opponents as inadmissible weakness or even betrayal, etc.

In other words, an extremely dangerous phenomenon for a society is the elevation of collective values to the rank of public ethics. This causes a moral decline and dehumanization of politics. For example, bolsheviks considered ethical "only what serves the cause of working class, which creates the society of communists." As a result, they neglected the common values of the mankind and provoked the bloody bacchanalia of the civil war. During Stalin's period, snitching against friends and relatives was officially supported by the soviet authorities. Also recall the extremely cruel, barbaric treatment of the political opponents in Pol Pot's Cambodia, Mao's China and some other countries. As reasonably noted by Father A. Men,[16] the relativization of morality, the pretentiousness and impenetrability of collective standards for more general ethical values inevitably lead to violence and "the pluralism of skulls."

The fixation of basic ethical principles in the system of legal regulation and also the development of special structures in state authorities to control the ethical behavior of public politicians and officials (e.g., restriction of gifts, prevention of nepotism, etc.) are of crucial importance. Another considerable aspect is to organize public control of state authorities (in form of mass media and non-governmental organizations reporting of corruption, false, and so on).

[16]Alexander V. Men, (1935–1990), was a Russian Orthodox priest, theologian, Biblical scholar and writer.

In each country, such a political course must be implemented together with a proper moral climate in which neither the leader nor ordinary people would shift the burden of responsibility to certain public or political institutions (family, party, organization). Only the ethical independence of an individual can serve as a foundation for the raising of politically conscientious citizens and maintaining morality as a source for a human-oriented political system of state power.

Chapter 2
Static Case of the Golden Rule

Celui qui croit pouvoir trouver en soi-même de quoi se passer de tout le monde se trompe fort; mais celui qui croit qu'on ne peut se passer de lui se trompe encore davantage.
—La Rochefoucauld[1]

In this chapter, the concept of Berge equilibrium is introduced as a mathematical model of the Golden Rule. This concept was suggested by the Russian mathematician K. Vaisman in 1994. The Berge–Pareto equilibrium is formalized and sufficient conditions for the existence of such an equilibrium are established. As an application, the existence in the class of mixed strategies is proved.

2.1 What is the Content of the Golden Rule?

Virtue is its own reward.
—English proverb

In the religious-ethical foundations, most nations are guided by the same strategy of behavior, embodied in the demands of the so-called Golden Rule (see Chap. 1). It will hopefully become an established ethical rule for the behavior of the mankind. The well-known statement of the Golden Rule declares, "Behave to others as you would like them to behave to you" (from a lecture delivered by Academician A. Guseinov, Director of RAS Institute of Philosophy, during The IX Moscow Science Festival on October 10, 2014 at Moscow State University). It originates from the

[1]French "He who thinks he has the power to content the world greatly deceives himself, but he who thinks that the world cannot be content with him deceives himself yet more." François de La Rochefoucauld (1613–1680) was a French classical writer; a quote from *Réflexions ou Sentences et Maximes morales* (1665).

© Springer Nature Switzerland AG 2020

M. E. Salukvadze, V. I. Zhukovskiy, *The Berge Equilibrium: A Game-Theoretic Framework for the Golden Rule of Ethics*, Static & Dynamic Game Theory: Foundations & Applications, https://doi.org/10.1007/978-3-030-25546-6_2

New Testament, see the Gospel of Luke, Chapter 6:31, precepting that "And as ye would that men should do to you, do ye also to them likewise."

Prior to formalizing the concept of Berge equilibrium that matches the Golden Rule, we will present the concept of Nash equilibrium as a standard approach to resolve conflicts. In fact, a critical discussion of the latter has led to the Berge equilibrium, the new solution of noncooperative games that is cultivated in this book.

2.2 Main Notions

Suum cuique.[2]

Nowadays, when the world shudders at the possibility of escalating military conflicts, the Golden Rule becomes more relevant than ever. Indeed, the Golden Rule is a possible way to avoid wars and blood-letting. The modern science of warfare relies mostly on the concept of *Nash equilibrium*. In this section, the definition of a Nash equilibrium is given, preceded by background material from mathematical theory of noncooperative games.

2.2.1 Preliminaries

Non multa sed multum.[3]

Some general notions from the mathematical theory of noncooperative games that will be needed in the text are presented.

Which mythical means were used by Pygmalion to revivify Galatea? We do not know the true answer, but Pygmalion surely was an operations researcher by vocation: at some moment of time his creation became alive. This idea underlines creative activities in any field, including mathematical modeling. To build an integral entity from a set of odd parts means "to revivify" it in an appropriate sense:

"She has not yet been born:
she is music and word,
and therefore the un-torn,
fabric of what is stirred." (Mandelshtam[4])

[2]Latin "To each according to its own merits; to each his/her own." This phrase appeared in philosophical dialogs and treatises *On Duties* 1, 5, 14, and *Tusculan Disputations*, Vol. 22, by Marcus Tullius Cicero (102–43 BC), a Roman statesman, lawyer, scholar, and writer.

[3]Latin "Not many, but much," meaning not quantity but quality. This phrase belongs to Plinius the Younger (62–114 A.D.); see *Letters*, VII, 9.

This section is devoted to the revivification of a conflict.

What is a conflict? Looking beyond the common (somewhat criminal) meaning of this word, we will use the following notion from [189, p. 333]: *"conceptually a conflict is any phenomenon that can be considered in terms of its participants, their actions, the outcomes yielded by these actions as well as in terms of parties interested in one way or another in these outcomes, including the nature of these interests."* As a matter of fact, game theory suggests mathematical models of optimal decision-making under conflict. The logical foundation of game theory is a formalization of three fundamental ingredients, namely, the features of a conflict, decision-making rules and the optimality of solutions. In this book, we study "rigid" conflicts only, in which each party is guided by his own reasons according to his perception and hence pursues individual goals, *l'esprit les intérêst du clocher.*[5]

The branch of game theory dealing with such rigid conflicts is known as the theory of noncooperative games. The noncooperative games described in Chap. 2 possess a series of peculiarities. Let us illustrate them using two simple examples.

Example 2.1.1 Imagine several competing companies (firms) that supply the same product in the market. Product price (hence, the profit of each firm) depends on the total quantity of products supplied in the market. The goal of each firm is to maximize its profit by choosing an appropriate quantity of supply.

Example 2.1.2 The economic potential of an individual country can be assessed by a special indicator—a function that depends on controllable factors (taxation, financial and economic policy, industrial and agricultural development, foreign supplies, investments, credits, etc.) and also on uncontrollable factors (climate changes and environmental disasters, anthropogenic accidents, suddenly sparked wars, etc.). Each country seeks to achieve a maximal economic potential through a reasonable choice of the controllable factors with a proper consideration of the existing economic relations with other countries.

These examples elucidate well the character of noncooperative games.

The *differentia specifica*[6] of such games are the following.

First, the decision-making process involves several parties (decision makers, e.g., sellers or governments), which are often called **players** in game theory. Note that a priori they are competitors: *quilibet (quisque) fortunae suae faber.*[7]

Second, each player has an individual goal (profit or economic potential maximization) and the goals are bound to each other: *tout s'enchaine, tout se lie dans ce*

---

[4]Osip E. Mandelshtam, (1891–1938), was a major Russian poet, prose writer, and literary essayist.

[5]French, meaning narrow-mindedness and a lack of understanding or even interest in the world beyond one's own town's boundaries.

[6]Latin, meaning a feature by which two subclasses of the same class of named objects can be distinguished.

[7]Latin "Every man is the artisan of his own fortune." This phrase goes back to Appius Claudius Caecus (4–3 centuries BC), an outstanding statesman, legal expert and author of early Rome who was one of the first notable personalities in Roman history.

monde.[8] A dazzling success of one party may turn out to be a disaster for another party.

Third, each player uses his own tools for achieving his goal (for sellers, the quantity of products supplied; for a country, the controllable factors in Example 2.1.2); in game theory, the controllable factors of each player are called his ***strategies***, while a specific strategy chosen by a player is his decision or action in a noncooperative game.

Let us list three important circumstances.

First, quantitative analysis in any field requires an appropriate mathematical model; this fully applies to noncooperative games. In the course of mathematical modeling, a researcher inevitably faces the risks of going too deep into details ("not see the wood for the trees") and presenting the phenomenon under study in a rough outline ("throwing out the baby with the bathwater"). The mathematical model of a noncooperative game often includes the following elements:

- the set of players;
- for each player, the set of his strategies;
- for each player, a scalar functional defined on the set of players' strategies. The value of this functional is the degree to which a given player achieves his goal under given strategies. In game theory, the functional is called the ***payoff function*** (or utility function) of a given player.

Second, "many intricate phenomena become clear naturally if treated in terms of game theory." [21, p. 97]. Following these *ex cathedra*[9] pronouncements by Russian game theory maître N. Vorobiev, we are employing the framework of noncooperative games in this book.

A series of conventional requirements have been established for a game-theoretic model (of course, including a sufficient adequacy to the conflict under consideration) as follows.

First, the model must incorporate all interested parties of the conflict (*players*).

Second, the model must specify possible actions of all parties (the *strategies* of players).

Third, the model must describe the interests of all parties (for each player and each admissible collection of actions chosen by all players, the model must assign a value called the *payoff* of that player).

The *main challenges of game theory* [24] are

(1°) the design of optimality principles;
(2°) the proof of existence of *optimal actions* for players;
(3°) the calculation of optimal actions.

Different game-theoretic concepts of optimality often reflect intuitive ideas of *profitability, stability* and *equitability*, rarely with an appropriate axiomatic

[8]French, meaning that all things in the worlds are interconnected.

[9]Latin "From the chair," used with regard to statements made by people in positions of authority.

characterization. Therefore, in most cases the notion of optimality in game theory (an optimal solution of a game) is not unique, prior or absolute.

We will make the *normative*[10] approach to noncooperative games the cornerstone of this book: it will be established which behavior of the players should be considered optimal (rational or reasonable) [47, 48].

Depending on the feasibility of joint actions among the players (coordination of their individual actions), the games are classified as noncooperative, cooperative, and coalitional [50].

In the *noncooperative setup* of a game (simply called a noncooperative game, see above), each player chooses his action (strategy) in order to achieve the best individual result for himself without any coordination with other players: *chacun pour soi, chacun chez soi.*[11]

The *cooperative setup* of a game (cooperative game) is opposite to the noncooperative one. Here all players jointly choose their strategies in a coordinate way and, in some cases, even share the results (their payoffs). *Alle für einen, einer für alle.*[12]

Finally, in the *coalitional setup* of a game (coalitional game), all players are partitioned into pairwise disjoint groups (coalitions) so that the members of each coalition act cooperatively while all coalitions play a noncooperative game with each other.

2.2.2 Elements of the Mathematical Model

Ad Disputandum[13] Consider several subsystems that are interconnected with each other. In economics, these can be industrial enterprises or sectors, countries, sellers in a market, producers of every sort and kind with the same type of products, and other economic systems (called firms in [124, p. 28]). In ecology, industrial enterprises with the same purification and treatment facilities, competing populations of different species (e.g., predators and preys), epidemics propagation and control. In the mechanics of controlled systems, a group of controlled objects (aircrafts, missiles) that attempt to approach each other or to capture an evader.

Each subsystem is controlled by a supervisor (henceforth called a player), who undertakes certain actions for achieving his goal based on available information. In social and economic systems, the role of players is assigned to the general managers of industrial enterprises and business companies, the heads of states,

[10]There also exist other approaches to conflict analysis: *descriptive*, which is to find the resulting collections of players' actions (the so-called strategy profiles) in a given conflict; *constructive*, which is to implement the desired (e.g., optimal) strategies in a given conflict; *predictive*, which is to forecast the actual result (outcome) of a given conflict.

[11]French "Every man for himself, every man to himself."

[12]German "One for all and all for one."

[13]Latin "For discussion."

sellers (suppliers) and buyers (customers). In mechanical control systems, this role is played by the captains of ships or aircrafts and the chiefs of control centers.

Assume that, due to a priori conditions, the players have to follow the "Help yourself" slogan. This leads to the noncooperative setup of their interaction.

As an example, consider a simplified mathematical model of competition among N firms in a market.

Example 2.2.3 There are $N \geqslant 2$ competing firms (players) that supply an infinitely divisible good of the same type (flour, sugar, etc.) in a market. The cost of one unit of good for firm i is $c_i > 0,\ i \in \mathbb{N} = \{1, \dots, N\}$. Suppose the number of market participants is sufficiently small so that the prices for goods depend directly on the quantity supplied by each firm. More specifically, denote by K the total supply of goods in the market; then the price p of one good can be calculated as $p = \max\{a - Kb, 0\}$, where $a > 0$ gives the constant price of one good without any supply in the market, while $b > 0$ is the elasticity coefficient that characterizes the price drop in response to the supply of one unit of good. Here a natural assumption is that $c_i < a,\ i \in \mathbb{N}$, since otherwise the activity of firms makes no economic sense. In addition, the production capacities of the players are unlimited and they sell the goods at the price p.

Suppose the firms operate in stable (not extreme) conditions and hence their behavior is aimed at increasing profits. Denote by x_i the quantity of goods supplied by firm i $(i \in \mathbb{N})$. Then the total supply of goods in the market is given by

$$K = \sum_{i=1}^{N} x_i,$$

while the profit of firm i is described by the function

$$f_i(x) = px_i - c_i x_i \quad (i \in \mathbb{N}),$$

where (as before) p is the unit price.

Another reasonable hypothesis is that $a - Kb > 0$, since otherwise $p = 0$ and production yields no benefit for all firms (the profits become negative, $f_i(x) = -c_i x_i < 0,\ i \in \mathbf{N}$). In this case, the function

$$f_i(x) = \left[a - b\sum_{k=1}^{N} x_k\right] x_i - c_i x_i \quad (i \in \mathbb{N})$$

is the profit of firm i.

Therefore, in Example 2.2.3 the players are the competing firms and the action (strategy) of each player $i \in \mathbb{N}$ consists in choosing the quantity $x_i \in X_i = [0, +\infty)$ of its goods supplied in the market. Making its choice, each player i seeks to maximize its profit $f_i(x)$ (payoff) given the supplied quantities $x = (x_1, \dots, x_N)$ of all players.

Idée Générale Now, let us clarify the framework of the noncooperative games studied in this chapter.

For this purpose, we will answer Quintilian's questions "*Quis? Quid? Ubi? Quibus auxiliis? Cur? Quomodo? Quando?*"[14]

Quis? (Who?) In fact, a leading part in noncooperative games is assigned to the **players**. As mentioned earlier, players can be general managers of industrial enterprises and business companies, heads of states, sellers (suppliers) and buyers (customers), captains of ships or aircrafts and so on, i.e., those who have the right or authority to make decisions, give instructions and control their implementation (interestingly, some people considering themselves to be (fairly!) serious strongly object to such a game-theoretic interpretation of their activity). Each player has a corresponding serial number: $1, 2, \ldots, i, \ldots, N$. Denote by $\mathbb{N} = \{1, 2, \ldots, N\}$ the set of all players and *let the set* $\mathbb{N}$ *be finite*. Note that games with an infinite number of players (called non-atomic games) are also studied in game theory [160, 171]. Players may form groups, i.e., coalitions $\mathbb{K} \subseteq \mathbb{N}$. *A coalition is any subset* $\mathbb{K} = \{i_1, \ldots, i_k\}$ *of the player set* $\mathbb{N}$. In particular, possible coalitions are singletons (the noncoalitional setup of the game) and the whole set $\mathbb{N}$ (the cooperative setup of the game). A partition of the set $\mathbb{N}$ into pairwise disjoint subsets forming **N** in union is a *coalitional structure* of the game:

$$\mathcal{P} = \{\mathbb{K}_1, \mathbb{K}_2, \ldots, \mathbb{K}_l \mid \mathbb{K}_i \cap \mathbb{K}_j = \varnothing \ (i, j = 1, \ldots, l; i \neq j), \ \bigcup_{i=1}^{l} \mathbb{K}_i = \mathbb{N}\}.$$

For example, in the noncooperative three-player game ($\mathbb{N} = \{1, 2, 3\}$), there exist five possible coalitional structures, given by $\mathcal{P}_1 = \{\{1\}, \{2\}, \{3\}\}, \mathcal{P}_2 = \{\{1, 2\}, \{3\}\}$, $\mathcal{P}_3 = \{\{1\}, \{2, 3\}\}, \mathcal{P}_4 = \{\{1, 3\}, \{2\}\}, \mathcal{P}_5 = \{\{1, 2, 3\}\}$.

For a compact notation, we will sometimes consider only two-player games, letting $\mathbb{N} = \{1, 2\}$.

In Example 2.2.3, the players are the general managers of competing firms.

Quid? (What?) Each player chooses and then uses his *strategy. A strategy is understood as a rule that associates each state of the player's awareness with a certain action (behavior) from a set of admissible actions (behaviors) given this awareness.* For the head of a state, this is a direction of strategic development. In a sector composed of several industrial enterprises, a strategy of a general manager is the output of his enterprise, the price of products, the amount of raw materials and equipment purchased, supply contracts, investments, innovations and implementation of new technologies, payroll redistribution, penalties, bonuses and other incentive and punishment mechanisms. For a seller, a strategy is the price of

[14] Latin "Who? What? Where? Who helped? Why? How? When?"; a well-known system of seven questions for crime investigation suggested by Roman rhetorician Quintilian, Latin in full Marcus Fabius Quintilianus, (appr. 35–100 A.D.).

one good; for the captain of a ship, own course (rudder angle, the direction and magnitude of reactive force).

Thus, the action of each player consists in choosing and using his individual strategy, which gives an answer to the question *Quid?* Speaking formally, the strategy of player i in the game Γ_3 is x_i while the strategy set of this player is denoted by X_i.

Ubi? (Where?) Here the answer is short: in the conflict, more precisely put, in its mathematical model described by the noncooperative game. In Example 2.2.3, this is the market of goods.

Quibus Auxiliis? Quomodo? (Who helped? How?) Actually the players affect the conflict using their strategies, which is the answer to both questions.

In Example 2.2.3, the firms choose the quantities of their goods supplied in the market as their strategies. The resulting situation in the market is the strategy profile in the corresponding noncooperative game.

Cur? (Why?) The answer is: in order to assess the performance of each player. The noncooperative game (the mathematical model of a conflict adopted in our book) incorporates the *payoff function* of player i ($i \in \mathbb{N}$). The value of this function (called *payoff* or *outcome* in game theory) is a numerical assessment of the desired performance. In Example 2.2.3, the payoff function of player i has the form

$$f_i(x, y) = \left[a - b\sum_{k \in \mathbb{N}} x_k\right] x_i - c_i x_i.$$

It measures the profit of firm i in the single-stage game. The following circumstances should be taken into account while assessing the performance of each player in a noncooperative game.

First, the design of payoff functions (performance assessment criteria) is a rather difficult and at times subjective task: "*Nous ne désirerions guére de choses avec ardeur, si nous connaissions parfaitement ce que nous dèsirons.*"[15] [119, p. 55].

Sometimes, the goal consists in higher profit or lower cost; in other cases, in smaller environmental impact. Other goals are possible as well. As a rule, in a noncooperative game these criteria represent scalar functions defined on the set of all admissible strategy profiles. For the sake of definiteness, assume each player seeks to *increase* his payoff function as much as possible.

Second, in accordance with the noncoalitional setup of the game, the players act in an isolated way and do not form coalitions. Being guided by the *Suum cuique* slogan,[16] each player chooses his strategy by maximizing his own payoff.

[15] French "We would yearn for very few things if we clearly understood what we wanted." A quote from *Réflexions ou Sentences et Maximes Morales* by F. de La Rochefoucauld.

[16] Latin "To each his own," or "May all get their due"; also, see the epigraph to Sect. 2.2.1.

As a result, each player endeavors to implement his cherished goal: "*Chacun produit selon ses facultés et recoit selon ses besoins.*"[17]

Third, the decision-making process in the noncooperative game is organized as follows. Each player chooses and then uses his strategy, which yields a strategy profile of the game. The payoff function of each player is defined on the set of all admissible strategy profiles. The value of this function (**payoff**) is a numerical assessment of the player's performance.

In game theory, both terms are equivalent and widespread! Person = player.

At a conceptual level, during the decision-making process in the noncoalitional game player i chooses his strategy $x_i \in \mathrm{X}_i$ so that

first, this choice occurs simultaneously for all N players;
second, no agreements or coalitions among the players and no information exchange are allowed *during the game* [178, p. 1].

Quando? (When?) The answer to the last question of Quantilian's system is the shortest: at the time of decision-making in the conflict (within its mathematical model—the noncoalitional game) through an appropriate choice of strategies by the players.

In principle, a conflict can be treated as a certain controlled system, a "black box" in which the players input their strategies and receive their payoffs at the output. This is a standard approach to "instantaneous, single-period, static" noncooperative games in general game theory [23]. However, in most applications (particularly, in economics and the mechanics of controlled systems), the controlled system itself undergoes some changes with time, and the players are able to vary their strategies during the whole conflict. The games whose state evolves in time are called *dynamic*. Hopefully, our next book will be focused on the analysis of dynamic games.

2.2.3 Nash Equilibrium

Politica del campanile.[18]

A generally accepted solution concept for noncooperative games is the so-called Nash equilibrium.[19] Nash equilibrium is widely used in economics, military

[17] French "From each according to his ability, to each according to his needs."

[18] Italian, "The policy of his/her own bell tower." Used to describe narrow-mindness and commitment to local interests.

[19] John Forbes Nash, Jr. Born June 13, 1928, in Bluefield, West Virginia. Successfully graduated from the Carnegie Institute of Technology (now, Carnegie Mellon University) with bachelor's and master's degrees in mathematics. Richard Duffin, Nash's undergraduate advisor at the Carnegie Institute of Technology, gave him a brief characterization, "He is a mathematical genius." In 1948 Nash started his postgraduate study at Princeton University, where he was particularly influenced by International Economy, the faculty course of J. von Neumann, and by the famous book *Theory of Games and Economic Behavior* (1944), written by von Neumann together with O. Morgenstern.

science, policy and sociology. Almost each issue of modern journals on operations research, systems analysis or game theory contains papers involving the concept of Nash equilibrium.

Thus, let us consider a noncooperative three-player game described by

$$\Gamma_3 = \left\langle \{1, 2, 3\}, \{X_i\}_{i=1,2,3}, \{f_i(x)\}_{i=1,2,3} \right\rangle,$$

where each player $i = \{1, 2, 3\}$ chooses an individual *strategy* $x_i \in X_i \subseteq \mathbb{R}^{n_i}$ in order to increase his performance $f_i(x = (x_1, x_2, x_3))$, i.e., his *payoff* $f_i(x)$ in a current *strategy profile* $x = (x_1, x_2, x_3) \in \in X_1 \times X_2 \times X_3 = X$.

A *Nash equilibrium* is a pair $(x^e, f^e = (f_1(x^e), f_2(x^e), f_3(x^e)) \in X \times \mathbb{R}^3$ defined by the three equalities

$$\begin{aligned} f_1(x^e) &= \max_{x_1 \in X_1} f_1\left(x_1, x_2^e, x_3^e\right), \\ f_2(x^e) &= \max_{x_2 \in X_2} f_2\left(x_1^e, x_2, x_3^e\right), \\ f_3(x^e) &= \max_{x_3 \in X_3} f_3\left(x_1^e, x_2^e, x_3\right). \end{aligned} \tag{2.2.1}$$

Each player therefore acts selfishly, seeking to satisfy his individual ambitions regardless of the interests of the other players. As repeatedly mentioned earlier, this concept of equilibrium was suggested in 1949 by J. Nash, a Princeton University graduate at that time and a famous American mathematician and economist as we know him today. Moreover, 45 years later J. Nash, J. Harsanyi and R. Selten were awarded the Nobel Prize in Economic Sciences "for the pioneering analysis of equilibria in the theory of non-cooperative games." Let us note two important aspects. First, owing to his research in the field of game theory, by the end of the twentieth century J. Nash became a leading American apologist of the Cold War. Second, the Nash equilibrium had been so widely used in economics, sociology, and military science that during the period 1994–2012 the Nobel Committee awarded seven Nobel Prizes for different investigations that to a large degree stemmed from the concept of Nash equilibrium. However, the selfish character of NE prevents it from "paving the way" towards a peaceful resolution of conflicts.

In 1949 Nash presented his thesis on equilibrium solutions of noncooperative games; after 45 years—in 1994—he was awarded the Noble Prize in Economic Sciences for that research. From 1951 to 1959 worked at the Cambridge at Massachusetts Institute of Technology (MIT). In 1958 *Fortune* called Nash "America's brilliant young star of the 'new mathematics.'" In 1959 moved to California to work for the RAND Corporation and became a leading expert in the Cold War. Since 1959 suffered from a mental disorder (completely overcame the disease by 1980, to the great astonishment of doctors). Since 1980 again worked at Princeton University as a consulting professor. Was killed in a car crash on May, 24, 2015, at the age of 86. Throughout the world, Nash is well-known through R. Howard's movie *A Beautiful Mind* (2001, featuring R. Crowe) based on S. Nasar's book *Beautiful Mind: The Life of Mathematical Genius and Nobel Laureate John Nash*. The movie received four Oscars and the Golden Globe.

2.2.4 Berge Equilibrium

Que jamais le mérite avec lui ne perd rien, Et que, mieux que du mal, il se souvient du bien.[20]

Almost all notions in the modern theory of measure and integral go back to Lebesgue's works, and introduction of these notions was in some sense a turning point of transition from the mathematics of the 19th century to the science of the 20th century.
—Vilenkin[21]

A peaceful resolution of conflicts can be achieved using Berge equilibrium (BE). This concept appeared in 1994 in Russia, following a critical analysis of C. Berge's book [202]. Interestingly, Berge wrote his book as a visiting professor at Princeton University, simultaneously with Nash, who also worked there under support of the Alfred P. Sloan Foundation.

A Berge equilibrium is a pair $(x^{\mathrm{B}}, f^{\mathrm{B}} = (f_1(x^{\mathrm{B}}), f_2(x^{\mathrm{B}}), f_3(x^{\mathrm{B}})))$ defined by the equalities

$$\begin{aligned} f_1\left(x^{\mathrm{B}}\right) &= \max_{(x_2,x_3)\in \mathrm{X}_2\times \mathrm{X}_3} f_1\left(x_1^{\mathrm{B}}, x_2, x_3\right), \\ f_2\left(x^{\mathrm{B}}\right) &= \max_{(x_1,x_3)\in \mathrm{X}_1\times \mathrm{X}_3} f_2\left(x_1, x_2^{\mathrm{B}}, x_3\right), \\ f_3\left(x^{\mathrm{B}}\right) &= \max_{(x_1,x_2)\in \mathrm{X}_1\times \mathrm{X}_2} f_1\left(x_1, x_2, x_3^{\mathrm{B}}\right). \end{aligned} \tag{2.2.2}$$

Equilibria (2.2.2) and (2.2.1) exhibit the following fundamental difference. In (2.2.1), each player directs all efforts to increase his *individual* payoff (the value of his payoff function) as much as possible. The antipode of (2.2.1) is (2.2.2), where each player strives to maximize the payoffs of the other players, ignoring his individual interests. Such an altruistic approach is intrinsic to kindred relations and occurs in religious communities. The elements of such altruism can be found in charity, sponsorship, and so on. The concept of Berge equilibrium also provides a solution to the Tucker problem in the well-known Prisoner's Dilemma (see Example 2.6.1 below). Due to (2.2.2), an application of this equilibrium concept eliminates armed clashes and murderous wars. This is an absolute advantage of Berge equilibrium.

As a matter of fact, the Berge equilibrium had an unenviable fate. The publication of the book [202] in 1957 initiated a sharp response of Shubik [269, p. 821] ("...no attention has been paid to applications to economics...the book will be of a little direct interest to economists..."). Most likely, such a negative review

[20]"And will not let true merit miss its due, Remembering always rather good than evil." A quote from *Tartuffe*, Scene VII, a famous theatrical comedy by Molière (1622–1873).

[21]Naum Ya. Vilenkin, (1920–1991), was a Soviet mathematician and student of A. G. Kurosh, who contributed to general algebra, topology, real-variable theory and functional analysis. A quote from *Kvant*, 1975, no. 8, p. 2.

in combination with Shubik's authority in scientific community pushed away the Western experts in game theory and economics from the book [202]. In Russia, after its translation in 1961, the book was analyzed in depth (Russian researchers were not acquainted with Shubik's review!) and the concept of Berge equilibrium was suggested on the basis of an appropriate modification of the notion of Nash equilibrium. The difference between Berge equilibrium and Nash equilibrium is that the former postulates stable payoffs against the deviations of all players and also reassigns the "ownership" of the payoff function (in the definition of a Nash equilibrium, the strategies of a separate player and all other players are interchanged). Note that the book [202] did not actually introduce the definition of Berge equilibrium, but it inevitably comes to mind while studying the results of Chaps. 1 and 5 of [202].

Subsequently, the Berge equilibrium was rigorously defined in 1994–1995 by K. Vaisman in his papers and dissertation [11, 13, 302], under the scientific supervision of V. Zhukovskiy. This concept was immediately applied in [280, 281] for noncooperative linear-quadratic positional games under uncertainty. Unfortunately, Vaisman's sudden death at the age of 35 suspended further research on Berge equilibrium in Russia. At that time, however, the concept of Berge equilibrium was "exported from Russia" by Algerian postgraduates of V. Zhukovskiy Radjef [266] and Larbani [248]. Later on, it was actively used by Western researchers (e.g., see the survey [255] with over 50 references and also the recent review [131, pp. 53–56] published in Ukraine). As shown by these and more than 100 subsequent publications, most of research works are dedicated to the properties of Berge equilibrium, the specific features and modifications of this concept, and relations to Nash equilibrium. It seems that an incipient theory of Berge equilibrium will soon emerge as a rigorous mathematical theory. Hopefully, an intensive accumulation of facts will be replaced by the stage of evolutionary internal development.

This chapter reveals the internal instability of the set of Berge equilibria. To eliminate this negative feature, we suggest a method to construct a Berge equilibrium that is Pareto-maximal with respect to all other Berge equilibria. The method reduces to a saddle point calculation for an auxiliary zero-sum two-player game that is effectively designed using the original noncooperative game. As a supplement, we prove the existence of such a (Pareto refined) Berge equilibrium in mixed strategies under standard assumptions of mathematical game theory, i.e., compact strategy sets and continuous payoff functions of the players.

2.3 Compactness of the Set X^B

> The notion of infinity is our greatest friend;
> it is also the greatest enemy of our peace in mind.
> —Pierpont[22]

[22]James P. Pierpont, (1866–1938), was an American mathematician. Known for research in the field of real and complex variable functions.

| It is shown that the set of Berge equilibria is closed and bounded.

Thus, we consider the mathematical model of a conflict in the form of a noncooperative N-player game, $N \geqslant 2$, described by an ordered triplet

$$\Gamma = \langle\{\mathbb{N}\}, \{X_i\}_{i\in\mathbb{N}}, \{f_i(x)\}_{i\in\mathbb{N}}\rangle. \tag{2.3.1}$$

Here $\mathbb{N} = \{1, 2, \dots, N\}$ denotes the set of players; each of the N players, forming no coalitions with other players, chooses his strategy (action) $x_i \in X_i \subseteq \mathbb{R}^{n_i}$ (throughout the book, the symbol $\mathbb{R}^k$, $k \geqslant 1$, stands for the k-dimensional Euclidean space whose elements are ordered sets of k real numbers in the form of columns, with the standard scalar product and the Euclidean norm); such a choice yields a *strategy profile* $x = (x_1, \dots, x_N) \in X = \prod_{i\in\mathbb{N}} X_i \subseteq \mathbb{R}^n$ $(n = \sum_{i\in\mathbb{N}} n_i)$; a payoff function $f_i(x)$ defined on the set X numerically assesses the performance of player i $(i \in \mathbb{N})$; let $(x\|z_i) = (x_1, \dots, x_{i-1}, z_i, x_{i+1}, \dots, x_N)$ and $f = (f_1, \dots, f_N)$.

A pair $(x^B, f^B) = ((x_1^B, \dots, x_N^B), (f_1(x^B), \dots, f_N(x^B))) \in X \times \mathbb{R}^N$ is called *a Berge equilibrium* in game (2.3.1) if

$$\max_{x\in X} f_i\left(x\|x_i^B\right) = f_i\left(x^B\right) \quad (i \in \mathbb{N}). \tag{2.3.2}$$

In the sequel, we will consider mostly the strategy profiles x^B from such pairs, also calling them *Berge equilibria* in game (2.3.1).

Property 2.3.1 If in the game Γ the sets X_i are closed and bounded, i.e., $X_i \in \operatorname{comp} \mathbb{R}^{n_i}$, and the payoff functions $f_i(\cdot)$ are continuous, $f_i(\cdot) \in \in C(X)$ $(i \in \mathbb{N})$, then the set X^B of all Berge equilibria in the game Γ is compact in X (possibly, empty) and $f(X^B) \in \operatorname{comp} \mathbb{R}^N$.

Proof Since $X^B \subseteq X$ and $X \in \operatorname{comp} \mathbb{R}^n$, then X^B is bounded. Thus, if we can show that X^B is closed, then $X^B \in \operatorname{comp} \mathbb{R}^n$. Let us prove the closedness of X^B by contradiction. Assume that, for a infinite sequence $\{x^{(k)}\}_{k=0}^{\infty}$, $x^{(k)} \in X^B$, there exist a subsequence $\{x^{(k_r)}\}_{r=0}^{\infty}$ and a strategy profile $x^* \in X$ such that, *first*, $\lim_{r\to\infty} x^{(k_r)} = x^*$ and, *second*, $x^* \notin X^B$.

Since $x^* \notin X^B$, there exist a strategy profile $\bar{x} \in X$ and a number $j \in \mathbb{N}$ such that $f_j(\bar{x}\|x_j^*) > f_j(x^*)$, where $x^* = (x_1^*, \dots, x_j^*, \dots, x_N^*)$ and, as before, $(\bar{x}\|x_j^*) = (\bar{x}_1, \dots, \bar{x}_{j-1}, x_j^*, \bar{x}_{j+1}, \dots, \bar{x}_N)$.

Owing to the continuity of $f_j(\bar{x}\|x_j)$ and $f_j(x)$ in $x \in X$ and the convergence $\lim_{r\to\infty} x^{(k_r)} = x^*$, there exists an integer $M > 0$ such that, for $r \geqslant M$, $f_j(\bar{x}\|x_j^{(k_r)}) > f_i(x^{(k_r)})$. This strict inequality contradicts $f_j(x\|x_j^B) \leqslant f_j(x^B)$ $\forall x \in X$, and the conclusion follows. ∎

Corollary 2.3.1 *Let the hypotheses of Property 2.3.1 be valid and the set* $X^B \neq \emptyset$ *in the game* Γ*. Then there exists a Berge equilibrium that is Pareto-maximal with respect to all other equilibria* $x^B \in X^B$ *in this game.*

Indeed, since the set X^B is compact, $f_i(\cdot) \in C(X)$ $(i \in \mathbb{N})$ and the hypotheses of Property 2.3.1 hold, the N-criteria choice problem

$$\left\langle X^B, \{f_i(x)\}_{i\in\mathbb{N}}\right\rangle$$

has a Pareto-maximal alternative $x^{\mathrm{B}} \in X^{\mathrm{B}}$ [152, p. 149]. In other words, for every $x \in X^{\mathrm{B}}$, the system of N inequalities

$$f_i(x) \geqslant f_i(x^{\mathrm{B}}) \quad (i \in \mathbb{N}),$$

with at least one strict inequality, is inconsistent.

2.4 Internal Instability of the Set X^B

It is easier to stop the Sun and move the Earth
than to decrease the sum of angles in a triangle,
to make parallels converge, and to drop perpendiculars
to the same line from a far distance.
—Kagan[23]

It is found that there may exist two Berge equilibria, in one of which each player has a strictly greater payoff than in the other.

Property 2.4.1 The set X^{B} of all Berge equilibria can be internally unstable, i.e., in the game Γ there may exist two Berge equilibria $x^{(1)}$ and $x^{(2)}$ such that, for all $i \in \mathbb{N}$,

$$f_i\left(x^{(1)}\right) > f_i\left(x^{(2)}\right).$$

Example 2.4.1 Consider a noncooperative two-player game ($N = 2$) of the form

$$\Gamma_2 = \left\langle \{1, 2\},\ \{X_i = [-1, +1]\}_{i=1,2},\ \left\{f_1(x) = -x_2^2 + 2x_1x_2,\ f_2(x) = -x_1^2 + 2x_1x_2\right\}\right\rangle.$$

In this game, the strategy profiles are $x = (x_1, x_2) \in [-1, +1]^2$, the strategy sets of both players coincide, $X_i = [-1, +1]$ $(i = 1, 2)$, while the Berge equilibrium $x^{\mathrm{B}} = (x_1^{\mathrm{B}}, x_2^{\mathrm{B}})$ is defined by the inequalities

$$\begin{aligned} &-x_2^2 + 2x_1^{\mathrm{B}}x_2 \leqslant -\left(x_2^{\mathrm{B}}\right)^2 + 2x_1^{\mathrm{B}}x_2^{\mathrm{B}}, \\ &-x_1^2 + 2x_1x_2^{\mathrm{B}} \leqslant -\left(x_1^{\mathrm{B}}\right)^2 + 2x_1^{\mathrm{B}}x_2^{\mathrm{B}} \quad \forall\, x_i \in [-1, +1] \quad (i = 1, 2), \end{aligned}$$

[23]Veniamin F. Kagan, (1860–1953), was a Russian and Soviet mathematician. A quote translated into English from *Kvant*, 1975, no. 6, p. 16.

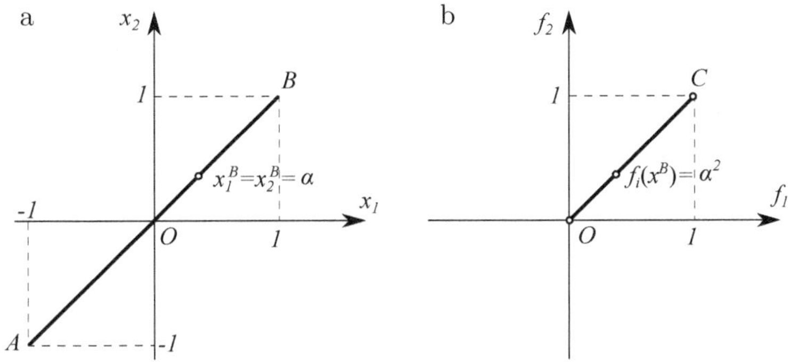

Fig. 2.1 (**a**) Set of Berge equilibria. (**b**) Payoffs in Berge equilibria

or

$$-\left(x_2 - x_1^{\mathrm{B}}\right)^2 \leqslant -\left(x_2^{\mathrm{B}} - x_1^{\mathrm{B}}\right)^2, \ -\left(x_1 - x_2^{\mathrm{B}}\right)^2 \leqslant -(x_1^{\mathrm{B}} - x_2^{\mathrm{B}})^2 \quad \forall x_1, x_2 \in [-1, +1]$$

(these inequalities follow from (2.3.2)). Hence, $x_1^{\mathrm{B}} = x_2^{\mathrm{B}} = \alpha$ for all $\alpha = \mathrm{const} \in [-1, +1]$ (see Fig. 2.1a), and then $f_i^{\mathrm{B}} = f_i(x^{\mathrm{B}}) = \alpha^2$ for all $\alpha = \mathrm{const} \in [-1, +1]$ (see Fig. 2.1b).

Thus, we have established that, *first*, there may exist a continuum of Berge equilibria (in Example 2.4.1, the set $\mathrm{X}^{\mathrm{B}} = AB$ as illustrated by Fig. 2.1a) and, *second*, the set X^{B} is internally unstable, since $f_i(0, 0) = 0 < f_i(1, 1) = 1$ $(i = 1, 2)$ (see Fig. 2.1b).

Hence, in the game Γ the players should use the Berge equilibrium that is Pareto-maximal with respect to all other Berge equilibria. We introduce the following definition for further exposition.

Definition 2.4.1 A strategy profile $x^* \in \mathrm{X}$ is called a Berge–Pareto equilibrium *(BPE)* in the game Γ if

(1) x^* is a Berge equilibrium in Γ (x^* satisfies conditions (2.3.2));
(2) x^* is a Pareto-maximal alternative in the N-criteria choice problem

$$\Gamma_{\mathrm{c}} = \left\langle \mathrm{X}^{\mathrm{B}}, \{f_i(x)\}_{i \in \mathbb{N}} \right\rangle,$$

i.e., for any alternatives $x \in \mathrm{X}^{\mathrm{B}}$, the system of inequalities

$$f_i(x) \geqslant f_i(x^*) \ \ (i \in \mathbb{N}),$$

with at least one strict inequality, is inconsistent.

In Example 2.4.1, we have two BPE, $x^{(1)} = (-1; -1)$ and $x^{(2)} = (+1; +1)$, with the same payoffs $f_i(x^{(1)}) = f_i(x^{(2)}) = 1$ $(i = 1, 2)$.

Remark 2.4.1 If $X^B \neq \emptyset$, $X_i \in \text{comp}\,\mathbb{R}^{n_i}$, and $f_i(\cdot) \in C(X)$ $(i \in \mathbb{N})$, then Definition 2.4.1 relies on Corollary 2.3.1, stating that the set of BPE is nonempty under the two requirements above.

Interestingly, the set of Nash equilibria in the game Γ is also internally unstable (this is demonstrated by Example 2.4.1 with the change $x_1 \leftrightarrow x_2$).

In the forthcoming sections, we will establish sufficient conditions for the existence of a BPE, which are reduced to a saddle point calculation for an auxiliary zero-sum two-player game that is effectively designed using the original noncooperative game.

2.5 No Guaranteed Individual Rationality of the Set X^B

Among the splendid generalizations effected by modern mathematics,
there is none more brilliant or more inspiring or more fruitful,
and none more commensurate with the limitless immensity of being itself,
than that which produced the great concept designated ...
hyperspace or multidimensional space.
—Keyser[24]

A Nash equilibrium has the property of individual rationality, whereas a Berge equilibrium generally does not, as illustrated by an example in this section. It is also established that there may exist a Berge equilibrium in which at least one player obtains a smaller payoff than the maximin.

Another negative property of a Berge equilibrium is the following.

Property 2.5.1 A Berge equilibrium may not satisfy the individual rationality conditions, as opposed to the Nash equilibrium x^e in the game Γ_2 (under the assumptions $X_i \in \text{comp}\,\mathbb{R}^{n_i}$ and $f_i(\cdot) \in C(X)$ $(i \in \mathbb{N})$, the game Γ_2 (the game Γ with $\mathbb{N} = \{1, 2\}$) satisfies the inequalities

$$f_1(x^e) \geqslant \max_{x_1 \in X_1} \min_{x_2 \in X_2} f_1(x_1, x_2), \quad f_2(x^e) \geqslant \max_{x_2 \in X_2} \min_{x_1 \in X_1} f_2(x_1, x_2),$$

known as the individual rationality conditions).

Example 2.5.1 Consider a noncooperative two-player game of the form

$$\Gamma_2' = \langle \{1, 2\}, \{X_1 = (-\infty, +\infty), X_2 = [-1, +1]\}, \{f_1(x) = \\ = -4x_1^2 + 2x_1x_2 + x_2^2,\ f_2(x) = -(x_1 - 1)^2 + 5\}\rangle,$$

[24]Cassius Jackson Keyser, (1862–1947), was an American mathematician of pronounced philosophical inclinations. A quote from *On Mathematics and Mathematicians*, R.E. Moritz, Ed., New York: Dover, 1958, pp. 360–361.

where $x = (x_1, x_2)$. A Berge equilibrium $x^B = (x_1^B, x_2^B)$ in the game Γ_2' is defined by the two equalities

$$\max_{x_2 \in X_2} f_1\left(x_1^B, x_2\right) = f_1\left(x^B\right), \quad \max_{x_1 \in X_1} f_2\left(x_1, x_2^B\right) = f_2\left(x^B\right).$$

The second equality holds only for the strategy $x_1^B = 1$. Due to the strong convexity of $f_1(x)$ in x_2 (which follows from the fact that $\left.\frac{\partial^2 f_1(x_1^B, x_2)}{\partial x_2^2}\right|_{x_2} = 2 > 0$), the maximum of the function

$$f_1\left(x_1^B, x_2\right) = -4 + 2x_2 + x_2^2$$

is achieved on the boundary of X_2, more specifically, at the point $x_2^B = 1$. Thus, the game Γ_2' has the Berge equilibrium $x^B = (1, 1)$, and the corresponding payoff is $f_1(x^B) = f_1(1, 1) = -1$.

Now, find $\max_{x_1 \in X_1} \min_{x_2 \in X_2} f_1(x_1, x_2)$ in two steps as follows. In the *first* step, construct a scalar function $x_2(x_1)$ that implements the *inner minimum*:

$$\min_{x_2 \in X_2} f_1(x_1, x_2) = f_1(x_1, x_2(x_1)) \quad \forall x_1 \in \mathbf{X}_1.$$

By the strong convexity of $f_1(x_1, x_2)$ in x_2,

$$\left.\frac{\partial f_1(x_1, x_2)}{\partial x_2}\right|_{x_2(x_1)} = 2x_1 + 2x_2(x_1) = 0,$$

yielding the unique solution $x_2(x_1) = -x_1$ and $f_1[x_1] = f_1(x_1, x_2(x_1)) = -5x_1^2$.

In the *second* step, construct *the outer maximum*, i.e., find

$$\max_{x_1 \in X_1} f_1[x_1] = \max_{x_1 \in \mathbb{R}} f_1(x_1, x_2(x_1)) = \max_{x_1 \in \mathbb{R}} \left[-5x_1^2\right] = 0.$$

Consequently,

$$f_1(x^B) = -1 < 0 = \max_{x_1 \in \mathbb{R}} f_1(x_1, x_2(x_1)) = \max_{x_1 \in \mathbb{R}} \min_{x_2 \in [-1, +1]} f_1(x_1, x_2),$$

which shows that the individual rationality property may fail for a Berge equilibrium.

Remark 2.5.1 Individual rationality is a requirement for a "good" solution in both noncooperative and cooperative games: each player can guarantee the maximin individually, i.e., by his own maximin strategy, regardless of the behavior of the other players [173]. However, in a series of applications (especially for the linear-quadratic setups of the game), the maximin often does not exist. Such games were studied in the books [52, pp. 95–97, 110–116, 120] and [93, pp. 124–131].

In the case where game (2.3.1) has maximins, Vaisman suggested to incorporate the individual rationality property into the definition of a Berge equilibrium. Such equilibria are called *Berge–Vaisman equilibria.*

2.6 Two-Player Game

You don't have to be a mathematician
to have a feel for numbers.
—Nash[25]

The specific features of Berge equilibria in two-player games are identified.

Non-antagonistic Case Consider a special case of game (2.3.1) with two players, i.e., the game Γ in which $\mathbb{N} = \{1, 2\}$. Then a Berge equilibrium $x^{\mathrm{B}} = (x_1^{\mathrm{B}}, x_2^{\mathrm{B}})$ is defined by the equalities

$$f_1\left(x^{\mathrm{B}}\right) = \max_{x_2 \in \mathrm{X}_2} f_1\left(x_1^{\mathrm{B}}, x_2\right), \quad f_2\left(x^{\mathrm{B}}\right) = \max_{x_1 \in \mathrm{X}_1} f_2\left(x_1, x_2^{\mathrm{B}}\right).$$

Recall that a Nash equilibrium x^{e} in this two-player game is given by the conditions

$$f_1\left(x^{\mathrm{e}}\right) = \max_{x_1 \in \mathrm{X}_1} f_1\left(x_1, x_2^{\mathrm{e}}\right), \quad f_2\left(x^{\mathrm{e}}\right) = \max_{x_2 \in \mathrm{X}_2} f_2\left(x_1^{\mathrm{e}}, x_2\right).$$

A direct comparison of these independent formulas leads to the following result.

Property 2.6.1 A Berge equilibrium in game (2.3.1) with $\mathbb{N} = \{1, 2\}$ coincides with a Nash equilibrium if both players interchange their payoff functions and then apply the concept of Nash equilibrium to solve the modified game.

Remark 2.6.1 In view of Property 2.6.1, a special theoretical study of Berge equilibrium in game (2.3.1) with $\mathbb{N} = \{1, 2\}$ seems unreasonable, despite careful attempts by a number of researchers. In fact, all results concerning Nash equilibrium *in a two-player game* are automatically transferred to the Berge equilibrium setting (of course, with an appropriate "interchange" of the payoff functions, as described by Property 2.6.1).

Let us proceed with an example of a two-player matrix game in which the players have *higher payoffs in a Berge equilibrium than in a Nash equilibrium* (in the setting of game, this is an analog of the Prisoner's Dilemma).

Also note the following interesting fact for game (2.3.1) with $\mathbb{N} = \{1, 2\}$, $f_i(x) = x_1^T A_i x_1 + x_2^T B_i x_2$, the strategies $x_1 \in \mathbb{R}^{n_1}$ and $x_2 \in \mathbb{R}^{n_2}$, where the matrices A_i and B_i of compatible dimensions are square, constant and symmetric, $A_1 > 0$, $B_1 < 0$, $A_2 < 0$, and $B_2 > 0$ (the notation $A > 0$ $(<)$ stands for the positive (negative) definiteness of the quadratic form $x^T Ax$): in this game, there exist no

[25]From a PBS interview with John F. Nash.

Nash equilibria, while the strategy profile $(0_{n_1}, 0_{n_2})$ forms a Berge equilibrium (as before, 0_k denotes the zero vector of dimension k).

Example 2.6.1 Consider the bimatrix game in which player 1 has two strategies, i.e., chooses between rows 1 and 2. Accordingly, the strategies of player 2 are represented by columns 1 and 2. For example, the choice of the strategy profile $(1, 2)$ means that the payoffs of players 1 and 2 are 4 and 7, respectively.

Player 1

Player 2	Row 1	Row 2
Column 1	Berge 6 6	4 7
Column 2	7 4	Nash 5 5

According to the above definitions, in this bimatrix game the strategy profiles $(2, 2)$ and $(1, 1)$ are a Nash and Berge equilibrium, respectively. As $6 > 5$, *the payoffs of both players in the Berge equilibrium are strictly greater than their counterparts in the Nash equilibrium.* The same result occurs in the Prisoner's Dilemma, a well-known bimatrix game. Note that the paper [255] gave some examples of 2×2 bimatrix games in which the payoffs in a Nash equilibrium are greater than or equal to those in a Berge equilibrium.

Antagonistic Case To conclude this section, consider the antagonistic case of game (2.3.1), which arises for Γ with $\mathbb{N} = \{1, 2\}$ and $f_2(x) = -f_1(x) = f(x)$. In other words, consider an ordered triplet

$$\Gamma_{\mathrm{a}} = \left\langle \{1, 2\}, \{\mathrm{X}_i\}_{i=1,2}, f(x) \right\rangle.$$

A conventional solution of the game Γ_{a} is the saddle point $x^0 = (x_1^0, x_2^0) \in \mathrm{X}_1 \times \mathrm{X}_2$, which is formalized here by the chain of inequalities

$$f\left(x_1^0, x_2\right) \leqslant f\left(x^0\right) \leqslant f\left(x_1, x_2^0\right) \quad \forall\, x_i \in \mathrm{X}_i \ (i = 1, 2). \tag{2.6.1}$$

Property 2.6.2 For the antagonistic case Γ_{a} of the game Γ, the Berge equilibrium $(x_1^{\mathrm{B}}, x_2^{\mathrm{B}})$ matches the saddle point (x_1^0, x_2^0) defined by (2.6.1).

The proof of this property follows immediately from the inequalities

$$f_1\left(x_1^{\mathrm{B}}, x_2\right) \leqslant f_1\left(x^{\mathrm{B}}\right), \quad f_2\left(x_1, x_2^{\mathrm{B}}\right) \leqslant f_2\left(x^{\mathrm{B}}\right) \quad \forall\, x_i \in \mathrm{X}_i \ (i = 1, 2)$$

and the identity $f(x) = f_2(x) = -f_1(x)\ \forall\, x \in \mathrm{X}$. ■

2.7 Comparison of Nash and Berge Equilibria

There are only two kinds of certain knowledge:
Awareness of our own existence and the truths of mathematics.
—d'Alembert[26]

A detailed comparison of Berge and Nash equilibria is made.

NE	BE
Stability	
against a unilateral deviation of a single player, since the inequality $$f_i(x_1^e,\dots,x_{i-1}^e,x_i,x_{i+1}^e,\dots,x_N^e)\leqslant f_i(x^e)$$ holding $\forall x_i\in \mathrm{X}_i$ $(i\in\mathbb{N})$ implies that the payoff of the deviating player i is not greater than in the NE.	*against the deviations of the coalition of all players except player* i, since the inequality $$f_i(x_1,\dots,x_{i-1},x_i^{\mathrm{B}},x_{i+1},\dots,x_N)\leqslant f_i(x^{\mathrm{B}})$$ holding $\forall x_j\in \mathrm{X}_j$ $(j\in\mathbb{N}\setminus\{i\},\ i\in\mathbb{N})$ implies that the payoff of each player i under such a deviation of the coalition of the other $N-1$ players from the BE is not greater than in the BE.
Individual rationality (IR)	
Here and in the sequel, $x_{\mathbb{N}\setminus\{i\}}=(x_1,\dots,x_{i-1},x_{i+1},\dots,x_N)\in\mathbf{X}_{\mathbb{N}\setminus\{i\}}=\prod_{j\in\mathbb{N}\setminus\{i\}}\mathbf{X}_j$. If x^{e} exists and $f_i^g=\max\limits_{x_i\in\mathbf{X}_i}\ \min\limits_{x_{\mathbb{N}\setminus\{i\}}\in\mathbf{X}_{\mathbb{N}\setminus\{i\}}} f(x\|x_i)=$ $=\min\limits_{x_{\mathbb{N}\setminus\{i\}}\in\mathbf{X}_{\mathbb{N}\setminus\{i\}}} f_i(x\|x_i^g)$ $(i\in\mathbb{N})$, then $f_i(x^{\mathrm{e}})\geqslant f_i^g$ $(i\in\mathbb{N})$, i.e., NE satisfies the IR condition.	Generally speaking, fails (see Property 2.5.1, Example 2.5.1, and Remark 2.5.1).
Internal instability	
The set of NE is internally unstable (see proof in [54]).	The set of BE is internally unstable (see Property 2.4.1 and Example 2.4.1).
To eliminate this drawback both for the NE *and* BE, *Pareto maximality with respect to* *the other equilibria of a given type is required.*	
Saddle point (SP) in the game $\Gamma_a=\langle\{1,2\},\{\mathbf{X}_i\}_{i=1,2},\{f_1(x_1,x_2),\ f_2(x_1,x_2)=-f_1(x_1,x_2)\}\rangle$ is a special case of NE and BE.	
NE coincides with the SP $(x_1^{\mathrm{e}},x_2^{\mathrm{e}})$ of the form $\max\limits_{x_1\in\mathbf{X}_1} f_1(x_1,x_2^{\mathrm{e}})=$ $=f_1(x_1^{\mathrm{e}},x_2^{\mathrm{e}})=\min\limits_{x_2\in\mathbf{X}_2} f_1(x_1^{\mathrm{e}},x_2)$.	The BE coincides with the SP $(x_1^{\mathrm{B}},x_2^{\mathrm{B}})$ of the form $\max\limits_{x_2\in\mathbf{X}_2} f_1(x_1^{\mathrm{B}},x_2)=$ $=f_1(x_1^{\mathrm{B}},x_2^{\mathrm{B}})=\min\limits_{x_1\in\mathbf{X}_1} f_1(x_1,x_2^{\mathrm{B}})$.

[26]Jean Le Rond d'Alembert, (1717–1783), was a French mathematician, philosopher, and writer.

Conclusions

The main difficulty in many modern developments of mathematics
is not to learn new ideas but to forget old ones.
—Sawyer[27]

Nash equilibria have three undisputable *advantages*, namely, they are stable, coincide with the saddle point (containing this generally accepted concept as a special case), and satisfy the individual rationality condition. The first and second advantages are shared also by the Berge equilibria.

At the same time, Nash equilibria suffer from several drawbacks, namely, the internal instability of the set of NE and selfishness (each player seeks to increase his individual payoff, as by definition $\forall i \in \mathbb{N}$: $f_i(x^{\mathrm{e}}) = \max_{x_i \in \mathrm{X}_i} f_i(x^{\mathrm{e}} \| x_i)$).

Internal instability is intrinsic to the set of BE too. This negative feature can be eliminated by requiring Pareto maximality for the NE and BE. The selfish nature of NE is eliminated using the altruistic orientation of BE ("help the others if you seek for their help"). This constitutes a clear merit of BE as a way of *benevolent conflict resolution*.

2.8 Sufficient Conditions

Mathematics as an expression of the human mind reflects the active will,
the contemplative reason, and the desire for aesthetic perfection.
Its basic elements are logic and intuition, analysis and construction,
generality and individuality. Though different traditions may emphasize
different aspects, it is only the interplay of these antithetic forces
and the struggle for their synthesis that constitute the life, usefulness,
and supreme value of mathematical science.
—Courant[28]

2.8.1 Continuity of the Maximum Function of a Finite Number of Continuous Functions

I turn with terror and horror from this lamentable scourge
of continuous functions with no derivatives.[29]

[27] Walter Warwick Sawyer, (1911–2008), was a British mathematician, mathematics educator and author, who popularized mathematics on several continents.

[28] Richard Courant, (1888–1972), was a German-born American mathematician, educator and scientific organizer who made significant advances in the calculus of variations. A quote from *The Australian Mathematics Teacher*, vols. 39–40, Australian Association of Mathematics Teachers, 1983, p. 3.

[29] From a letter of French mathematician Charles Hermite, (1822–1901), to Dutch mathematician Thomas Joannes Stieltjes, (1856–1894), written in 1893.

An auxiliary result from operations research is described, which will prove fruitful for the ensuing theoretical developments.

Consider $N+1$ scalar functions $\varphi_i(x,z) = f_i(x\|z_i) - f_i(z)$ $(i \in \mathbb{N})$ and $\varphi_{N+1}(x,z) = \sum\limits_{i\in\mathbb{N}} [f_i(x) - f_i(z)]$ that are defined on the Cartesian product $\mathrm{X} \times \mathrm{Z}$; from this point on, all strategy profiles $x = (x_1,\ldots,x_N) \in \mathrm{X} = \prod_{i\in\mathbb{N}} \mathrm{X}_i \subset \mathbb{R}^n$ $(n = \sum_{i\in\mathbb{N}} n_i)$, and also $x_i, z_i \in \mathrm{X}_i$ $(i \in \mathbb{N})$, $z = (z_1,\ldots,z_N) \in \mathrm{Z} = \mathrm{X} \subset \mathbb{R}^n$ (recall that $(x\|z_i) = (x_1,\ldots,x_{i-1},z_i,x_{i+1},\ldots,x_N)$).

Lemma 2.8.1 *If the $N+1$ scalar functions $\varphi_j(x,z)$ $(j = 1,\ldots,N,N+1)$ are continuous on $\mathrm{X} \times \mathrm{Z}$ while the sets X and Z are compact* ($\mathrm{X}, \mathrm{Z} \in \operatorname{comp} \mathbb{R}^n$), *then the function*

$$\varphi(x,z) = \max_{j=1,\ldots,N+1} \varphi_j(z,z) \qquad (2.8.1)$$

is also continuous on $\mathrm{X} \times \mathrm{Z}$.

The proof of a more general result can be found in many textbooks on operations research, e.g., [136, p. 54]; it was included even in textbooks on convex analysis [46, p. 146]. Note that function (2.8.1) is called *the Germeier convolution* of the functions $\varphi_j(x,z)$ $(j=1,\ldots,N+1)$.

Finally, note that our choice of Hermite's quote on "terror and horror" refers to the fact that although each of the functions $\varphi_j(x,z)$ can be differentiable, in general this is not necessarily true for the function $\varphi(x,z)$ defined by (2.8.1).

2.8.2 Reduction to Saddle Point Design

The result presented in this section is the pinnacle of our book.

Thus, using the payoff functions $f_i(x)$ of game (2.3.1), construct the Germeier convolution

$$\varphi(x,z) = \max\left\{[f_i(x\|z_i) - f_i(z)\ (i \in \mathbb{N})], \left(\sum_{j\in\mathbb{N}} f_j(x) - \sum_{j\in\mathbb{N}} f_j(z)\right)\right\}, \qquad (2.8.2)$$

with the domain of definition $\mathrm{X} \times (\mathrm{Z} = \mathrm{X})$.

A saddle point $(x^0, z^{\mathrm{B}}) \in \mathrm{X} \times \mathrm{Z}$ of the scalar function $\varphi(x,z)$ in the zero-sum two-player game

$$\Gamma^{\mathrm{a}} = \langle \mathrm{X}, \mathrm{Z} = \mathrm{X}, \varphi(x,z) \rangle$$

is defined by the chain of inequalities

$$\varphi\left(x, z^{\mathrm{B}}\right) \leqslant \varphi\left(x^0, z^{\mathrm{B}}\right) \leqslant \varphi\left(x^0, z\right) \quad \forall\, x, z \in \mathrm{X}. \tag{2.8.3}$$

Theorem 2.8.1 *If in the zero-sum two-player game* Γ^{a}, *there exists a saddle point* (x^0, z^{B}), *then the minimax strategy* z^{B} *is a Berge–Pareto equilibrium in noncooperative game* (2.3.1).

Proof In view of (2.8.2), the first inequality in (2.8.3) with $z = x^0$ shows that $\varphi(x^0, x^0) = 0$. By (2.8.3) and transitivity, for all $x \in \mathrm{X}$,

$$\varphi\left(x, z^{\mathrm{B}}\right) = \max\left\{\left(f_i\left(x \| z_i^{\mathrm{B}}\right) - f_i\left(z^{\mathrm{B}}\right)\right)\ (i \in \mathbb{N}), \left(\sum_{j \in \mathbb{N}} f_j(x) - \sum_{j \in \mathbb{N}} f_j\left(z^{\mathrm{B}}\right)\right)\right\} \leqslant 0.$$

Hence, for each $i \in \mathbb{N}$ and all $x \in \mathrm{X}$,

$$f_i\left(x \| z_i^{\mathrm{B}}\right) - f_i\left(z^{\mathrm{B}}\right) \leqslant 0, \quad \sum_{j \in \mathbb{N}} f_j(x) \leqslant \sum_{j \in \mathbb{N}} f_j\left(z^{\mathrm{B}}\right).$$

Hence, for all $x \in \mathrm{X}$ we have

$$f_i\left(x \| z_i^{\mathrm{B}}\right) \leqslant f_i\left(z^{\mathrm{B}}\right) \quad (i \in \mathbb{N}), \quad \max_{x \in X^B} \sum_{j \in \mathbb{N}} f_j(x) = \sum_{j \in \mathbb{N}} f_j\left(z^{\mathrm{B}}\right). \tag{2.8.4}$$

Since the first N inequalities in (2.8.4) hold for all $x \in \mathrm{X}$, the strategy profile $x^{\mathrm{B}} = z^{\mathrm{B}}$ satisfies the Berge equilibrium requirements (2.3.2) in the game Γ. The last equality in (2.8.4) where $x \in \mathrm{X}^{\mathrm{B}}$ (the set of Berge equilibria) is a sufficient condition [152, p. 71] for $x^{\mathrm{B}} = z^{\mathrm{B}}$ to be a Pareto-maximal alternative in the N-criteria choice problem $\langle \mathrm{X}^{\mathrm{B}}, \{f_i(x)\}_{i \in \mathbb{N}} \rangle$. Thus, by Definition 2.4.1, the resulting strategy profile $z^{\mathrm{B}} \in \mathrm{X}$ is a Berge–Pareto equilibrium in game (2.3.1). ■

Remark 2.8.1 Theorem 2.8.1 suggests the following design method of a Berge–Pareto equilibrium in the noncooperative game (2.3.1):
first, construct the function $\varphi(x, z)$ using formula (2.8.2);
second, find the saddle point (x^0, z^{B}) of the function $\varphi(x, z)$ from the chain of inequalities (2.8.3).

Then the resulting strategy profile $z^{\mathrm{B}} \in \mathrm{X}$ is a Berge–Pareto equilibrium in game (2.3.1).

2.8.3 Germeier Convolution

The world of curves has a richer texture than the world of points.
It has been left for the twentieth century to penetrate into this full richness.
—Wiener[30]

Let us associate with game (2.3.1) the N-criteria choice problem

$$\Gamma_\nu = \langle \mathrm{X}, \{f_i(x)\}_{i\in\mathbb{N}}\rangle,$$

where $\mathrm{X} = \prod_{i\in\mathbb{N}} \mathrm{X}_i$ is the set of all admissible alternatives and $f(x) = (f_1(x), \ldots, f_N(x))$ is the N-dimensional vector criterion. In the problem Γ_ν, a decision maker (DM) seeks to choose an alternative $x \in \mathrm{X}$ in order to maximize the values of all N criteria (objective functions) $f_1(x), \ldots, f_N(x)$.

2.8.3.1 Necessary and Sufficient Conditions

In Definition 2.4.1 we have used the following notion of a vector optimum for the problem Γ_ν.

Definition 2.8.1 An alternative $x^{\mathrm{P}} \in \mathrm{X}$ is called Pareto-maximal in the problem Γ_ν if, for any $x \in \mathrm{X}$, the combined inequalities

$$f_i(x) \geqslant f_i(x^{\mathrm{P}}) \quad (i \in \mathbb{N}),$$

with at least one strict inequality, are inconsistent. An alternative $x^{\mathrm{S}} \in \mathrm{X}$ is called Slater-maximal in the problem Γ_ν if, for any $x \in \mathrm{X}$, the combined strict inequalities

$$f_i(x) > f_i(x^{\mathrm{S}}) \quad (i \in \mathbb{N})$$

are inconsistent.

In this section, we are dealing with the Germeier convolution

$$\max_{i\in\mathbb{N}} \mu_i f_i(x) = \varphi(x), \tag{2.8.5}$$

where the constants μ_i belong to the set M of positive vectors from $\mathbb{R}^N$ (sometimes, with the unit sum of their components).

[30]Norbert Wiener, (1894–1964), was an outstanding American mathematician and philosopher, the father of cybernetics. A quote from his book *I Am a Mathematician: the Later Life of a Prodigy*, MIT Press, 1964.

Note that if $f_i(x) = -\psi_i(x)$, then formula (2.8.5) yields the standard Germeier convolution (with $\varphi(x) = -\psi(x)$) given by

$$\psi(x) = \min_{i \in \mathbb{N}} \mu_i \psi_i(x), \tag{2.8.6}$$

since

$$\max_{i \in \mathbb{N}} \mu_i f_i(x) = -\min_{i \in \mathbb{N}} \mu_i \psi_i(x).$$

Most applications employ Germeier convolutions of two types:

$$\psi(x) = \min_{i \in \mathbb{N}} \frac{\psi_i(x)}{a_i},$$

where $a_i = \text{const} > 0$ are convolution parameters, $i = 1, \dots, N$, and

$$\psi(x) = \min_{i \in \mathbb{N}} \mu_i \psi_i(x),$$

where $\mu_i = const > 0$ are convolution parameters, $i = 1, \dots, N$. Clearly, the transition from the first form to the second can be performed by the change of variables $\mu_i = \frac{1}{a_i}$.

The following results were obtained in multicriteria choice theory.

Germeier's theorem ([152, p. 66]). *Consider the* N*-criteria choice problem*

$$\Gamma_\nu = \langle \mathrm{X}, \{f_i(x)\}_{i \in \mathbb{N}} \rangle,$$

and assume that the objective functions $f_i(x)$ *are positive for all* $x \in \mathrm{X}$ *and* $i \in \mathbb{N}$.

An alternative $x^{\mathrm{S}} \in \mathrm{X}$ *is Slater-maximal in* Γ_ν *if and only if there exists a vector* $\mu = (\mu_1, \dots, \mu_N) \in \mathrm{M}$ *such that*

$$\max_{x \in \mathrm{X}} \min_{i \in \mathbb{N}} \mu_i f_i(x) = \min_{i \in \mathbb{N}} \mu_i f_i(x^{\mathrm{S}}). \tag{2.8.7}$$

For the Slater-maximal alternatives $x^{\mathrm{S}} \in \mathrm{X}$, let $\mu = \mu^{\mathrm{S}} = (\mu_1^{\mathrm{S}}, \dots, \mu_N^{\mathrm{S}})$, where

$$\mu_i^{\mathrm{S}} = \frac{\lambda^0}{f_i(x^{\mathrm{S}})} \quad (i \in \mathbb{N}), \quad \lambda^0 = \frac{1}{\sum\limits_{i \in \mathbb{N}} \frac{1}{f_i(x^{\mathrm{S}})}},$$

which leads to

$$\max_{x \in \mathrm{X}} \min_{i \in \mathbb{N}} \mu_i^{\mathrm{S}} f_i(x^S) = \lambda^0.$$

Recall that M is the set of positive vectors $\mu = (\mu_1, \ldots, \mu_N) \in \mathbb{R}^N$ (possibly with the unit sum of components). The next result is a useful generalization of Germeier's theorem.

Corollary 2.8.1 ([152, p. 67]). *Suppose* $x^{\mathrm{S}} \in \mathrm{X}$ *and* $\zeta_i(y)$ $(i \in \mathbb{N})$ *are increasing functions of the variable* $y \in \mathbb{R}$ *that satisfy*

$$\zeta_1(f_1(x^{\mathrm{S}})) = \zeta_2(f_2(x^{\mathrm{S}})) = \cdots = \zeta_N(f_N(x^{\mathrm{S}})).$$

An alternative x^{S} *is Slater-maximal in the multicriteria choice problem* Γ_ν *if and only if*

$$\zeta_1(f_1(x^{\mathrm{S}})) = \max_{x\in\mathrm{X}} \min_{i\in\mathbb{N}} \zeta_i(f_i(x)).$$

Corollary 2.8.2 ([152, p. 68]). *An alternative* x^{S} *is Slater-maximal in the multicriteria choice problem* Γ_ν *if and only if*

$$\max_{x\in\mathrm{X}} \min_{i\in\mathbb{N}} \left[f_i(x) - f_i(x^{\mathrm{S}}) \right] = 0.$$

Finally, a Pareto-maximal alternative x^{P} *in the problem* Γ_ν *has the following property.*

Proposition 2.8.1 ([152, p. 72]). *Let* $x^{\mathrm{P}} \in \mathrm{X}$ *and* $f_i(x^{\mathrm{P}}) > 0$ $(i \in \mathbb{N})$. *An alternative* x^{P} *is Pareto-maximal in the multicriteria choice problem* Γ_ν *if and only if there exists a vector*

$$\mu = (\mu_1, \ldots, \mu_N) \in M = \{\mu \mid \mu_i = \mathrm{const} > 0 \ (i \in \mathbb{N}), \ \sum_{i\in\mathbb{N}} \mu_i = 1\}$$

such that $f(x^{\mathrm{P}})$ *yields the maximum point of the function* $\sum_{i\in\mathbb{N}} f_i(x)$ *on the set*

$$\left\{ f(\mathrm{X}) = \bigcup_{x\in\mathrm{X}} f(x) \mid \min_{i\in\mathbb{N}} \mu_i f_i(x^{\mathrm{P}}) \geqslant \max_{x\in\mathrm{X}} \min_{i\in\mathbb{N}} \mu_i f_i(x) \right\}.$$

2.8.3.2 Geometrical Interpretation

Consider the Germeier convolution in the case of two criteria in the choice problem Γ_ν, i.e., $f(x) = (f_1(x), f_2(x))$. Assume that at some point $A = (f_1(x^A), f_2(x^A))$ one has

$$\mu_1 f_1(x^A) = \mu_2 f_2(x^A) = 1$$

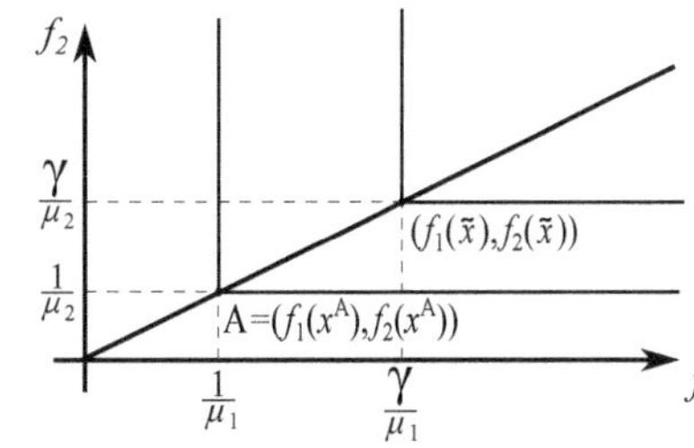

Fig. 2.2 Contour lines of the function $\min\limits_{i\in\{1,2\}} \mu_i f_i(x)$

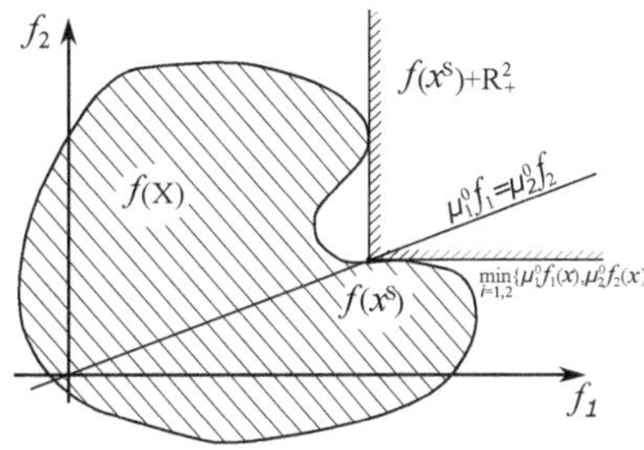

Fig. 2.3 Geometrical interpretation of Germeier's theorem

for some parameter values, i.e., $f_1(x^A) = \frac{1}{\mu_1}$ and $f_2(x^A) = \frac{1}{\mu_2}$ (see Fig. 2.2). Then the Germeier convolution takes the form

$$\min_{i=1,2} f_i(x^A) = 1.$$

Then the following relations hold on the rays originating from the point A parallel to the coordinate axes:

1. $\mu_1 f_1(x) \geqslant 1$ and $\mu_1 f_2(x) = 1$ on the horizontal ray, or
2. $\mu_1 f_1(x) = 1$ and $\mu_1 f_2(x) \geqslant 1$ on the vertical ray.

Hence, $\min_{i\in\mathbb{N}} \mu_i f_i(x) = 1$ on these rays. Consequently, the contour lines of the Germeier convolution coincide with the boundaries of the cone $\{f(x^A) + \mathbb{R}^2_+\}$, where $\mathbb{R}^2_+ = \{f = (f_1, f_2) \mid f_i \geqslant 0\ (i = 1, 2)\}$. In the same way, the contour lines of $\min_{i\in\{1;2\}} \mu_i f_i(x) = \gamma$ are defined by the vertical and horizontal rays originating from the point $f(\widetilde{x}) = (f_1(\widetilde{x}), f_2(\widetilde{x}))$, where $\mu_1 f_1(\widetilde{x}) = \mu_2 f_2(\widetilde{x}) = \gamma$. In other words, the contour lines of the Germeier convolution $\min_{i\in\{1;2\}} \mu_i f_i(x) = \gamma$ form the boundaries of the cone $\{f(\widetilde{x}) + \mathbb{R}^2_+\}$, where $f(\widetilde{x}) = \gamma f(x^A)$.

In the general case of N criteria, the level surfaces form the boundaries of the cone $\{f(\widetilde{x})+\mathbb{R}^N_+\}$, where $f(\widetilde{x})$ is any point satisfying the relation $\min_{i\in\mathbb{N}} \mu_i f_i(x) = \gamma = \text{const} > 0\ (i \in \mathbb{N})$. Therefore, the level surfaces of the Germeier convolution are the boundaries of the cone of points that dominate its vertex.

It is geometrically obvious that x^{S} is a Slater-maximal alternative in the bicriteria choice problem Γ_ν $(N = 2)$ if and only if the interior of the orthant $\mathbb{R}^N_+$ shifted to the point $f(x^{\mathrm{S}})$ does not intersect $f(\mathrm{X})$ (see Fig. 2.3).

2.9 Mixed Extension of a Noncooperative Game

> The value of pure existence proofs consists precisely in that the individual construction is eliminated by them and that many different constructions are subsumed under one fundamental idea, so that only what is essential to the proof stands out clearly; brevity and economy of thought are the *raison d'être* of existence proofs... To prohibit existence proofs... is tantamount to relinquishing the science of mathematics altogether.
>
> —Hilbert[31]

2.9.1 Mixed Strategies and Mixed Extension of a Game

> The theory of probabilities is at bottom nothing but common sense reduced to calculus; it enables us to appreciate with exactness that which accurate minds feel with a sort of instinct for which ofttimes they are unable to account... It teaches us to avoid the illusions which often mislead us; ... there is no science more worthy of our contemplations nor a more useful one for admission to our system of public education.
>
> —Laplace[32]

The mixed extension of a game that includes mixed strategies and profiles as well as expected payoffs is formalized.

Let us, consider the noncooperative N-player game (2.3.1). For each compact set $\mathrm{X}_i \subset \mathbb{R}^{n_i}$ $(i \in \mathbb{N})$, consider the Borel σ-algebra $\mathcal{B}(\mathrm{X}_i)$, i.e., the minimal σ-algebra that contains all closed subsets of the compact set X_i (recall that a σ-algebra is closed under taking complements and unions of countable collections of sets).

Assuming that there exist no Berge–Pareto equilibria x^{B} (see Definition 2.4.1) in the class of pure strategies $x_i \in \mathrm{X}_i$ $(i \in \mathbb{N})$, we will extend the set X_i of pure strategies x_i to the mixed ones, using the approach of Borel [204], von Neumann [261], and Nash [257] and their followers [192, 194, 195, 197, 198]. Next, the idea is to establish the existence of (properly formalized) mixed strategy profiles in game (2.3.1) that satisfy the requirements of a Berge–Pareto equilibrium (an analog of Definition 2.4.1).

Thus, we use the Borel σ-algebras $\mathcal{B}(\mathrm{X}_i)$ for the compact sets X_i $(i \in \mathbb{N})$ and the Borel σ-algebra $\mathcal{B}(\mathrm{X})$ for the set of strategy profiles $\mathrm{X} = \prod_{i \in \mathbb{N}} \mathrm{X}_i$, so that $\mathcal{B}(\mathrm{X})$

---

[31]David Hilbert, (1862–1943), was a German mathematician who axiomatized geometry and contributed substantially to the establishment of the formalistic foundations of mathematics. Recognized as one of the most influential and universal mathematicians of the 19th and early 20th centuries. A quote from the book *Hilbert* by C.B. Reid, Springer, 1996.

[32]Pierre-Simon, marquis de Laplace, (1749–1827), was a French scholar who made important contributions to the development of mathematics, statistics, physics and astronomy. An English translation of a quote from his book *Théorie Analytique des Probabilités*, 1795.

contains all Cartesian products of arbitrary elements of the Borel σ-algebras $\mathcal{B}(\mathrm{X}_i)$ $(i \in \mathbb{N})$.

In accordance with mathematical game theory, *a mixed strategy* $\nu_i(\cdot)$ of player i will be identified with *a probability measure on the compact set* X_i. By the definition in [122, p. 271] and notations in [108, p. 284], a probability measure is *a nonnegative* scalar function $\nu_i(\cdot)$ defined on the Borel σ-algebra $\mathcal{B}(\mathrm{X}_i)$ of all subsets of the compact set $\mathrm{X}_i \subset \mathbb{R}^{n_i}$ that satisfies the following conditions:

1. $\nu_i(\bigcup_k Q_k^{(i)}) = \bigcup_k \nu_i(Q_k^{(i)})$ for any sequence $\{Q_k^{(i)}\}_{k=1}^{\infty}$ of pairwise disjoint elements from $\mathcal{B}(\mathrm{X}_i)$ (*countable additivity*);
2. $\nu_i(\mathrm{X}_i) = 1$ (*normalization*), which yields $\nu_i(Q^{(i)}) \leqslant 1$ for all $Q^{(i)} \in \mathcal{B}(\mathrm{X}_i)$.

Let $\{\nu_i\}$ denote the set of all mixed strategies of player i $(i \in \mathbb{N})$.

Also note that the product measures $\nu(dx) = \nu_1(dx_1)\cdots\nu_N(dx_N)$, see the definitions in [122, p. 370] (and the notations in [108, p. 123]), are probability measures on the strategy profile set X. Let $\{\nu\}$ be the set of such probability measures (strategy profiles). Once again, we emphasize that in the design of the product measure $\nu(dx)$ the role of a σ-algebra of subsets of the set $\mathrm{X}_1 \times \cdots \times \mathrm{X}_N = \mathrm{X}$ is played by the smallest σ-algebra $\mathcal{B}(\mathrm{X})$ that contains all Cartesian products $Q^{(1)} \times \cdots \times Q^{(N)}$, where $Q^{(i)} \in \mathcal{B}(\mathrm{X}_\mathrm{i})$ $(i \in \mathbb{N})$. The well-known properties of probability measures [41, p. 288], [122, p. 254] imply that the sets of all possible measures $\nu_i(dx_i)$ $(i \in \mathbb{N})$ and $\nu(dx)$ are *weakly closed and weakly compact* ([122, pp. 212, 254], [180, pp. 48, 49]). As applied, e.g., to $\{\nu\}$, this means that from any infinite sequence $\{\nu^{(k)}\}$ $(k = 1, 2, \ldots)$ one can extract a subsequence $\{\nu^{(k_j)}\}$ $(j = 1, 2, \ldots)$ that *weakly converges* to a measure $\nu^{(0)}(\cdot) \in \{\nu\}$. In other words, for any continuous scalar function $\varphi(x)$ on X, we have

$$\lim_{j\to\infty} \int_{\mathrm{X}} \varphi(x)\nu^{(k_j)}(dx) = \int_{\mathrm{X}} \varphi(x)\nu^{(0)}(dx)$$

and $\nu^{(0)}(\cdot) \in \{\nu\}$. Owing to the continuity of $\varphi(x)$, the integrals $\int_{\mathrm{X}} \varphi(x)\nu(dx)$ (the expectations) are well-defined; by Fubini's theorem,

$$\int_{\mathrm{X}} \varphi(x)\nu(dx) = \int_{\mathrm{X}_1} \cdots \int_{\mathrm{X}_N} \varphi(x)\nu_N(dx_N)\cdots\nu_1(dx_1),$$

and the order of integration can be interchanged.

Let us associate with game (2.3.1) in pure strategies *its mixed extension*

$$\langle \mathbb{N}, \{\nu_i\}_{i\in\mathbb{N}}, \{f_i[\nu] = \int_{\mathrm{X}} f_i[x]\nu(dx)\}_{i\in\mathbb{N}} \rangle, \tag{2.9.1}$$

where, like in (2.3.1), $\mathbb{N}$ is the set of players while $\{\nu_i\}$ is the set of mixed strategies $\nu_i(\cdot)$ of player i. In game (2.9.1), each conflicting party $i \in \mathbb{N}$ chooses its mixed strategy $\nu_i(\cdot) \in \{\nu_i\}$, thereby forming a mixed strategy profile $\nu(\cdot) \in \{\nu\}$; the payoff function of each player i, i.e., the expectation

$$f_i[\nu] = \int\limits_{\mathrm{X}} f_i[x]\nu(dx),$$

is defined on the set $\{\nu\}$.

For game (2.9.1), the notion of a Berge–Pareto equilibrium x^* (see Definition 2.4.1) has the following analog.

Definition 2.9.1 A mixed strategy profile $\nu^*(\cdot) \in \{\nu\}$ is called a Berge–Pareto equilibrium in the mixed extension (2.9.1) (equivalently, a Berge–Pareto equilibrium in mixed strategies in game (2.3.1)) if

first, the profile $\nu^*(\cdot)$ is a Berge equilibrium in game (2.9.1), i.e.,

$$\max_{\nu_{\mathbb{N}\setminus\{i\}}(\cdot)\in\{\nu_{\mathbb{N}\setminus\{i\}}\}} f_i\left(\nu \| \nu_i^*\right) = f_i(\nu^*) \quad (i \in \mathbb{N}), \tag{2.9.2}$$

and second, $\nu^*(\cdot)$ is a Pareto-maximal alternative in the N-criteria choice problem

$$\langle\{\nu^{\mathrm{B}}\}, \{f_i(\nu)\}_{i\in\mathbb{N}}\rangle,$$

i.e., for all $\nu(\cdot) \in \{\nu^B\}$, the system of inequalities

$$f_i(\nu) \geqslant f_i(\nu^*) \quad (i \in \mathbb{N}),$$

with at least one strict inequality, is inconsistent.

Here and in the sequel,

$\nu_{\mathbb{N}\setminus\{i\}}(dx_{\mathbb{N}\setminus\{i\}}) = \nu_1(dx_1)\cdots\nu_{i-1}(dx_{i-1})\nu_{i+1}(dx_{i+1})\cdots\nu_N(dx_N)$,
$(\nu\|\nu_i^*) = (\nu_1(dx_1)\cdots\nu_{i-1}(dx_{i-1})\nu_i^*(dx_i)\nu_{i+1}(dx_{i+1})\cdots\nu_N(dx_N))$,
$\nu^*(dx) = \nu_1^*(dx_1)\cdots\nu_N^*(dx_N)$, $\{\nu_{\mathbb{N}\setminus\{i\}}\} = \{\nu_{\mathbb{N}\setminus\{i\}}(\cdot)\}$; in addition, $\{\nu^{\mathrm{B}}(\cdot)\}$ denotes the set of Berge equilibria $\nu^{\mathrm{B}}(\cdot)$, i.e., the strategy profiles that satisfy (2.9.2) with ν^* replaced by ν^{B}. Let $\{\nu^*\}$ be the set of mixed strategy profiles in game (2.9.1) that are given by the two requirements of Definition 2.9.1.

The following sufficient condition for Pareto maximality is obvious, see the statement below.

Remark 2.9.1 A mixed strategy profile $\nu^*(\cdot) \in \{\nu\}$ is Pareto-maximal in the game $\tilde{\Gamma}_\nu = \langle\{\nu^{\mathrm{B}}\}, \{f_i(\nu)\}_{i\in\mathbb{N}}\rangle$ if

$$\max_{\nu(\cdot)\in\{\nu^{\mathrm{B}}\}} \sum_{i\in\mathbb{N}} f_i(\nu) = \sum_{i\in\mathbb{N}} f_i(\nu^*).$$

2.9.2 Préambule

Proposition 2.9.1 *In game* (2.3.1)*, suppose the sets* X_i *are compact, the payoff functions* $f_i(x)$ *are continuous on* $\mathrm{X} = \mathrm{X}_1 \times \cdots \times \mathrm{X}_N$*, and the set of mixed strategy Berge equilibria* $\{\nu^{\mathrm{B}}\}$ *that satisfy* (2.9.2) *with* ν^* *replaced by* ν^{B} *is nonempty.*

Then $\{\nu^{\mathrm{B}}\}$ *is a weakly compact subset of the set of mixed strategy profiles* $\{\nu\}$ *in game* (2.9.1)*.*

Proof To establish the weak compactness of the set $\{\nu^{\mathrm{B}}\}$, take an arbitrary scalar function $\psi(x)$ that is continuous on the compact set X and an infinite sequence of mixed strategy profiles

$$\nu^{(k)}(\cdot) \in \{\nu^{\mathrm{B}}\} \quad (k = 1, 2, \ldots) \tag{2.9.3}$$

in game (2.9.1). Inclusion (2.9.3) (hence, $\{\nu^{\mathrm{B}}\} \subset \{\nu\}$) implies $\{\nu^{(k)}(\cdot)\} \subset \{\nu\}$. As mentioned earlier, the set $\{\nu\}$ is weakly compact; hence, there exist a subsequence $\{\nu^{(k_j)}(\cdot)\}$ and a measure $\nu^{(0)}(\cdot) \in \{\nu\}$ such that

$$\lim_{j\to\infty} \int\limits_{\mathrm{X}} \psi(x)\nu^{(k_j)}(dx) = \int\limits_{\mathrm{X}} \psi(x)\nu^{(0)}(dx).$$

We will show that $\nu^{(0)}(\cdot) \in \{\nu^{\mathrm{B}}(\cdot)\}$ *by contradiction*. Assume that $\nu^{(0)}(\cdot)$ does not belong to $\{\nu^B\}$. Then for sufficiently large j, one can find a number $i \in \mathbb{N}$ and a strategy profile $\bar{\nu}(\cdot) \in \{\nu\}$ such that

$$f_i\left[\bar{\nu}\|\nu_i^{(k_j)}\right] > f_i\left[\nu^{(k_j)}\right],$$

which clearly contradicts the inclusion $\{\nu^{(k_j)}(\cdot)\} \in \{\nu^{\mathrm{B}}\}$.

Thus, we have proved the requisite weak compactness. ■

Corollary 2.9.1 *In a similar fashion, one can prove the compactness (closedness and boundedness) of the set*

$$f\left[\left\{\nu^B\right\}\right] = \bigcup_{\nu(\cdot)\in\{\nu^B\}} f[\nu], \text{ where } f = (f_1, \ldots, f_N),$$

in the criteria space $\mathbb{R}^N$.

Proposition 2.9.2 *If in game* (2.9.1) *the sets* $\mathrm{X}_i \in \operatorname{comp} \mathbb{R}^{n_i}$ *and* $f_i(\cdot) \in C(\mathrm{X})$ $(i \in \mathbb{N})$*, then the function*

$$\varphi(x, z) = \max_{r=1,\ldots,N+1} \varphi_r(x, z) \tag{2.9.4}$$

satisfies the inequality

$$\max_{r=1,\dots,N+1} \int\limits_{X\times X} \varphi_r(x,z)\mu(dx)\nu(dz) \leqslant \int\limits_{X\times X} \max_{r=1,\dots,N+1} \varphi_r(x,z)\mu(dx)\nu(dz) \tag{2.9.5}$$

for any $\mu(\cdot) \in \{\nu\}$, $\nu(\cdot) \in \{\nu\}$; *recall that*

$$\begin{aligned} \varphi_i(x,z) &= f_i(x\|z_i) - f_i(z) \quad (i \in \mathbb{N}), \\ \varphi_{N+1}(x,z) &= \sum_{i\in\mathbb{N}} [f_i(x) - f_i(z)] . \end{aligned} \tag{2.9.6}$$

Proof Indeed, from (2.9.4) we have $N+1$ inequalities of the form

$$\varphi_r(x,z) \leqslant \max_{j=1,\dots,N+1} \varphi_j(x,z) \quad (r = 1,\dots,N+1)$$

for each $x, z \in X$. Integrating both sides of these inequalities with respect to an arbitrary product measure $\mu(dx)\nu(dz)$ yields

$$\varphi_r(\mu,\nu) = \int\limits_{X\times X} \varphi_r(x,z)\mu(dx)\nu(dz) \leqslant \int\limits_{X\times X} \max_{j=1,\dots,N+1} \varphi_j(x,z)\mu(dx)\nu(dz)$$

for all $\mu(\cdot)\in\{\nu\}$, $\nu(\cdot)\in\{\nu\}$ and each $r = 1,\dots,N+1$. Consequently,

$$\max_{r=1,\dots,N+1} \varphi_r(\mu,\nu) = \max_{r=1,\dots,N+1} \int\limits_{X\times X} \varphi_r(x,z)\mu(dx)\nu(dz) \leqslant$$

$$\leqslant \int\limits_{X\times X} \max_{j=1,\dots,N+1} \varphi_j(x,z)\mu(dx)\nu(dz) \quad \forall\, \mu(\cdot) \in \{\nu\},\ \nu(\cdot) \in \{\nu\},$$

which proves (2.9.5). ∎

Remark 2.9.2 In fact, formula (2.9.5) generalizes the well-known property of maximization: the maximum of a sum does not exceed the sum of the maxima.

2.9.3 Existence Theorem

Good mathematicians see analogies.
Great mathematicians see analogies between analogies.
—Banach[33]

[33]Stefan Banach, (1892–1945), was a Polish mathematician who founded modern functional analysis and helped contributed to the development of the theory of topological vector spaces. Generally considered one of the most important and influential mathematicians of the twentieth century.

The central result of Chap. 2—the existence of a Berge–Pareto equilibrium in mixed strategies—is established.

Theorem 2.9.1 *If in game* (2.3.1) *the sets* $\mathrm{X}_i \in \operatorname{cocomp} \mathbb{R}^{n_i}$ *and* $f_i(\cdot) \in C(\mathrm{X})$ $(i \in \mathbb{N})$*, then there exists a Berge–Pareto equilibrium in mixed strategies.*

Proof Consider the auxiliary zero-sum two-player game

$$\Gamma^{\mathrm{a}} = \langle \{1, 2\}, \{\mathrm{X}, \mathrm{Z} = \mathrm{X}\}, \varphi(x, z) \rangle.$$

In the game Γ^{a}, the set X of strategies x chosen by player 1 (which seeks to maximize $\varphi(x, z)$) coincides with the set of strategy profiles of game (2.3.1); the set Z of strategies z chosen by player 2 (which seeks to minimize $\varphi(x, z)$) coincides with the same set X. A solution of the game Γ^{a} is *a saddle point* $(x^0, z^{\mathrm{B}}) \in \mathrm{X} \times \mathrm{X}$; for all $x \in \mathrm{X}$ and each $z \in \mathrm{X}$, it satisfies the chain of inequalities

$$\varphi\left(x, z^{\mathrm{B}}\right) \leqslant \varphi\left(x^0, z^{\mathrm{B}}\right) \leqslant \varphi\left(x^0, z\right).$$

Now, associate with the game Γ^a its mixed extension

$$\tilde{\Gamma}^{\mathrm{a}} = \langle \{1, 2\}, \{\mu\}, \{\nu\}, \varphi(\mu, \nu) \rangle,$$

where $\{\nu\}$ and $\{\mu\} = \{\nu\}$ denote the sets of mixed strategies $\nu(\cdot)$ and $\mu(\cdot)$ of players 1 and 2, respectively. The payoff function of player 1 is the expectation

$$\varphi(\mu, \nu) = \int\limits_{\mathrm{X}\times\mathrm{X}} \varphi(x, y)\mu(dx)\nu(dz).$$

The solution of the game $\tilde{\Gamma}^{\mathrm{a}}$ (the mixed extension of the game Γ^{a}) is also *a saddle point* (μ^0, ν^*) defined by the two inequalities

$$\varphi\left(\mu, \nu^*\right) \leqslant \varphi\left(\mu^0, \nu^*\right) \leqslant \varphi\left(\mu^0, \nu\right) \tag{2.9.7}$$

for any $\nu(\cdot) \in \{\nu\}$ and $\mu(\cdot) \in \{\nu\}$.

Sometimes, this pair (μ^0, ν^*) is called *the solution of the game* Γ^{a} *in mixed strategies.*

In 1952, Gliksberg [30] established the existence of a mixed strategy Nash equilibrium for a noncooperative game of $N \geqslant 2$ players. Applying this existence result to the zero-sum two-player game Γ^{a} as a special case, we obtain the following statement. In the game Γ^{a}, let the set $\mathrm{X} \subset \mathbb{R}^n$ be nonempty and compact and let the payoff function $\varphi(x, z)$ of player 1 be continuous on $\mathrm{X} \times \mathrm{X}$ (note that the continuity of $\varphi(x, z)$ is assumed in Lemma 2.8.1). Then the game Γ^{a} has a solution (μ^0, ν^*) defined by (2.9.7), i.e., there exists a saddle point in mixed strategies.

In view of (2.9.4), inequalities (2.9.7) can be written as

$$\begin{aligned} &\int\limits_{X\times X} \max_{j=1,\dots,N+1} \varphi_j(x, z)\mu(dx)\nu^*(dz) \\ &\quad\leqslant \int\limits_{X\times X} \max_{j=1,\dots,N+1} \varphi_j(x, z)\mu^0(dx)\nu^*(dz) \\ &\quad\leqslant \int\limits_{X\times X} \max_{j=1,\dots,N+1} \varphi_j(x, z)\mu^0(dx)\nu(dz) \end{aligned} \tag{2.9.8}$$

for all $\nu(\cdot) \in \{\nu\}$ and $\mu(\cdot) \in \{\nu\}$. Using the measure $\nu_i(dz_i) = \mu_i^0(dx_i)$ ($i \in \mathbb{N}$) (so that $\nu(dz) = \mu^0(dx)$) in the expression

$$\varphi(\mu^0, \nu) = \int\limits_{X\times X} \max_{j=1,\dots,N+1} \varphi_j(x, z)\mu^0(dx)\nu(dz),$$

we obtain $\varphi(\mu^0, \mu^0) = 0$ due to (2.9.6). Similarly, $\varphi(\nu^*, \nu^*) = 0$, and it follows from (2.9.7) that

$$\varphi(\mu^0, \nu^*) = 0. \tag{2.9.9}$$

The condition $\varphi(\mu^0, \mu^0) = 0$ and the chain of inequalities (2.9.7) give, by transitivity,

$$\varphi(\mu, \nu^*) = \int\limits_{X\times X} \max_{j=1,\dots,N+1} \varphi_j(x, z)\mu(dx)\nu^*(dz) \leqslant 0 \quad \forall\, \mu(\cdot) \in \{\nu\}.$$

In accordance with Proposition 2.9.2, then we have

$$0 \geqslant \int\limits_{X\times X} \max_{j=1,\dots,N+1} \varphi_j(x, z)\mu(dx)\nu^*(dz) \geqslant \max_{j=1,\dots,N+1} \int\limits_{X\times X} \varphi_j(x, z)\mu(dx)\nu^*(dz).$$

Therefore, for all $j = 1, \dots, N+1$,

$$\int\limits_{X\times X} \varphi_j(x, z)\mu(dx)\nu^*(dz) \leqslant 0 \quad \forall\, \mu(\cdot) \in \{\nu\}. \tag{2.9.10}$$

Consider two cases as follows.

Case I $(j = 1, \dots, N)$. Here, by (2.9.10), (2.9.6) and the normalization of $\nu(\cdot)$, we arrive at

$$
\begin{aligned}
0 \geqslant \int\limits_{X\times X} \varphi_i(x,z)\mu(dx)\nu^*(dz) &= \int\limits_{X\times X} [f_i(x\|z_i) - f_i(z)]\mu(dx)\nu^*(dz) \\
&= \int\limits_{X\times X_i} f_i(x\|z_i)\mu(dx)\nu_i^*(dz) - \int\limits_{X} f_i(z)\mu(dz)\cdot\int\limits_{X}\nu^*(dz) \\
&= f_i(\mu\|\nu_i^*) - f_i(\nu^*) \quad \forall\,\mu(\cdot)\in\{\nu\},\ i\in\mathbb{N}.
\end{aligned}
$$

By Definition 2.9.1, $\nu^*(\cdot)$ is a Berge equilibrium in mixed strategies in game (2.3.1).

Case II $(j = N+1)$. Again, using (2.9.10), (2.9.6) and the normalization of $\nu(\cdot)$ and $\mu(\cdot)$, we have

$$
\begin{aligned}
0 \geqslant \int\limits_{X\times X}\left[\sum_{r\in\mathbb{N}} f_r(x) - \sum_{r\in\mathbb{N}} f_r(z)\right]\mu(dx)\nu^*(dz) &= \int\limits_{X}\sum_{r\in\mathbb{N}} f_r(x)\mu(dx)\cdot\int\limits_{X}\nu^*(dz) \\
- \int\limits_{X}\mu(dx)\int\limits_{X}\sum_{r\in\mathbb{N}} f_r(z)\nu^*(dz) &= \sum_{r\in\mathbb{N}} f_r(\mu) - \sum_{r\in\mathbb{N}} f_r(\nu^*) \quad \forall\,\mu(\cdot)\in\{\nu^B\}.
\end{aligned}
$$

In accordance with Remark 2.9.1, the mixed strategy profile $\nu^*(\cdot) \in \{\nu\}$ of game (2.3.1) is a Pareto-maximal alternative in the multicriteria choice problem

$$
\tilde{\Gamma}_\nu = \left\langle \left\{\nu^{\mathrm{B}}\right\}, \{f_i(\nu)\}_{i\in\mathbb{N}}\right\rangle.
$$

Thus, we have proved that the mixed strategy profile $\nu^*(\cdot)$ in game (2.3.1) is a Berge equilibrium that satisfies Pareto maximality. Hence, by Definition 2.9.1, the mixed strategy profile $\nu^*(\cdot)$ is a Berge–Pareto equilibrium in game (2.3.1). ■

2.10 Linear-Quadratic Two-Player Game

Verba docent, exempla trahunt.[34]

Readers who studied Lyapunov's stability theory surely remember algebraic coefficient criteria. The whole idea of such criteria is to establish the stability of unperturbed motion without solving a system of differential equations, by using the signs of coefficients and/or their relationships. In this section of the book, we are endeavoring to propose a similar approach to equilibrium choice in noncooperative

[34]Latin "Words instruct, illustrations lead."

linear-quadratic two-player games. More specifically, our approach allows one to decide about the existence of a Nash equilibrium and/or a Berge equilibrium in these games based on the sign of quadratic forms in the payoff functions of players.

2.10.1 Preliminaries

Consider a noncooperative linear-quadratic two-player game described by

$$\Gamma_2 = \langle \{1, 2\}, \{\mathrm{X}_i = \mathbb{R}^{n_i}\}_{i=1,2}, \{f_i(x_1, x_2)\}_{i=1,2} \rangle.$$

A distinctive feature of Γ_2 is the absence of constraints on the strategy sets X_i: the strategies of player i can be any column vectors of dimension n_i, i.e., elements of the n_i-dimensional Euclidean space $\mathbb{R}^{n_i}$ with the standard Euclidean norm $\|\cdot\|$ and the scalar product. Let the payoff function of player i $(i = 1, 2)$ have the form

$$f_i(x_1, x_2) = x_1' A_i x_1 + 2x_1' B_i x_2 + x_2' C_i x_2 + 2a_i' x_1 + 2c_i' x_2, \tag{2.10.1}$$

where A_i and C_i are constant symmetric matrices, B_i is a constant rectangular matrix, and a_i and c_i are constant vectors, all of compatible dimensions; $'$ denotes transposition; det A denotes, the determinant of a matrix A. Henceforth, $A < 0$ $(>, \geqslant)$ means that the quadratic form $z'Az$ is negative definite (positive definite, positive semidefinite, respectively). We will adopt the following rules of differentiation of bilinear quadratic forms with respect to the vector argument [19, 27]:

$$\left[\frac{\partial}{\partial x_1}\left[x_1' B_i x_2\right] = B_i x_2\right] \Rightarrow \left[\frac{\partial}{\partial x_2}\left[x_1' B_i x_2\right] = B_i' x_1 \wedge \frac{\partial}{\partial x_1}\left[x_1' A_i x_1\right]\right.$$
$$\left. = 2A_i x_1 \wedge \frac{\partial}{\partial x_1}\left[a_1' x_1\right] = a_1\right],$$
$$\frac{\partial^2}{\partial x_i^2}\left[x_1' A_i x_1\right] = 2A_i. \tag{2.10.2}$$

For a scalar function $\Psi(x)$ of a k-dimensional vector argument x, *sufficient conditions* for

$$\max_{x \in \mathbb{R}^k} \Psi(x) = \Psi(x^*)$$

are

$$
\begin{aligned}
&(1)\quad \left.\frac{\partial \Psi(x)}{\partial x}\right|_{x=x^*} = \operatorname{grad} \Psi(x)|_{x=x^*} = 0_k, \\
&(2)\quad \left.\frac{\partial^2 \Psi(x)}{\partial x^2}\right|_{x=x^*} < 0,
\end{aligned} \tag{2.10.3}
$$

where 0_k denotes a zero column vector of dimension k.

2.10.2 Berge Equilibrium

For the payoff functions (2.10.1), relations (2.10.3) yield the following sufficient condition for the existence of a Berge equilibrium in the game Γ_2.

Proposition 2.10.1 *Assume that in the game* Γ_2

$$A_2 < 0, \quad C_1 < 0, \tag{2.10.4}$$

and

$$\det\left[C_1 - B_1' A_2^{-1} B_2\right] \neq 0. \tag{2.10.5}$$

Then the Berge equilibrium $x^{\mathrm{B}} = (x_1^{\mathrm{B}}, x_2^{\mathrm{B}})$ *is given by*

$$
\begin{aligned}
x_1^{\mathrm{B}} &= -A_2^{-1} B_2 \left[C_1 - B_1' A_2^{-1} B_2\right]^{-1} \left(B_1' A_2^{-1} a_2 - c_1\right) - A_2^{-1} a_2, \\
x_2^{\mathrm{B}} &= \left[C_1 - B_1' A_2^{-1} B_2\right]^{-1} \left(B_1' A_2^{-1} a_2 - c_1\right).
\end{aligned} \tag{2.10.6}
$$

Proof By definition, a strategy profile $(x_1^{\mathrm{B}}, x_2^{\mathrm{B}}) = x^{\mathrm{B}}$ is a Berge equilibrium in the game Γ_2 if and only if

$$
\begin{aligned}
\max_{x_2 \in \mathbb{R}^{n_2}} f_1\left(x_1^{\mathrm{B}}, x_2\right) &= f_1\left(x^{\mathrm{B}}\right), \\
\max_{x_1 \in \mathbb{R}^{n_1}} f_2\left(x_1, x_2^{\mathrm{B}}\right) &= f_2\left(x^{\mathrm{B}}\right).
\end{aligned} \tag{2.10.7}
$$

In view of (2.10.3) and(2.10.1), sufficient conditions for the first equality in (2.10.7) to hold can be written as

$$\left.\frac{\partial f_1\left(x_1^{\mathrm{B}}, x_2\right)}{\partial x_2}\right|_{x_2=x_2^{\mathrm{B}}} = 2B_1' x_1^{\mathrm{B}} + 2C_1 x_2^{\mathrm{B}} + 2c_1 = 0_{n_2},$$

$$\left.\frac{\partial^2 f_1\left(x_1^{\mathrm{B}}, x_2\right)}{\partial x_2^2}\right|_{x_2=x_2^{\mathrm{B}}} = 2C_1.$$

Similarly, for the second equality in (2.10.7) we obtain

$$\left.\frac{\partial f_2\left(x_1, x_2^{\mathrm{B}}\right)}{\partial x_1}\right|_{x_1=x_1^{\mathrm{B}}} = 2A_2 x_1^{\mathrm{B}} + 2B_2 x_2^{\mathrm{B}} + 2a_1 = 0_{n_1},$$

$$\left.\frac{\partial^2 f_2\left(x_1, x_2^{\mathrm{B}}\right)}{\partial x_1^2}\right|_{x_1=x_1^{\mathrm{B}}} = 2A_2.$$

In accordance with (2.10.4), the matrices C_1 and A_2 are negative definite and hence the Berge equilibrium $(x_1^{\mathrm{B}}, x_2^{\mathrm{B}})$ in the game Γ_2 satisfies the linear nonhomogeneous system of matrix equations

$$\begin{cases} A_2 x_1^{\mathrm{B}} + B_2 x_2^{\mathrm{B}} = -a_2, \\ B_1' x_1^{\mathrm{B}} + C_1 x_2^{\mathrm{B}} = -c_1. \end{cases} \tag{2.10.8}$$

Using the chain of implications $[A_2 < 0] \Rightarrow [\det A_2 \neq 0] \Rightarrow \left[\exists A_2^{-1}\right]$, we multiply the first equation in (2.10.8) on the left by A_2^{-1} to get

$$x_1^{\mathrm{B}} = -A_2^{-1} B_2 x_2^{\mathrm{B}} - A_2^{-1} a_2. \tag{2.10.9}$$

Substituting this expression into the second equation of system (2.10.8), one obtains

$$\left[C_1 - B_1' A_2^{-1} B_2\right] x_2^{\mathrm{B}} = -c_1 + B_1' A_2^{-1} a_2, \tag{2.10.10}$$

or

$$x_2^{\mathrm{B}} = \left[C_1 - B_1' A_2^{-1} B_2\right]^{-1} \left(B_1' A_2^{-1} a_2 - c_1\right). \tag{2.10.11}$$

Here we used the fact that

$$\left[\det\left[C_1 - B_1' A_2^{-1} B_2\right] \neq 0\right] \Rightarrow \left[\exists\left(C_1 - B_1' A_2^{-1} B_2\right)^{-1}\right];$$

formula (2.10.11) is easily derived upon multiplying both sides of (2.10.10) on the left by the matrix $\left[C_1 - B_1' A_2^{-1} B_2\right]^{-1}$. Finally, using the resulting expression for x_2^B in (2.10.9), we arrive at the first equality of system (2.10.6).

In the same fashion, it is possible to solve system (2.10.8) by multiplying the second equation by C_1^{-1}. This leads to

Proposition 2.10.2 *Assume inequalities* (2.10.4) *and*

$$\det\left[A_2 - B_2 C_1^{-1} B_1'\right] \neq 0 \tag{2.10.12}$$

hold in the game Γ_2. *Then the Berge equilibrium* $x^{\mathrm{B}} = (x_1^{\mathrm{B}}, x_2^{\mathrm{B}})$ *has the form*

$$x_1^{\mathrm{B}} = \left[A_2 - B_2 C_1^{-1} B_1'\right]^{-1} (B_2 C_1^{-1} c_1 - a_2),$$
$$x_2^{\mathrm{B}} = -C_1^{-1} B_1' \left[A_2 - B_2 C_1^{-1} B_1'\right]^{-1} \left(B_2 C_1^{-1} c_1 - a_2\right) - C_1^{-1} c_1.$$

Remark 2.10.1 System (2.10.8) has a unique solution for $A_2 < 0$ and $C_1 < 0$. The two explicit forms of this solution presented above are equivalent and can be reduced to each other.

2.10.3 Nash Equilibrium

In this section, we derive similar results for the Nash equilibrium in the game Γ_2. Instead of (2.10.7), we consider a Nash equilibrium $x^{\mathrm{e}} = (x_1^{\mathrm{e}}, x_2^{\mathrm{e}})$ defined by the two equalities

$$\max_{x_1 \in \mathbb{R}^{n_1}} f_1\left(x_1, x_2^{\mathrm{e}}\right) = f_1\left(x^{\mathrm{e}}\right), \qquad \max_{x_2 \in \mathbb{R}^{n_2}} f_2\left(x_1^{\mathrm{e}}, x_2\right) = f_2\left(x^{\mathrm{e}}\right). \tag{2.10.13}$$

The sufficient conditions for implementing (2.10.13) take the form

$$\operatorname{grad}_{x_1} f_1\left(x_1, x_2^{\mathrm{e}}\right)\big|_{x_1 = x_1^{\mathrm{e}}} = \left.\frac{\partial f_1\left(x_1, x_2^{\mathrm{e}}\right)}{\partial x_1}\right|_{x_1 = x_1^{\mathrm{e}}} = 2A_1 x_1^{\mathrm{e}} + 2B_1 x_2^{\mathrm{e}} + 2a_1 = 0_{n_1},$$

$$\operatorname{grad}_{x_2} f_2\left(x_1^{\mathrm{e}}, x_2\right)\big|_{x_2 = x_2^{\mathrm{e}}} = \left.\frac{\partial f_2\left(x_1^{\mathrm{e}}, x_2\right)}{\partial x_2}\right|_{x_2 = x_2^{\mathrm{e}}} = 2B_2' x_1^{\mathrm{e}} + 2C_2 x_2^{\mathrm{e}} + 2c_2 = 0_{n_2},$$

$$\left.\frac{\partial^2 f_1\left(x_1, x_2^{\mathrm{e}}\right)}{\partial x_1^2}\right|_{x_1 = x_1^{\mathrm{e}}} = 2A_1 < 0,$$

$$\left.\frac{\partial^2 f_2\left(x_1^{\mathrm{e}}, x_2\right)}{\partial x_2^2}\right|_{x_2 = x_2^{\mathrm{e}}} = 2C_2 < 0.$$

The first two conditions give the linear nonhomogeneous system of matrix equations

$$\begin{cases} A_1 x_1^e + B_1 x_2^e = -a_1, \\ B_2' x_1^e + C_2 x^e B_2 = -c_2. \end{cases}$$

As in Propositions 2.10.1 and 2.10.2, the conditions $A_1 < 0$ and $C_2 < 0$ allow us to establish the following results. The *onus probandi*[35] is left to the reader.

Proposition 2.10.3 *Assume the inequalities*

$$A_1 < 0, \quad C_2 < 0, \tag{2.10.14}$$

and

$$\det\left[C_2 - B_2' A_1^{-1} B_1\right] \neq 0 \tag{2.10.15}$$

hold in the game Γ_2. *Then the Nash equilibrium* $x^e = (x_1^e, x_2^e)$ *has the form*

$$x_1^e = -A_1^{-1} B_1 \left[C_2 - B_2' A_1^{-1} B_1\right]^{-1} (B_2' A_1^{-1} a_1 - c_2) - A_1^{-1} a_1,$$
$$x_2^e = \left[C_2 - B_2' A_1^{-1} B_1\right]^{-1} (B_2' A_1^{-1} a_1 - c_2).$$

Proposition 2.10.4 *Assume inequalities* (2.10.14) *and*

$$\det\left[A_1 - B_1 C_2^{-1} B_2'\right] \neq 0 \tag{2.10.16}$$

are satisfied in the game Γ_2. *Then the Nash equilibrium* $x^e = (x_1^e, x_2^e)$ *has the form*

$$x_1^e = \left[A_1 - B_1 C_2^{-1} B_2'\right]^{-1} \left(B_1 C_2^{-1} c_2 - a_1\right),$$
$$x_2^e = -C_2^{-1} B_2' \left[A_1 - B_1 C_2^{-1} B_2'\right]^{-1} \left(B_1 C_2^{-1} c_2 - a_1\right) - C_2^{-1} c_2.$$

2.10.4 Auxiliary Lemma

Despite the *negativa non probantur*[36] principle of Roman law, we will rigorously obtain a result useful for discarding the games without any Berge and/or Nash equilibria.

[35] Latin "The burden of proof."

[36] Latin "Negative statements are not proved."

Lemma 2.10.1 *In the game* Γ_2 *with* $A_1 > 0$, *there exists no* $\bar{x}_1$ *such that, for each fixed* x_2,

$$\max_{x_1 \in \mathbb{R}^{n_1}} f_1(x_1, x_2) = f_1(\bar{x}_1, x_2). \tag{2.10.17}$$

In other words, the payoff function f_1 *of the first player is not maximized in this game.*

Proof Let us "freeze" a certain strategy $x_2 \in \mathbb{R}^{n_2}$ of the second player. Then the payoff function of the first player can be written in the form

$$f_1(x_1, x_2) = x_1' A_1 x_1 + 2x_1' \varphi(x_2) + \psi(x_2),$$

where the column vector $\varphi(x_2)$ of dimension n_1 and the scalar function $\psi(x_2)$ depend on the frozen value x_2 only.

By the hypothesis of Lemma 2.10.1, the symmetric matrix A_1 is positive definite. In this case, the characteristic equation $\det[A_1 - E_{n_1}\lambda] = 0$ (where E_{n_1} denotes an identity matrix of dimensions $n_1 \times n_1$) has n_1 positive real roots owing to symmetry and, in addition,

$$x_1' A_1 x_1 \geqslant \lambda^* \|x_1\|^2 \quad \forall x_1 \in \mathbb{R}^{n_1}, \tag{2.10.18}$$

where $\lambda^* > 0$ is the smallest root among them. Thus, maximum (2.10.17) is not achieved if, for any large value $m > 0$, there exists a strategy $x_1(m, x_2) \in \mathbb{R}^{n_1}$ such that

$$f_1(x_1(m, x_2), x_2) > m.$$

Under (2.10.18), this inequality holds if

$$\lambda^* \|x_1(m, x_2)\|^2 + 2x_1'(m, x_2)\varphi(x_2) + \psi(x_2) > m. \tag{2.10.19}$$

Let us construct a solution $x_1(m, x_2)$ of inequality (2.10.19) in the form

$$x_1(m, x_2) = \beta e_{n_1}, \tag{2.10.20}$$

where the constant $\beta > 0$ will be specified below and e_1's the n_1-dimensional vector with all components equal to 1.

Substituting (2.10.20) into (2.10.19) yields the following inequality for β:

$$\lambda^* \beta^2 n_1 + 2\beta \left(e_{n_1}, \varphi(x_2)\right) + \psi(x_2) - m > 0.$$

Hence, for any constant

$$\beta > \beta_+ = \frac{\left|\left(e_{n_1}, \varphi(x_2)\right)\right| + \sqrt{\left(e_{n_1}, \varphi(x_2)\right)^2 + \lambda^* n_1 |\psi(x_2) - m|}}{\lambda^* n_1}$$

and corresponding strategy $x_1(m, x_2) = \beta e_{n_1}$ of the first player, we have

$$f_1(x_1(m, x_2), x_2) > m.$$

Thus, maximum (2.10.17) does not exist. ∎

Remark 2.10.2 In this case, the game Γ_2 with a matrix $A_1 > 0$ admits no Nash equilibria. In combination with Proposition 2.10.1, this shows that

1. the game Γ_2 with matrices $A_1 > 0$ or (and) $C_2 > 0$ and also $A_2 < 0$, $C_1 < 0$ that satisfies condition (2.10.5) has no Nash equilibria, but does have a Berge equilibrium defined by (2.10.6).
   In a similar way, it is easy to obtain the following.
2. If $A_2 < 0$, $C_1 < 0$, condition (2.10.5) or (2.10.12) holds and also $A_1 > 0$ or (and) $C_2 > 0$, then the game Γ_2 has a Berge equilibrium only.
3. If $A_1 < 0$, $C_2 < 0$, condition (2.10.15) or (2.10.16) holds and also $A_2 > 0$ or (and) $C_1 > 0$, then the game Γ_2 has a Nash equilibrium only.
4. If $A_1 > 0$ or (and) $C_2 > 0$ and also $A_2 > 0$ or (and) $C_1 > 0$, then the game Γ_2 has none of these equilibria.
5. If $A_2 < 0$, $C_1 < 0$, $A_1 < 0$, $C_2 < 0$ and also conditions (2.10.5) or (2.10.12) and (2.10.15) or (2.10.16) hold, then the game Γ_2 has both types of equilibrium.

2.10.5 Concluding Remarks

Thus, we have considered the noncooperative linear-quadratic two-player game without constraints ($\mathrm{X}_i = \mathbb{R}^{n_i}$, $i = 1, 2$) and with the payoff functions

$$f_1(x_1, x_2) = x_1' A_1 x_1 + 2x_1' B_1 x_2 + x_2' C_1 x_2 + 2a_1' x_1 + 2c_1' x_2,$$
$$f_2(x_1, x_2) = x_1' A_2 x_1 + 2x_1' B_2 x_2 + x_2' C_2 x_2 + 2a_2' x_1 + 2c_2' x_2.$$

Here $'$ denotes transposition; A_i and C_i are constant symmetric matrices of dimensions $n_1 \times n_1$ and $n_2 \times n_2$, respectively; B_i is a constant rectangular matrix of dimensions $n_1 \times n_2$; finally, a_i and c_i are constant vectors of dimensions n_1 and n_2, respectively ($i = 1, 2$).

Based on Propositions 2.10.1–2.10.4, we introduce the following *coefficient criteria* for the existence of Nash and Berge equilibria in the game Γ_2. *Par acquit de conscience,*[37] they are presented in form of Table 2.1.

How should one use Table 2.1? Just follow the three simple steps indicated below.

[37] French "For our peace of mind."

Table 2.1 Coefficient criteria of equilibria

					BE	NE	
Only one of the equilibria exists							
$A_1 > 0$	$A_2 < 0$	$C_1 < 0$		(2.10.5)	$\exists$	$\neg\exists$	$\forall\, C_2, B_i, a_i, c_i$
	$A_2 < 0$	$C_1 < 0$	$C_2 > 0$	(2.10.12)	$\exists$	$\neg\exists$	$\forall\, A_1, B_i, a_i, c_i$
$A_1 < 0$	$A_2 > 0$		$C_2 < 0$	(2.10.15)	$\neg\exists$	$\exists$	$\forall\, C_1, B_i, a_i, c_i$
$A_1 < 0$		$C_1 > 0$	$C_2 < 0$	(2.10.16)	$\neg\exists$	$\exists$	$\forall\, A_2, B_i, a_i, c_i$
None of the equilibria exists							
$A_1 > 0$	$A_2 > 0$				$\neg\exists$	$\neg\exists$	$\forall\, B_i, C_i, a_i, c_i$
$A_1 > 0$		$C_1 > 0$			$\neg\exists$	$\neg\exists$	$\forall\, A_2, C_2, B_i, a_i, c_i$
	$A_2 > 0$		$C_2 > 0$		$\neg\exists$	$\neg\exists$	$\forall\, A_1, C_1, B_i, a_i, c_i$
		$C_1 > 0$	$C_2 > 0$		$\neg\exists$	$\neg\exists$	$\forall\, A_1, A_2, B_i, a_i, c_i$
Both equilibria exist							
$A_1 < 0$	$A_2 < 0$	$C_1 < 0$	$C_2 < 0$	(2.10.5) and (2.10.15)	$\exists$	$\exists$	$\forall\, B_i, a_i, c_i$
$A_1 < 0$	$A_2 < 0$	$C_1 < 0$	$C_2 < 0$	(2.10.12) (2.10.16)	$\exists$	$\exists$	$\forall\, B_i, a_i, c_i$

Step 1. First, check the signs of the quadratic forms with the matrices A_1, A_2, C_1, and C_2. For example, suppose $A_1 < 0$, $C_2 < 0$ (both matrices are negative definite), while $A_2 > 0$ (i.e., A_2 is positive definite).

Step 2. Find the corresponding row in Table 2.1 (in our case, the conditions $A_1 < 0$, $C_2 < 0$ and $A_2 > 0$ are in row 3); then verify the nondegeneracy of the matrix (2.10.15) in column 5, i.e., the condition $\det\left[C_2 - B_2' A_1^{-1} B_1\right] \neq 0$.

Step 3. As shown in columns 6 and 7 of Table 2.1, the game Γ_2 with these matrices has no Berge equilibria, but has a Nash equilibrium for any matrices C_1, B_i and any vectors a_i, b_i of compatible dimensions. The explicit form of this Nash equilibrium is given in Proposition 2.10.3.

Chapter 3
The Golden Rule Under Uncertainty

Dubia plus torquent mala.[1]

As an English proverb goes, "Between the cup and lip a morsel may slip." This chapter is devoted to the Golden Rule under uncertainty, which accompanies every concept of equilibrium (in particular, Berge equilibrium).

3.1 Uncertainty and Types of Uncertainty

L'homme propose et dieu dispose.[2]

The harm and good of action are conditioned by
a totality of the circumstances.
—Kozma Prutkov[3]

What is uncertainty? How does uncertainty appear in economic and mechanical systems, sociology and decision-making? These questions are discussed below.

[1]Latin "Doubtful ills plague us worst." A quote from *Agamemnon* 480, by Seneca the Younger. In full Lucius Annaeus Seneca, (c. 4 B.C.–65 A.D.), was a Roman philosopher, statesman, orator, and tragedian.

[2]French "Man proposes but God disposes." This proverb emphasizes an influence of various contingencies on one's own plans, intentions, or even life.

[3]An English translation of a quote from [168, p. 230].

© Springer Nature Switzerland AG 2020

M. E. Salukvadze, V. I. Zhukovskiy, *The Berge Equilibrium: A Game-Theoretic Framework for the Golden Rule of Ethics*, Static & Dynamic Game Theory: Foundations & Applications, https://doi.org/10.1007/978-3-030-25546-6_3

3.1.1 Conceptual Meaning of Uncertainty

The following situation seems common for almost everybody: one needs to reach a place of employment from home. First of all, a person in this situation (henceforth called passenger) has to decide which means of transportation to use (subway, bus, tramcar, suburban electric train, etc.). Choosing a means of transportation (strategy), the passenger inevitably encounters incomplete and/or inaccurate information: delays or breakdowns of vehicles, sudden changes of schedule, strikes of drivers, weather fluctuations, crashes on routes, and so on. As noted by O. Holmes, "The longing for certainty... is in every human mind. But certainty is generally illusion."[4] At best the passenger knows the ranges of variation of these factors, without any probabilistic appraisals. Nevertheless, he/she has to make a decision! As a matter of fact, **the incomplete and/or inaccurate information about the conditions under which his/her strategy will be implemented** results in its inherent *uncertainty*. The uncertainty is caused by the embarrassment of choice.[5] We end this section by quoting Napoleon Bonaparte: "If the art of war were nothing but the art of avoiding risks, glory would become the prey of mediocre minds... I have made all the calculations; fate will do the rest."[6]

3.1.2 Uncertainty in Economic Systems

The following types of uncertainty are common in economic systems [25, 117, 118, 123, 125, 126, 129, 130, 175]:

1. uncertainty in economic indicators;
2. uncertainty about future disturbances, endogenous and exogenous;
3. uncertainty induced by mathematical modeling.

Pliny the Elder was used to say, "In these matters the only certainty is that there is nothing certain."[7] Among the sources and causes of uncertainty, we are identifying

---

[4]Oliver Wendell Holmes, Jr., by name The Great Dissenter, (1841–1935), was a justice of the United States Supreme Court, U.S. legal historian and philosopher who advocated judicial restraint.

[5]A house owner (H) asked a heating engineer (E) how much firewood will be required for a winter season. The latter requested information about the area of the house, the number of rooms, the location of windows, the number of fireplaces and also a mass of other technical details.

E: You will need from three to nineteen cubic meters of firewood.

H: Why is the answer so inaccurate?

E: Everything depends on how severe the coming winter will be. See [98, p. 41].

[6]Napoleon I, French in full Napoléon Bonaparte, (1769–1821), was a French general, first consul (1799–1804), and emperor of the French (1804–1814/1815).

[7]Gaius Plinius Secundus, (23–79 A.D.), well-known as Pliny the Elder, was a Roman writer, natural philosopher and scientist.

pure economic and also political factors. The latter include such unforeseen events as

- military conflicts and bans on exports and imports dictated by wartime (closure of borders, military operations in a country, migration, etc.);
- disposition of immovable and movable property (in particular, financial assets) on political grounds;
- inefficient economic policy and related ethnical and regional problems, polarization of different social groups.

An economic system, e.g., a firm, is often subject to sudden influence that is difficult to predict, namely, *exogenous disturbances* in the form of

- forces of nature (earthquakes, floods, storms, hurricanes, and other natural phenomena such as cold, ice, hail, thunder, drought, etc.);
- various accidents (fires, blasts, emissions of atomic and heat power plants, etc.);
- product price fluctuations caused by demand-supply dynamics, the varying number and range of supplies, purchase price fluctuations, the disruption of supplies;
- bad faith, low qualification or incompetence of economic partners, counteractions of rivals, acts of terrorism or racketeering;
- emergence or implementation of new technologies (investments made in technological progress and the resulting economic effects are often separated in time and therefore can be predicted on a long-term basis only);

as well as *endogenous disturbances* in the form of

- breakdown and failure of industrial equipment;
- unplanned additional cost and the losses of materials or energy during product storage and transportation;
- industrial accidents and employee illness;
- mistakes in personnel management;
- incorrect marketing or pricing policies (no sales, old stocks);
- mistakes in planning and product design;
- innovations suggested by employees.

New technologies and also anthropogenic and weather changes may cause uncertainty in *ecological systems.* In this context, we also mention epidemics among biological species and sudden pollution of their habitats [32, 147, 183].

3.1.3 Uncertainty in Mechanical Control Systems

In mechanical control systems, le vague[8] can be induced by exogenous disturbances, which lead to uncertainty in the forces affecting these systems [1, 107, 108, 153, 154, 169, 170, 174]. Atmospheric phenomena such as puff and varying air

[8]French "Uncertainty."

density can be sources of exogeneous disturbance. Incomplete information can be also a consequence of control program errors. Other disturbing factors include inaccurate initial data, the spread of characteristics and design parameters of a moving body, as well as gravitational and other perturbations. A primary cause of incomplete information in mechanical control systems consists in the inherent noises of measurement channels, which yield inaccurate motion parameters of the systems.

Information delays associated with finite periods of time needed to acquire and process measurement data also cause uncertainty in mechanical control systems.

3.1.4 Uncertainty in Decision-Making

As a matter of fact, uncertainty occurs *in decision-making too.*

First, in the course of mathematical modeling, since it often seems impossible to consider the whole variety of *constraints* on the uncontrolled and controlled parameters of the process under study within the current level and methods of science [6, 15–17, 132, 133, 135, 177, 178].

Second, in the understanding of all goals to be achieved by a controlled process: in many cases these goals are unclear or ambiguous, and their formalization has a subjective character defined by a player [7, 17, 139–142, 151].

Third, relationships between the process variables in the form of differential and/or algebraic equations may be inadequate for the process itself [9, 10, 143–146].

3.1.5 Classification of Uncontrolled Factors

In accordance with operations research [28], the strategies are the factors *controlled* by a player, i.e., chosen at his own discretion. Also, there exist *uncontrolled* factors [295, 296] affecting the outcome, which are not at the player's disposal (e.g., environmental conditions). Obviously, players should have some information about the values of uncontrolled factors.

Based on the awareness of players, operations research [28] divides the uncontrolled factors into three groups: *fixed, random*, and *uncertain.*

The fixed uncontrolled factors are the ones that have precisely known (given) values; e.g., a share sale is transacted if the buyers are informed about the exact price quotations. In this example the price quotations act as an uncontrolled factor.

The random uncontrolled factors are represented by random variables obeying given probability distributions.

Finally, the uncertain uncontrolled factors (hereinafter referred to as *uncertainty*) are deterministic or random variables with given value ranges or given classes of admissible probability distributions.

Among the above-mentioned groups, of crucial importance are the random and uncertain uncontrolled factors. In fact, the fixed uncontrolled factors do not differ from the other parameters of a mathematical model: their values are given and not varied at the wish of players. The random factors and uncertainty are also not affected by the players, but they take unknown values. As a rule, the random factors have a given probability distribution. In other words, if a random factor takes a finite set of values $y_1, \ldots, y_k$, then the players know the probabilities $p_1, \ldots, p_k$ associated with these values. For a random factor described by a continuous random variable, one deals with a given probability density function $p(x)$. In both cases, the optimization criteria (payoffs functions) are defined in terms of expectation.

Even less information is available about uncertainty. Whenever it represents a deterministic variable, we will assume that there is a given domain Y of its admissible values and consider the values $y \in Y$ only. If uncertainty is a random variable, then by assumption it belongs to a given class of admissible probability distributions.

Modern publications on economics distinguish three types of uncertainty as follows:

- **interval uncertainty**, for which the only available information consists of the ranges of admissible values (any probabilistic characteristics are absent for some reason). This type of uncertainty will be studied in our book;
- **random uncertainty**, as discussed above;
- **fuzzy uncertainty**, which is ruled by fuzzy mathematics, an intensively developing branch [99] founded by L. A. Zadeh.

3.1.6 Classification of Uncertainty

Using different sources of uncertainty, it is possible to suggest four groups of uncertainty [297–300], namely,

1^0. uncertainty caused by the purposeful actions of other persons who are not players;
2^0. uncertainty reflecting the fuzzy knowledge of all players about their goals;
3^0. uncertainty occurring due to an insufficient exploration of processes or characteristics;
4^0. uncertainty arising in the course of data acquisition, processing and transfer.

Let us discuss each group in detail.

1^0. Real control systems (especially economic, ecological, and social ones) often operate under conflict. In such systems, uncertainty is connected with the actions of conflicting parties, which are pursuing individual goals. Uncertainties of this type are called *strategic* [28] and cover any uncertainty caused by the actions of such goal-oriented parties actually not representing players. For example, the operation of an economic object can be influenced by other enterprises and firms, regardless of their economic relations with this object (say, an import product put in a market). These relations are incorporated into a mathematical model using several parameters with given ranges of variation (as the only information available to the players), e.g., the minimal and maximal quantity of products released in the market by an importer. The specific values of these parameters depend on the specific actions of other enterprises, i.e., the importer.

In this case, the parameters themselves constitute the uncertainty. Besides, this type of uncertainty also covers some exogenous disturbances such as the disruption and variation of the quantity (range) of supply, demand fluctuations for the products supplied by a given enterprise, the emergence of new technologies, etc.

2^0. A special status is assigned to the uncertainty that reflects the player's understanding of his goals. Roughly speaking, this uncertainty is not a controlled factor because each player chooses goals at his wish. However, if a player is unable to make choices or has some doubts, the resulting situation resembles the case of uncontrolled factors. For further analysis, we will assume that such a situation can be described by a set of criteria $f_1(x),\ldots, f_N(x)$, each maximized by a given player without a clear view of a single criterion. As demonstrated below, this player operates under the same conditions as uncontrolled factors. A similar state of affairs occurs if the player's criterion depends on the uncertainty taking a finite set of values: substituting these values into the criterion, we obtain a vector criterion with the same number of components as the number of uncertainty values.

Of course, an immediate issue is to design a uniform scalar criterion that would reflect the "desires" associated with all the elements of the vector criterion (the criteria convolution problem). The most widespread methods to convolute the criteria $f_1(x),\ldots, f_N(x)$ are (a) the weighted sum $\sum_{i=1}^{N} \alpha_i f_i(x)$ and (b) the weighted minimum $\min\limits_{1\leqslant i\leqslant N} \alpha_i f_i(x)$. In both cases, the weight coefficients must often satisfy the normalization conditions $\sum_{i=1}^{N} \alpha_i = 1$, where $\alpha_i > 0$ $(i = 1,\ldots, N)$. These coefficients can be used to transform the results into a universal measuring scale. Inaccurate knowledge about the player's goal is encoded by the uncertain values of α_i $(i = 1,\ldots, N)$.

However, such an approach, *first,* does not eliminate the existing uncertainty (yielding the uncertain parameters α_i) and, *second,* can be used if the uncertainty takes a finite set of values. If this set is infinite or even has the cardinality of the continuum, then the approach is called into question.

Finally, the relationship between the criterion values and uncertainty can be determined by different factors such as weather conditions, anthropogenic changes, a sudden appearance of competitors, price fluctuations in the market, and other exogenous disturbances [301–305].

3^0. An increasing amount of information, and consequently a rising number of studied objects (in particular, their gradual complication), is also increasing the existing uncertainty due to an insufficient exploration of processes and characteristics, compelling us *emere catullum in sacco*.[9]

[9]Latin "To buy a cat in the sack." Meaning to buy something sight unseen or without knowing anything about the object.

The growing uncertainty describes well the fact that, in the course of development, any fundamental or applied science[10] is posing many more problems than it actually solves. Decision-making based on incomplete data can be interpreted as conflict with nature. Note that this source of uncertainty has a subjective character in some sense. Indeed, such uncertainty depends on accumulated experience, the completeness of modern scientific knowledge, and access to new information. For example, flight missions to Mars are intended to eliminate blanks in what is known about this planet and will surely lead to new unexpected problems. The same applies to the appearance of new technologies.

4^0. Data acquisition, processing and transfer directly involve computers for different calculations. In practice, we have to be content with approximate solutions, reconciling ourselves with the element of uncertainty in the solutions. Rough information occurs as the result of many factors—computational errors, inaccurate data transfer as well as the limited precision of numerical representations and measurements, to name a few.

Solutions obtained by a numerical method are always approximate. There exist several sources of errors for numerical solutions, such as disagreement between a mathematical model and the real phenomenon,[11] inaccurate initial data, and imprecision of numerical methods (e.g., roundoff errors for arithmetical and other operations).

Even hand calculation [179] involves the roundoff effect, which is associated with a finite number of decimals used for different operations. This problem is equally important for computer systems and people.

There are several reasons explaining this situation.

First, the amount of computational job that can be performed manually is considerably smaller compared with that of modern computer systems.

Second, hand calculation allows us to observe roundoff effects and undertake necessary measures for avoiding mistakes.

Third, hand calculation often employs variable-length numbers, which are adjusted to eliminate rough errors; by contrast, computer calculation deals with floating-point numbers of fixed length.

Fourth, hand calculation allows us to estimate the maximal error induced by rounding. Such estimation is very costly for computer calculation, requiring the use of statistical estimates.

Practical calculations have led to several popular methods to use computer systems for error detection and estimation. The latter is vital: prior to writing programs for a computer system, one needs to assess the expected accuracy.

---

[10] *In jeder besonderen Naturlehre nur so viel eigentliche Wissenschaft angetroffen werden könne, als darin Mathematik anzutreffen ist.* (German "In every department of physical science there is only so much science, properly so-called, as there is mathematics.") A quote from *Metaphysische Anfangsgründe der Naturwissenschaft* (Metaphysical Foundations of Natural Science) by Immanuel Kant (1724–1804), an outstanding German philosopher.

[11] "If my husband would ever meet a woman on the street who looked the women in his paintings, he would fall over in a dead faint." —Mrs. Picasso.

Perhaps, the simplest and most successful approach to the roundoff problem is to define the range of admissible values. Then each quantity can be described by two values, i.e., the maximal and minimal ones. In a certain sense, each quantity is replaced by a range that covers its exact value. Different operations on quantities correspond to new ranges defined from the original ranges using appropriate rounding. Therefore, each stage of calculations has reliable limits for the correct value of a given quantity. These issues form the content of *interval analysis* [187, 256].

A direct application of interval methods in calculation processes allows us to impose limits on the solutions of problems with initial data belonging to given ranges. The resulting intervals also incorporate the roundoff errors caused by calculations. For precise initial data, these intervals contain the exact solution of an original problem and hence interval analysis gives the approximation and roundoff errors.

To pursue the path of two-sided estimation is a very promising approach, as it solves the issue of resulting errors. Two-sided estimation is proceeding with the so-called *interval arithmetics* [187], which operates with intervals instead of values. More specifically, it is assumed that initial data, intermediate calculations and final results belong to some intervals. Thus, a main element of interval calculus is an interval $[a, b]$ (also termed range) defined as a set of real values x such that $\{x \in \mathbb{R} | a \leqslant x \leqslant b\}$.

Generally, when a value x is specified for computer systems, it is assumed that x incorporates an error. In terms of interval analysis, this means that in a computer system a value x belongs to an interval.

With an interval algorithm used for solving a posed problem, we may construct an interval function that contains the exact solution. In this case, the accuracy of the resulting solution is taken into account and it is also possible to perform a prior analysis of roundoff errors.

Thus, we have presented a list of factors causing uncertainty in different systems, which does not claim to be exhaustive. But this brief discussion demonstrates that uncertainty should be considered for the elementary and difficult problems, particularly, for conflicts, in which the interests of many parties are clashing with one another and undergoing the influence of uncertain disturbances. Even in simple market problems these disturbances might not be neglectable. How can one account for them in noncooperative games under uncertainty (NGU), especially in *dynamic* (time-varying) *controlled systems*? A possible approach based on an appropriate modification of the principle of guaranteed result [28, 29] was developed for multicriteria choice problems in [295] and for conflicts in [51, 289]. An alternative framework using the principle of minimax regret [267, 268] is presented in the book [66] (though for the noncooperative setup only).

In the mathematical models of CGUs, the influence of several uncertain factors is assessed by the specific values $y_1, \ldots, y_m$ of corresponding scalar parameters. These values y_j $(j = 1, ..., m)$ describe for instance the quantity of imported products (put in the market), their unit price, the number of people suffered from an accident or fire, the delays of negotiated supplies, and so on. We will also adopt a column vector $y = (y_1, \dots, y_m)$, with a set of values denoted by $\mathrm{Y} \subset \mathbb{R}^m$.

Our book addresses uncertainties that cannot be described by statistical methods. This situation occurs at least in two cases as follows:

- the probabilistic characteristics of uncertainty exist in principle, but statistical data are not available (e.g., sudden anthropogenic accidents like the Chernobyl and Fukushima Daiichi nuclear disasters) or are very expensive to acquire;
- the uncertainty y does not have any probability distribution.

The uncertainty of the second type is well illustrated by the following example; for details see [18, p. 21]. For a clothing factory, production planning for a next year heavily affects future profits, which in turn depends on the length y of women's skirts. However, taking into account the vagaries of fashion and female logic dictating fashion trends, any probabilistic characteristics for the parameter y would be hardly expected. All one can do is to establish some obvious limits of length variations. In [18, p. 21], E. Ventsel' called such uncontrollable factors ill uncertainty due to an unpredictable character of their specific realization. This type of uncertainty will be considered below.

Once again, we emphasize that recent publications on competitive economics have identified three types of uncertainty, namely, interval uncertainty (studied in this book), random uncertainty (based on some probabilistic characteristics of a variable y distributed on a set Y), and fuzzy uncertainty (based on the concept of a fuzzy set introduced by Zadeh in [99]).

Thus, throughout this chapter it will be assumed that the players make their decisions using a value set Y of uncertain parameters y only, i.e., there exist no probability characteristics for y. Therefore, choosing their strategies, the players are expecting any realization of y from the set Y.

3.2 General Notions and Obtained Results

3.2.1 Saddle Point and Maximin

Maximin is the problem of finding the minimum amount of fabric required for sewing a maxi skirt.[12]

A single-criterion choice problem under uncertainty (SCPU) is described by a triplet

$$\langle\, \mathrm{X}_1, \mathrm{Y}, f_1(x_1, y)\,\rangle, \tag{3.2.1}$$

[12]A Russian translation from a humorous mathematical glossary in [34, p. 204].

where $\mathrm{X}_1 \subseteq \mathbb{R}^n$ denotes the set of alternatives x_1 selected by a decision maker (DM); $\mathrm{Y} \subseteq \mathbb{R}^m$ is the set of uncertain factors y; finally, $f_1(x_1, y)$ is an objective function defined on $\mathrm{X}_1 \times \mathrm{Y}$ that is maximized by the DM under any realization of $y \in \mathrm{Y}$.

For problem (3.2.1), game theory considers at least two types of solutions:

- *first*, the saddle point $(x_1^0, y^0) \in \mathrm{X}_1 \times \mathrm{Y}$, which is defined by the equalities

$$\max_{x_1 \in \mathrm{X}_1} f_1\left(x_1, y^0\right) = f_1\left(x_1^0, y^0\right) = \min_{y \in \mathrm{Y}} f_1\left(x_1^0, y\right); \tag{3.2.2}$$

- *second*, the maximin f_1^{g} and the maximin alternatives $x_1^{\mathrm{g}} \in \mathrm{X}_1$ suggested by A. Wald [282] in 1939, which are given by

$$f_1^{\mathrm{g}} = \min_{y \in \mathrm{Y}} f_1\left(x_1^{\mathrm{g}}, y\right) = \max_{x_1 \in \mathrm{X}_1} \min_{y \in \mathrm{Y}} f_1(x_1, y). \tag{3.2.3}$$

Remark 3.2.1 The chain of equalities (3.2.2) will be used below to formalize the guaranteed balanced equilibrium as a solution concept for the noncooperative N-player game under uncertainty (NGU), the first type of the guaranteed equilibria developed in this book.

3.2.2 Auxiliary Results from Operations Research, Theory of Multicriteria Choice and Game Theory

Some background material from operations research, theory of multicriteria choice and game theory (Nash and Berge equilibria) is provided.

Operations Research

Whilst we deliberate how to begin a thing,
it grows too late to begin it.
—Quintilian

Here we present some auxiliary results from operations research, multicriteria choice problems and noncooperative games. The following fact was established in [14, p. 160].

Proposition 3.2.1 *Assume that*

1^0. *the scalar function $F(x, y)$ is continuous on the product of compact sets $\mathrm{X} \subset \mathbb{R}^n$ and $\mathrm{Y} \subset \mathbb{R}^m$, where Y is also convex;*

2^0. *for each $x \in \mathrm{X}$, the function $F(x, y)$ is strictly convex in y on the set Y, i.e., for each $x \in \mathrm{X}$ and any $y^{(1)}, y^{(2)} \in \mathrm{Y}$,*

$$F\left(x, \alpha y^{(1)} + (1-\alpha) y^{(2)}\right) < \alpha F\left(x, y^{(1)}\right) + (1-\alpha) F\left(x, y^{(2)}\right)$$

for any $\alpha \in (0, 1)$.

Then the m-dimensional vector function $y(x) : \mathrm{X} \to \mathrm{Y}$ defined by

$$\min_{y \in \mathrm{Y}} F(x, y) = F(x, y(x)) \quad \forall x \in \mathrm{X} \tag{3.2.4}$$

is also continuous.

Theory of Multicriteria Choice

Vom Himmel fordert er
Die schönsten Sterne –
Und von der Erde
—Jede höchste Lust.[13]

We provide some background material from the theory of multicriteria choice that will be needed below. For two vectors $f^{(k)} = (f_1^{(k)}, \dots, f_N^{(k)})$ $(k = 1, 2)$, introduce the notations:

$$\begin{aligned}
&[f^{(1)} = f^{(2)}] \Longleftrightarrow \left[f_i^{(1)} = f_i^{(2)} \ (i \in \mathbb{N})\right];\\
&[f^{(1)} \neq f^{(2)}] \Longleftrightarrow \rceil(f^{(1)} = f^{(2)});\\
&[f^{(1)} \geqslant f^{(2)}] \Longleftrightarrow \left[f_i^{(1)} \geqslant f_i^{(2)} \ (i \in \mathbb{N})\right];\\
&[f^{(1)} \geq f^{(2)}] \Longleftrightarrow (f^{(1)} \geqslant f^{(2)}) \wedge (f^{(1)} \neq f^{(2)});\\
&[f^{(1)} \ngeq f^{(2)}] \Longleftrightarrow \rceil(f^{(1)} \geq f^{(2)});\\
&[f^{(1)} > f^{(2)}] \Longleftrightarrow \left[f_i^{(1)} > f_i^{(2)} \ (i \in \mathbb{N})\right];\\
&[f^{(1)} \not> f^{(2)}] \Longleftrightarrow \rceil(f^{(1)} > f^{(2)}).
\end{aligned} \tag{3.2.5}$$

In the sequel, an n-dimensional vector $x \in \mathrm{X}$ will be called an *alternative*, while an m-dimensional vector $y \in \mathrm{Y}$ will be called an uncertain factor, more specifically, a *pure uncertainty* if $y \in \mathrm{Y}$ and a *counter-situation* if $y(\cdot) \in \mathrm{Y}^{\mathrm{X}}$, where Y^{X} denotes the set of all m-dimensional vector functions $y(x)$ defined on the set X and taking values in the set Y. Further analysis will be confined to the counter-situations $y(\cdot) : \mathrm{Y} \to \mathrm{X}$ that are continuous on X, i.e., $y(\cdot) \in C(\mathrm{X}, \mathrm{Y})$.

Definition 3.2.1 For an N-criteria choice problem $\Gamma = \langle\, \mathrm{Y}, f(x, y) \,\rangle$ with a fixed alternative $x^* \in \mathrm{X}$,

(a) a pure uncertainty $y_S \in \mathrm{Y}$ is Slater minimal in Γ if

$$f(x^*, y) \not< f(x^*, y_S) \quad \forall\, y \in \mathrm{Y};$$

---

[13]German "The fairest stars from Heaven he requireth,
From Earth the highest raptures and the best."
A quote from *Faust*, Prologue in Heaven (Mephistopheles), by J. von Goethe. Johann Wolfgang von Goethe, (1749–1832), was a German poet, playwright, novelist, scientist, statesman, theatre director, critic, and amateur artist. Considered the greatest German literary figure of the modern era.

(b) a pure uncertainty $y_P \in Y$ is Pareto minimal in Γ if

$$f(x^*, y) \nleq f(x^*, y_P) \quad \forall\, y \in Y.$$

For an N-criteria choice problem $\Gamma(x) = \langle\, Y^X, f(x, y) \,\rangle$ that is defined for all $x \in X$,

(c) a counter-situation $y_S(x) \in Y^X$ is Slater minimal if, for each $x \in X$,

$$f(x, y) \nless f(x, y_S(x)) \quad \forall\, y \in Y;$$

(d) a counter-situation $y_P(x) \in Y^X$ is Pareto minimal if, for each $x \in X$,

$$f(x, y) \nleq f(x, y_P(x)) \quad \forall y \in Y.$$

Proposition 3.2.2

(a) *If in the problem* $\Gamma(x^*) = \langle\, Y, f(x^*, y) \,\rangle$ *the set* Y *is compact and the function* $f(x^*, y)$ *is continuous on* Y*, then the set* Y_S *of Slater-minimal pure uncertainties* y_S *is nonempty and compact* [152, p. 137].

(b) *The pure uncertainty* $y_S \in Y$ *that satisfies the condition*

$$\min_{y \in Y} \sum_{i \in \mathbb{N}} \alpha_i f_i(x^*, y) = \sum_{i \in \mathbf{N}} \alpha_i f_i(x^*, y_S) \tag{3.2.6}$$

for some $\alpha_i = \text{const} \geqslant 0$ *and* $\sum_{i \in \mathbb{N}} \alpha_i > 0$ *is Slater minimal in the problem* $\Gamma(x^*)$ [152, p. 68–69].

(c) *The pure strategy* $y_P \in Y$ *that satisfies*

$$\min_{y \in Y} \sum_{i \in \mathbb{N}} \alpha_i f_i(x^*, y) = \sum_{i \in \mathbb{N}} \alpha_i f_i(x^*, y_P) \tag{3.2.7}$$

for some $\alpha_i = \text{const} > 0$ $(i \in \mathbb{N})$ *is Pareto minimal in the problem* $\Gamma(x^*)$ [152, p. 71].

(d) *In addition, it follows from* (3.2.5) *that the set* $Y_S \supseteq Y_P$ *of Slater-minimal uncertainties is the set of the Pareto-minimal pure uncertainties* y_P *in the problem* $\Gamma(x^*)$.

Nash Equilibrium

> On ne peut pas savoir tout, il faut se contenter de comprendre.[14]

[14]French "To know everything is impossible, so one should be content with his/her own comprehension." An English translation of a quote from *Notes on the Personality of Belinskii* by Ivan A. Goncharov, (1812–1891), a Russian novelist.

Now, consider a noncooperative N-player game of the form

$$\langle\, \mathbb{N}, \{\mathrm{X}_i\}_{i\in\mathbb{N}}, \{f_i[x]\}_{i\in\mathbb{N}}\,\rangle, \tag{3.2.8}$$

where $\mathbb{N} = \{1, \ldots, N\}$ denotes the set of players and $\mathrm{X}_i \subseteq \mathbb{R}^{n_i}$ is the set of pure strategies x_i of player i ($i \in \mathbb{N}$).

In game (3.2.8), the players do not build any coalitions and each player i chooses his strategy $x_i \in \mathrm{X}_i$ simultaneously with the other players, which yields a strategy profile $x = (x_1, \ldots, x_N) \in \mathrm{X} = \prod_{i\in\mathbb{N}} \mathrm{X}_i$. A scalar payoff function $f_i[x]$ of player i is a priori defined on the set $\mathrm{X} \subseteq \mathbb{R}^n$ ($n = \sum_{i\in\mathbb{N}} n_i$); its value in a specific strategy profile gives the payoff of player i. At a conceptual level, each player i in game (3.2.8) seeks for choosing a strategy $x_i \in \mathrm{X}_i$ that would maximize his payoff in a specific strategy profile x.

In 1949, J. Nash formalized a solution of game (3.2.8), suggesting a strategy profile known today as Nash equilibrium; see [257].

Definition 3.2.2 A strategy profile $x^{\mathrm{e}} = (x_1^{\mathrm{e}}, \ldots, x_N^{\mathrm{e}}) \in \mathrm{X}$ is called a Nash equilibrium in game (3.2.8) if

$$\max_{x_i\in\mathrm{X}_i} f_i[x^{\mathrm{e}}||x_i] = f_i[x^{\mathrm{e}}] \quad (i \in \mathbb{N});$$

as before, $[x^{\mathrm{e}}||x_i] = [x_1^{\mathrm{e}}, \ldots, x_{i-1}^{\mathrm{e}}, x_i, x_{i+1}^{\mathrm{e}}, \ldots, x_N^{\mathrm{e}}]$.

Remark 3.2.2 In accordance with Definition 3.2.2, for compact sets X_i and continuous payoff functions $f_i[x]$ on X, the set X^{e} of all Nash equilibria in game (3.2.8) is a compact (possibly empty) subset of X [51, p. 174].

The next result was proved in [22, p. 93] using Brouwer's fixed-point theorem.

Theorem 3.2.1 *Consider game* (3.2.8) *under the assumptions that*

(1°) *the sets* X_i *are convex and compact;*
(2°) *each payoff function* $f_i[x]$ *is continuous on* X *and concave in the variable* x_i *for any fixed values of the other variables* ($i \in \mathbb{N}$).

Then there exists a Nash equilibrium in this game.

Now, consider a game (3.2.8) in which the sets X_i are compact and the payoff functions $f_i[x]$ are continuous on X. Associate with this game (3.2.8) its mixed extension

$$\langle\, \mathbb{N}, \{\nu_i\}_{i\in\mathbb{N}}, \{f_i[\nu]\}_{i\in\mathbb{N}}\,\rangle, \tag{3.2.9}$$

where $\mathbb{N}$ is the same as in (3.2.8); $\{\nu_i\}$ denotes the set of mixed strategies of player i, i.e., each $\nu_i(\cdot)$ represents a probability measure—a nonnegative scalar countably-additive function defined on the Borel σ-algebra of all subsets of the compact set X_i that is normalized by unity; $\nu(dx) = \nu_1(dx_1)\ldots\nu_N(dx_N)$ is the product measure; $\{\nu\}$ designates the set of all mixed strategy profiles $\nu(\cdot)$; finally, the payoff function of player i in game (3.2.9),

$$f_i[\nu] = \int\limits_{\mathrm{X}} f_i[x]\nu(dx) = \int\limits_{\mathrm{X}_1} \cdots \int\limits_{\mathrm{X}_N} f_i[x]\nu_N(dx_N)\cdots\nu_1(dx_1),$$

is defined as the expectation $f_i[x]$ for the payoff function of game (3.2.8) (using Fubini's theorem on switching the order of integration).

Definition 3.2.3 A mixed strategy profile $\nu^{\mathrm{e}}(\cdot) \in \{\nu\}$ is called a Nash equilibrium in game (3.2.9) if

$$\max_{\nu_i(\cdot)\in\{\nu_i\}} f_i[\nu^{\mathrm{e}}||\nu_i] = f_i[\nu^{\mathrm{e}}] \quad (i \in \mathbb{N}),$$

where $\nu^{\mathrm{e}}||\nu_i = \nu_1^{\mathrm{e}}(dx_1)\ldots\nu_{i-1}^{\mathrm{e}}(dx_{i-1})\nu_i(dx_i)\nu_{i+1}^{\mathrm{e}}(dx_{i+1})\cdots\nu_N^{\mathrm{e}}(dx_N)$ and $\nu^{\mathrm{e}}(dx) = \nu_1^{\mathrm{e}}(dx_1)\cdots\nu_N^{\mathrm{e}}(dx_N)$.

The following result was obtained in [22, p. 117–119] using Gliksberg's fixed-point theorem.

Theorem 3.2.2 *Consider game* (3.2.8) *under the assumptions that the sets* X_i *are convex and compact and the payoff functions* $f_i[x]$ *are continuous on* $\mathrm{X} = \prod_{i\in\mathbb{N}} \mathrm{X}_i$. *Then in this game there exists a mixed strategy Nash equilibrium.*

We conclude this section with an English translation of a remarkable quote from the book [10, p. 170]: "Intuition is not adapted to comprehend gaming opposition... Mixed strategies and Nash equilibrium are two revolutionary concepts that are described in each textbook, yet remain in the shadow of world view."

The next section introduces one possible concept of guaranteed equilibrium in a noncooperative game under uncertainty and establishes its existence in mixed strategies under standard assumptions of mathematical game theory.

Berge Equilibrium

As the call, so the echo.
—Russian proverb [127]

In 1994, V. Zhukovskiy and his postgraduate K. Vaisman formalized the Berge equilibrium as a solution concept for game (3.2.8); see the publications [11, 12, 302].

Definition 3.2.4 A strategy profile $x^{\mathrm{B}} = (x_1^{\mathrm{B}}, \ldots, x_N^{\mathrm{B}}) \in \mathrm{X}$ is called a Berge equilibrium in game (3.2.8) if

$$\max_{x\in\mathrm{X}} f_i\left[x||x_i^{\mathrm{B}}\right] = f_i\left[x^{\mathrm{B}}\right] \quad (i \in \mathbb{N}),$$

where $[x||x_i^{\mathrm{B}}] = [x_1, \ldots, x_{i-1}, x_i^{\mathrm{B}}, x_{i+1}, \ldots, x_N]$.

Theorem 3.2.1 and Property 2.6.1 directly imply

Proposition 3.2.3 *Consider game* (3.2.8) *with* $\mathbb{N} = \{1, 2\}$ *under the assumptions that, for each* $i = 1, 2$,

(1°) *the sets* X_i *are convex and compact;*
(2°) *the payoff functions* $f_i[x]$ $(i = 1, 2)$ *are continuous on* X, $f_1[x]$ *is concave in* x_2 *and* $f_2[x]$ *concave in* x_1 *for each fixed strategy of the other player.*

Then there exists a Berge equilibrium in this game.

Denote by X^B the set of all Berge equilibria in game (3.2.8). By Property 2.3.1, X^B is a (possibly, empty) compact set if the payoff functions $f_i[x]$ are continuous and the sets X_i $(i \in \mathbb{N})$ are compact.

Definition 3.2.5 A strategy profile $x^* \in X$ is called a Berge–Pareto equilibrium in game (3.2.8) if

first, x^* is a Berge equilibrium in (3.2.8), i.e.,

$$\max_{x_i \in X_i} f_i[x^* || x_i] = f_i[x^*] \quad (i \in \mathbb{N}),$$

and *second*, x^* is a Pareto-maximal alternative in the N-criteria choice problem

$$\langle X^B, \{f_i[x]\}_{i \in \mathbb{N}} \rangle,$$

i.e., for all $x \in X^B$, the system of inequalities

$$f_i[x] \geqslant f_i[x^*] \quad (i \in \mathbb{N}),$$

with at least one strict inequality, is inconsistent.

Now, let us pass to the mixed extension (2.9.1) of game (3.2.8) (see Sect. 2.9.1).

Definition 3.2.6 A mixed strategy profile $\nu^*(\cdot) \in \{\nu\}$ is called a Berge–Pareto equilibrium in mixed strategies in game (3.2.9) (equivalently, a Berge–Pareto equilibrium in the mixed extension of game (3.2.8)) if

first, $\nu^*(\cdot)$ is a Berge equilibrium in game (2.9.1), i.e., conditions (2.9.2) are satisfied,

and *second*, $\nu^*(\cdot)$ is a Pareto-maximal alternative in the N-criteria choice problem

$$\langle \{\nu^B\}, \{f_i[\nu]\}_{i \in \mathbb{N}} \rangle,$$

i.e., for all $\nu(\cdot) \in \{\nu^B\}$ the system of inequalities

$$f_i[\nu] \geqslant f_i[\nu^*] \quad (i \in \mathbb{N}),$$

with at least one strict inequality, is inconsistent.

The following result is a stronger analog of Theorem 3.2.2, which was proved in Sect. 2.9.3.

Theorem 3.2.3 *If in game* (3.2.8) *the sets* $\mathrm{X}_i \in \operatorname{comp} \mathbb{R}^{n_i}$ *and the functions* $f_i[\cdot] \in C(\mathrm{X})$ $(i \in \mathbb{N})$, *then this game possesses a Berge–Pareto equilibrium in mixed strategies.*

3.3 Balanced Equilibrium as an Analog of Saddle Point

Faber est suae quisque fortunae.[15]

3.3.1 Analogs of Saddle Point: The Idea and Formalization

Nothing obstructs seeing as much as a viewpoint.
—Don-Aminado[16]

The concept of a Slater-guaranteed balanced Berge equilibrium is formalized for the noncooperative N-player game under uncertainty.

As a matter of fact, the first type of equilibrium discussed below was suggested by V. Zhukovskiy in 1994 in the book [93, p. 233] for noncooperative games under uncertainty and later used by him for different types of equilibria [56] and also for cooperative games [52]. The whole idea is very simple: replace minimization in (3.2.2) by a vector minimum (in the sense of Slater, Pareto, Borwein, Geoffrion, or the A-minimum [295]) and replace maximization by an equilibrium design (in the sense of Nash, Berge, threats and counter-threats, or active equilibrium [54]). This approach was employed by K. Vaisman in a series of publications [280, 281]. One of his concepts will be presented below in Definition 3.3.1.

Consider a noncooperative N-player game with pure strategies and pure uncertainties, defined by

$$\langle\, \mathbb{N}, \{\mathrm{X}_i\}_{i\in\mathbf{N}}, \mathrm{Y}, \{f_i(x, y)\}_{i\in\mathbf{N}}\,\rangle. \tag{3.3.1}$$

In (3.3.1), $\mathbb{N} = \{1, \dots, N\}$ denotes the set of players; $\mathrm{X}_i \subseteq \mathbb{R}^{n_i}$ is the set of pure strategies x_i of player i; $\mathrm{Y} \subseteq \mathbb{R}^m$ gives the set of pure uncertainties y.

15Latin "Each man is the maker of his own fortune." This phrase appeared in *Letters to Ceasar* I by Gaius Sallustius Crispus, (86–35 B.C.), a Roman historian and politician. Considered as one of the great Latin literary stylists.

16Aminad P. Shpolyanskii, well-known in the Western world as Don–Aminado, (1888–1957), was a Russian émigré poet and satirist.

In this game, no coalitions are allowed and each player i chooses his strategy x_i simultaneously with the other players, which yields a strategy profile $x = (x_1, \ldots, x_N) \in \mathrm{X} = \prod_{i\in\mathbb{N}} \mathrm{X}_i$. Regardless of their choice, some pure uncertainty $y \in \mathrm{Y}$ arises in game (3.3.1). For each player i ($i \in \mathbb{N}$), a payoff function $f_i(x, y)$ is defined on all such pairs $(x, y) \in \mathrm{X} \times \mathrm{Y}$.

At a conceptual level, each player i in game (3.3.1) chooses a pure strategy $x_i \in \mathrm{X}_i$ in order to maximize his payoff $f_i(x, y)$ under any unpredictable realization of the pure uncertainty $y \in \mathrm{Y}$.

Definition 3.3.1 A pair $(\bar{x}^{\mathrm{B}}, \bar{f}^{\mathrm{S}}) \in \mathrm{X} \times \mathbb{R}^N$ is called a Slater-guaranteed balanced Berge equilibrium in game (3.3.1) if there exists an uncertain factor $y_{\mathrm{S}} \in \mathrm{Y}$ such that

(1°) the pure strategy profile x^{B} is a Berge equilibrium in the game

$$\langle\, \mathbb{N}, \{\mathrm{X}_i\}_{i\in\mathbb{N}}, \{f_i(x, y_{\mathrm{S}})\}_{i\in\mathbb{N}}\,\rangle \tag{3.3.2}$$

(which is obtained from (3.3.1) by setting $y = y_{\mathrm{S}}$), i.e., by Definition 3.2.4,

$$\max_{x\in\mathrm{X}} f_i\left(x\|x_i^{\mathrm{B}}, y_{\mathrm{S}}\right) = f_i\left(x^{\mathrm{B}}, y_{\mathrm{S}}\right) \quad (i \in \mathbb{N}); \tag{3.3.3}$$

(2°) the uncertain factor y_{S} is Slater minimal in the N-criteria choice problem

$$\langle\, \mathrm{Y}, \{f_i\left(x^{\mathrm{B}}, y\right)\}_{i\in\mathbb{N}}\,\rangle \tag{3.3.4}$$

(which is obtained from (3.3.1) by setting $x = x^{\mathrm{B}}$), i.e., by Definition 3.2.1,

$$f\left(x^{\mathrm{B}}, y\right) \not< f\left(x^{\mathrm{B}}, y_{\mathrm{S}}\right) \quad \forall y \in \mathrm{Y}; \tag{3.3.5}$$

(3°) the pair $(\bar{x}^{\mathrm{B}}, \bar{y}_{\mathrm{S}})$ is Slater-maximal in the N-criteria choice problem

$$\left\langle \left\{x^{\mathrm{B}}, y_{\mathrm{S}}\right\}, \{f_i(x, y)\}_{i\in\mathbb{N}} \right\rangle \tag{3.3.6}$$

(in which each element $(x^{\mathrm{B}}, y_{\mathrm{S}})$ of the set $\{x^{\mathrm{B}}, y_{\mathrm{S}}\}$ satisfies (3.3.3) and (3.3.5)), i.e., the vector

$$\bar{f}^{\mathrm{S}} = f\left(\bar{x}^{\mathrm{B}}, \bar{y}_{\mathrm{S}}\right) \not< f(x, y) \quad \forall (x, y) \in \left\{x^{\mathrm{B}}, y_{\mathrm{S}}\right\}. \tag{3.3.7}$$

In this case, x^{B} is called a Slater-guaranteeing profile in game (3.3.1) and $\bar{f}^{\mathrm{S}}$ is called a guaranteed vector payoff.

3.3.2 *Pro et contra of Balanced Equilibrium*[17]

Many intricate phenomena
are naturally clarified within
the framework of game theory.
—Vorobiev [24, p. 97].

The advantages of Slater-guaranteed balanced Berge equilibria are discussed.

Let us outline the *benefits* of this solution concept for the NGUs.

First, using their strategies from a profile $\bar{x}^{\mathrm{B}}$, the players are assured to obtain a guaranteed vector payoff $\bar{f}^{\mathrm{S}}$. In accordance with (3.3.5), for $x^{\mathrm{B}} = \bar{x}^{\mathrm{B}}$ the elements $f_i(\bar{x}^{\mathrm{B}}, y)$ $(i \in \mathbb{N})$ cannot be all simultaneously smaller than the corresponding elements $f_i(\bar{x}^{\mathrm{B}}, \bar{y}_{\mathrm{S}})$ $(i \in \mathbb{N})$, and by (3.3.7) this is the highest (Slater-maximal) guarantee among all the possible guarantees $f(x^{\mathrm{B}}, y_{\mathrm{S}})$ achieved on any pairs $(x^{\mathrm{B}}, y_{\mathrm{S}})$ that satisfy conditions 1° and 2° of Definition 3.3.1.

Second, the equilibrium $(\bar{x}^{\mathrm{B}}, \bar{f}^{\mathrm{S}})$ aims at "the maximum opposition to uncertainty," i.e., it is based on the principle of guaranteed result (which explains its "guaranteed" character).

Third, this solution concept is wide enough, since it contains main solution concepts from game theory (saddle point, Berge equilibrium) and theory of multicriteria choice (Slater optimum) as special cases. Note that we may also adopt other optimality principles (Pareto, Geoffrion, Borwein, cone optimality). Connections between such approaches were considered in [295].

Fourth, the notion of Slater-guaranteed equilibrium is well fitted for practical design and theoretical analysis (in particular, existence proofs). Indeed, introduce a dummy player with the set of strategies $y \in \mathrm{X}_{N+1} = \mathrm{Y}$ and the payoff function

$$\varphi_3(x, y) = -\sum_{i \in \mathbb{N}} \alpha_i f_i(x, y),$$

with some

$$\alpha_i = \text{const} \geqslant 0 \ (i \in \mathbb{N}) \ \wedge \sum_{i \in \mathbb{N}} \alpha_i > 0.$$

Add two other dummy players with the payoff functions

$$\varphi_1(x, z, y) = \max\{f_i(x \| z_i, y) - f_i(z, y) \ (i \in \mathbb{N}), \sum_{j \in \mathbb{N}} f_j(x, y) - \sum_{j \in \mathbb{N}} f_j(z, y)\}$$

[17] Latin "For and against."

and

$$\varphi_2(x, z, y) = -\varphi_1(x, z, y) = \varphi(x, z, y).$$

Let the *strategies* of player I be the profiles $x \in \mathrm{X}$ of game (3.3.1) while the strategies of player II be the profiles $z \in \mathrm{Z} = \mathrm{X}$ (of the same game (3.3.1)). As his strategy, player III chooses $y \in \mathrm{Y}$. Now consider the auxiliary three-player game

$$\langle\, \{\mathrm{I}, \mathrm{II}, \mathrm{III}\}, \{\mathrm{X}, \mathrm{Z}, \mathrm{Y}\}, \{\varphi_i(x, z, y)\}_{i=1,2,3} \,\rangle. \tag{3.3.8}$$

A Nash equilibrium $(x^{\mathrm{e}}, z^{\mathrm{e}}, y^{\mathrm{e}})$ in game (3.3.8) is given by the three conditions

$$\begin{aligned}
&\max_{x\in\mathrm{X}} \varphi_1(x, z^{\mathrm{e}}, y^{\mathrm{e}}) = \varphi_1(x^{\mathrm{e}}, z^{\mathrm{e}}, y^{\mathrm{e}}),\\
&\max_{z\in\mathrm{X}} \varphi_2(x^{\mathrm{e}}, z, y^{\mathrm{e}}) = \varphi_2(x^{\mathrm{e}}, z^{\mathrm{e}}, y^{\mathrm{e}}),\\
&\max_{y\in\mathrm{Y}} \varphi_3(x^{\mathrm{e}}, z^{\mathrm{e}}, y) = \varphi_3(x^{\mathrm{e}}, z^{\mathrm{e}}, y^{\mathrm{e}}).
\end{aligned} \tag{3.3.9}$$

Using the form of the functions $\varphi_i(x, z, y)$ $(i = 1, 2, 3)$, from the third equality one can see that $y^{\mathrm{e}} = y_{\mathrm{S}}$ and the pair $(x^{\mathrm{e}}, z^{\mathrm{e}})$ yields a saddle point of the zero-sum game

$$\langle\, \mathrm{X}, \mathrm{Z} = \mathrm{X}, \varphi(x, z, y_{\mathrm{S}}) = \varphi_1(x, z, y_{\mathrm{S}}) \,\rangle.$$

In combination with Theorem 2.8.1, this result implies the following. If there exists a Nash equilibrium in game (3.3.8), then $(z^{\mathrm{e}}, f^{\mathrm{S}} = f(x^{\mathrm{e}}, z^{\mathrm{e}}, y^{\mathrm{e}}))$ is a Slater-guaranteed balanced Berge equilibrium (condition 3° of Definition 3.3.1 becomes non-binding).

3.3.3 Games with Separated Payoff Functions

The simplest example is more convincing
than the most eloquent sermons.
—Seneca

The existence of a Slater-guaranteed balanced Berge equilibrium is established for the noncooperative two-player game under uncertainty with separated payoff functions that have a special concavity property.

Consider a particular case of game (3.3.1), described by

$$\langle\, \mathbb{N}, \{\mathrm{X}_i\}_{i\in\mathbb{N}}, \mathrm{Y}, \{f_i(x, y) = \varphi_i(x) + \psi_i(y)\}_{i\in\mathbb{N}} \,\rangle, \tag{3.3.10}$$

which differs from (3.3.1) only in the payoff functions $f_i(x, y) = \varphi_i(x) + \psi_i(y)$ $(i \in \mathbb{N})$. In other words, the payoff functions are split into two components associated with the strategy profiles $x \in \mathrm{X}$ and uncertain factors $y \in \mathrm{Y}$, respectively.

This separation of the functions $f_i(x, y)$ allows us to propose a constructive design method for a Slater-guaranteed balanced Berge equilibrium (see Definition 3.3.1), which proceeds from an independent analysis of the noncooperative N-player game

$$\Gamma_x = \langle\, \mathbb{N}, \{\mathrm{X}_i\}_{i\in\mathbb{N}}, \{\varphi_i(x)\}_{i\in\mathbb{N}}\,\rangle \tag{3.3.11}$$

and the N-criteria choice problem

$$\Gamma_y = \langle\, \mathrm{Y}, \{\psi_i(y)\}_{i\in\mathbb{N}}\,\rangle. \tag{3.3.12}$$

The ensuing exposition will use two N-dimensional vectors, $\varphi = (\varphi_1, \ldots, \varphi_N)$ and $\psi = (\psi_1, \ldots, \psi_N)$, as well as the following auxiliary and obvious statement.

Lemma 3.3.1 *For any constant N-dimensional vector $a = (a_1, \ldots, a_N)$,*
(*a*) *the system of inequalities*

$$\psi_i^{(1)} < \psi_i^{(2)} \quad (i \in \mathbb{N})$$

is inconsistent if and only if this is the case for the system of inequalities

$$\psi_i^{(1)} + a_i < \psi_i^{(2)} + a_i \quad (i \in \mathbb{N});$$

(*b*) *the following two systems of inequalities are equivalent:*

$$\left[\varphi_i^{(1)} \leqslant \varphi_i^{(2)} \quad (i \in \mathbb{N})\right] \Leftrightarrow \left[\varphi_i^{(1)} + a_i \leqslant \varphi_i^{(2)} + a_i \quad (i \in \mathbb{N})\right].$$

With Lemma 3.3.1, a Slater-guaranteed balanced Berge equilibrium in game (3.3.10) can be obtained by the following algorithm.

Step 1. For the N-criteria choice problem (3.3.12), construct the set $\mathrm{Y_S} \subseteq \mathrm{Y}$ of the Slater-minimal alternatives y_{S} and also the set of outcomes $\psi(\mathrm{Y_S}) = \bigcup_{y\in\mathrm{Y_S}} \psi(y)$, i.e., the system of inequalities

$$\psi_i(y) < \psi_i(y_S) \quad (i \in \mathbb{N})$$

must be inconsistent for any $y \in \mathrm{Y}$ and each $y_S \in \mathrm{Y_S}$ (then by Lemma 3.3.1a the system of inequalities

$$\varphi_i(x) + \psi_i(y) < \varphi_i(x) + \psi_i(y_S) \quad \forall x \in \mathrm{X}, y \in \mathrm{Y}\ (i \in \mathbb{N})$$

is also inconsistent, which gives condition 2° of Definition 3.3.1).

Step 2. For game (3.3.11), find the set $\mathrm{X}^{\mathrm{B}} \subseteq \mathrm{X}$ of all Berge equilibria $x^{\mathrm{B}} \in \mathrm{X}$ using the inequalities

$$\varphi_i(x||x_i^{\mathrm{B}}) \leqslant \varphi_i(x^{\mathrm{B}}) \quad \forall x \in \mathrm{X} \ (i \in \mathbb{N}),$$

and then construct the set $\varphi(\mathrm{X}^{\mathrm{B}}) = \bigcup_{x \in \mathrm{X}^{\mathrm{B}}} \varphi(x)$ (then by Lemma 3.3.1b the system of inequalities

$$\varphi_i(x||x_i^{\mathrm{B}}) + \psi_i(y_{\mathrm{S}}) \leqslant \varphi_i(x^{\mathrm{B}}) + \psi_i(y_{\mathrm{S}}) \quad \forall y_{\mathrm{S}} \in \mathrm{Y}, x \in \mathrm{X} \ (i \in \mathbb{N}),$$

holds, which matches condition 1° of Definition 3.3.1).

Step 3. Construct the sum of sets

$$\begin{aligned} \varphi(\mathrm{X}^{\mathrm{B}}) + \psi(\mathrm{Y}_{\mathrm{S}}) &= \left(\varphi(\mathrm{X}^{\mathrm{B}}) + \psi(y_{\mathrm{S}}) \mid y_{\mathrm{S}} \in \mathrm{Y}_{\mathrm{S}}\right) \\ &= \left(\varphi(x^{\mathrm{B}}) + \psi(\mathrm{Y}_{\mathrm{S}}) \mid x^{\mathrm{B}} \in \mathrm{X}^{\mathrm{B}}\right) \\ &= \left(\varphi(x^{\mathrm{B}}) + \psi(y_{\mathrm{S}}) \mid x^{\mathrm{B}} \in \mathrm{X}^{\mathrm{B}},\ y_{\mathrm{S}} \in \mathrm{Y}_{\mathrm{S}}\right). \end{aligned}$$

Step 4. Find the Slater-maximal alternative $(\bar{x}^{\mathrm{B}}, \bar{y}_{\mathrm{S}})$ in the N-criteria choice problem

$$\langle\, \mathrm{X}^{\mathrm{B}} \times \mathrm{Y}_{\mathrm{S}}, \{\varphi_i(x) + \psi_i(y)\}_{i \in \mathbb{N}} \,\rangle,$$

i.e., calculate $(\bar{x}^{\mathrm{B}}, \bar{y}_{\mathrm{S}})$ as follows: for all $x^{\mathrm{B}} \in \mathrm{X}^{\mathrm{B}}$ and all $y_{\mathrm{S}} \in \mathrm{Y}_{\mathrm{S}}$, the system of inequalities

$$\varphi_i(\bar{x}^{\mathrm{B}}) + \psi_i(\bar{y}_{\mathrm{S}}) < \varphi_i(x^{\mathrm{B}}) + \psi_i(y_{\mathrm{S}}) \quad (i \in \mathbb{N})$$

is inconsistent, which satisfies condition 3° of Definition 3.3.1.

The resulting strategy profile $(\bar{x}^{\mathrm{B}}, \varphi(\bar{x}^{\mathrm{B}}) + \psi(\bar{y}_{\mathrm{S}}))$ is a Slater-guaranteed balanced Berge equilibrium in game (3.3.10).

The suggested algorithm leads to the following existence theorem of a Slater-guaranteed balanced Berge equilibrium in game (3.3.10).

Theorem 3.3.1 *Consider game* (3.3.10) *with* $\mathbb{N} = \{1, 2\}$ *under the assumptions that*

(1) the sets X_i *and* Y *are compact and* X_i *are also convex;*
(2) the scalar functions $\varphi_i(x)$ *and* $\psi_i(y)$ *are continuous on* $\mathrm{X} = \prod_{i \in \{1,2\}} \mathrm{X}_i$ *and* Y*, respectively;*
(3) the functions $\varphi_i(x)$ *are concave in* x_j $(i, j = 1, 2;\ j \neq i)$ *for any fixed values of the other variables* $(i \in \{1, 2\})$*.*

Then there exists a Slater-guaranteed balanced Berge equilibrium.

Proof For proving this result, we follow the four steps of the above-mentioned algorithm.

Step 1. In problem (3.3.12), the set $\mathrm{Y_S}$ is a nonempty and compact (see Theorem 3.3.1) and hence (by the continuity of $\psi_i(y)$ on Y) $\psi(\mathrm{Y_S})$ is also a compact subset of $\mathbb{R}^N$ $(N = 2)$.

Step 2. In game (3.3.11), the set $\mathrm{X^B}$ of all Berge equilibria is a nonempty and compact (see Theorem 3.2.1 and Property 2.3.1). Then the set $\varphi(\mathrm{X}^B) = \bigcup_{x \in \mathrm{X}^B} \varphi(x)$ is also compact because the components of the N-dimensional vector function $\varphi(x)$ are continuous on X.

Step 3. From Steps 1 and 2 of this proof it follows that the product $\mathrm{X^B} \times \mathrm{Y_S}$ and the sum $\varphi(\mathrm{X^B}) + \psi(\mathrm{Y_S})$ are also compact sets.

Step 4. Consider the bicriteria choice problem

$$\langle\, \mathrm{X^B} \times \mathrm{Y_S}, \{\varphi_i(x) + \psi_i(y)\}_{i \in \mathbb{N}} \,\rangle. \tag{3.3.13}$$

The set $\mathrm{X^B} \times \mathrm{Y_S}$ is compact and the components of the N-dimensional vector function $\varphi(x) + \psi(y)$ are continuous on $\mathrm{X^B} \times \mathrm{Y_S}$. Therefore, there exists a Slater-maximal alternative $(\bar{x}^{\mathrm{B}}, \bar{y}_{\mathrm{S}}) \in \mathrm{X^B} \times \mathrm{Y_S}$ in problem (3.3.13), i.e., for any $(x^{\mathrm{B}}, y_{\mathrm{S}}) \in \mathrm{X^B} \times \mathrm{Y_S}$ the system of inequalities

$$\varphi_i(x^{\mathrm{B}}) + \psi_i(y_{\mathrm{S}}) \;>\; \varphi_i(\bar{x}^{\mathrm{B}}) + \psi_i(\bar{y}_{\mathrm{S}}) \quad (i \in \mathbb{N})$$

is inconsistent.

The resulting pair

$$\left(\bar{x}^{\mathrm{B}}, \bar{f}^{\mathrm{S}} = f\left(\bar{x}^{\mathrm{B}}, \bar{y}_{\mathrm{S}}\right) = \varphi_i\left(\bar{x}^{\mathrm{B}}\right) + \psi_i\left(\bar{y}_{\mathrm{S}}\right)\right)$$

is a Slater-guaranteed balanced Berge equilibrium in game (3.3.10). ■

Example 3.3.1 Consider a noncooperative two-player game under uncertainty with separated payoff functions given by

$$\langle\, \{1, 2\}, \{\mathrm{X}_i = [-1, 1]\}_{i=1,2}, \mathrm{Y}, \{f_i(x, y) = -x_j^2 + 2x_1x_2 + y_i\}_{i,j=1,2;i \neq j} \,\rangle, \tag{3.3.14}$$

in which $x = (x_1, x_2)$, $y = (y_1, y_2)$, and $\mathrm{Y} = \{y = (y_1, y_2) \mid y_1^2 + y_2^2 \leqslant 1\}$. We will construct a Slater-guaranteed balanced Berge equilibrium in this game using the suggested algorithm. In accordance with the latter, extract from (3.3.14) the noncooperative two-player game

$$\Gamma_x = \langle \{1, 2\}, \{\mathrm{X}_i = [-1, 1]\}_{i=1,2}, \{\varphi_i(x) = -x_j^2 + 2x_1x_2\}_{i,j=1,2;i \neq j} \rangle \tag{3.3.15}$$

and also the bicriteria choice problem

$$\Gamma_y = \langle\, \mathrm{Y}, \{\psi_i(y) = y_i\}_{i=1,2}\, \rangle, \tag{3.3.16}$$

where $\mathrm{Y} = \{y = (y_1, y_2) \mid y_1^2 + y_2^2 \leqslant 1\}$.

Step 1. The set Y represents a disc with center $(0, 0)$ and radius $R = 1$ in the space $\mathbb{R}^2$, and it coincides with the shaded set $\psi(\mathrm{Y})$ in Fig. 3.1. Then the Slater minima in problem (3.3.16) are the points lying on the circumference in the third quadrant; see the solid arc in Fig. 3.2. This set can be described as

$$\psi(\mathrm{Y_S}) = \Big\{ y_\mathrm{S} = \Big(y_1^{(\mathrm{S})}, y_2^{(\mathrm{S})}\Big) \mid y_1^{(\mathrm{S})} = -R\cos\beta,\, y_2^{(\mathrm{S})} = -R\sin\beta \quad \forall\beta \in [0, \pi/2]\Big\}.$$

Step 2. Game (3.3.15) was studied in [68, pp. 177–178]. The set of all Berge equilibria (Fig. 3.3) is

$$\mathrm{X^B} = \{(\alpha, \alpha) \mid \forall\alpha = \mathrm{const} \in [-1, 1]\},$$

and the corresponding payoffs (Fig. 3.4) are

$$\varphi(\mathrm{X^B}) = \Big\{(\alpha^2, \alpha^2) \mid \forall\alpha = \mathrm{const} \in [-1, 1]\Big\} = OC.$$

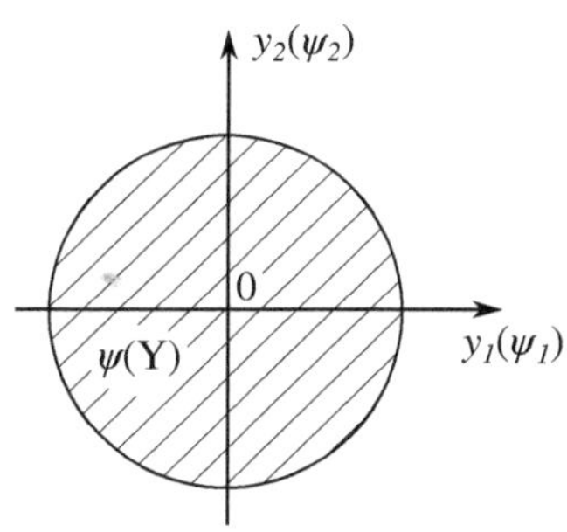

Fig. 3.1 Set Y

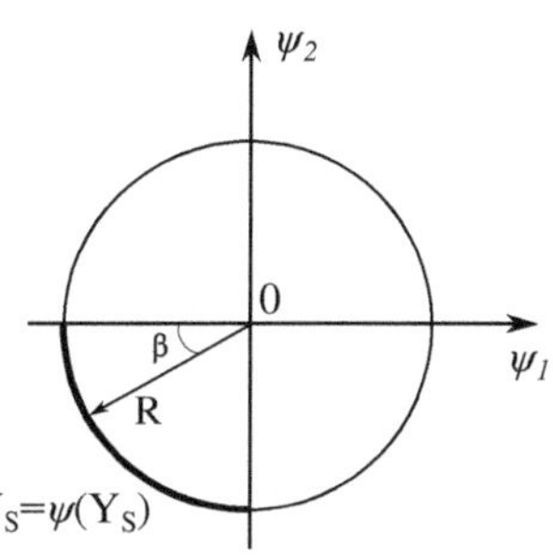

Fig. 3.2 Slater minima

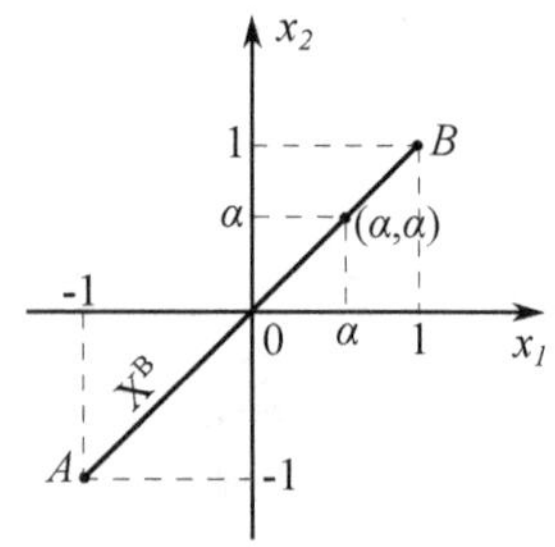

Fig. 3.3 Berge equilibria

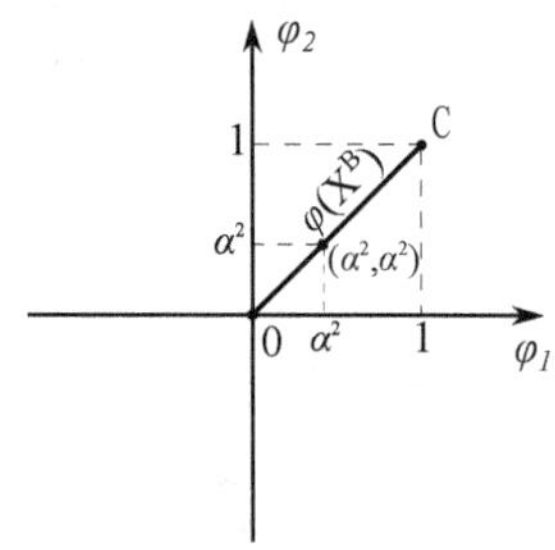

Fig. 3.4 Corresponding payoffs

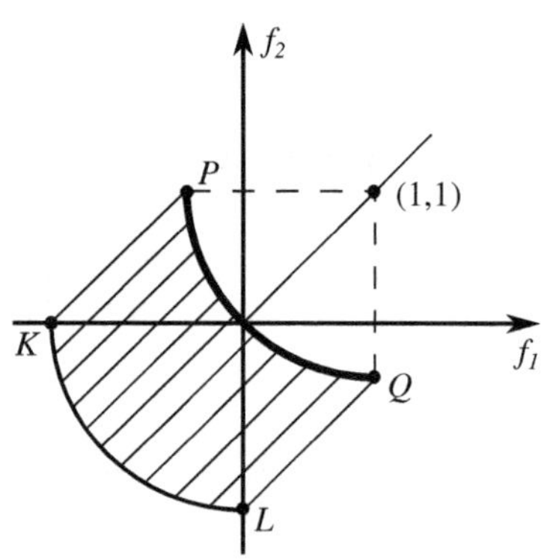

Fig. 3.5 Set $\varphi(\mathrm{X}^B) + \psi(\mathrm{Y}_S)$

Thus, every point (α, α) of the bisecting segment AB is a Berge equilibrium in game (3.3.15). The corresponding payoffs $\varphi(\mathrm{X}^{\mathrm{B}})$ form the segment OC, as illustrated in Fig. 3.4.

Step 3. Then

$$\varphi(\mathrm{X}^B) + \psi(\mathrm{Y}_S) = \{OC + \psi(\mathrm{Y}_S)\} = OC + \{y_S \mid \forall \beta \in [0, \pi/2]\} = KPQL$$

(see the shaded domain in Fig. 3.5).

Step 4. The Slater minima of the set $KPQL$ make up a quarter of the circumference (the bold arc PQ in Fig. 3.5), i.e.,

$$PQ = \{1 - \cos\beta, 1 - \sin\beta \mid \beta \in [0, \pi/2]\}.$$

Each pair $((1, 1), \ (1 - \cos\beta, 1 - \sin\beta))$ with any $\beta \in [0, \pi/2]$ is a Slater-guaranteed balanced Berge equilibrium in game (3.3.14).

Thus, the suggested algorithm dictates both players to choose $x_1^B = x_2^B = 1$ (the Slater-maximal Berge equilibrium $B = (1, 1)$ in game (3.3.14), see Fig. 3.4). In this case, the players obtain the guaranteed vector payoff $(1 - \cos\beta, 1 - \sin\beta) = \bar{f}^B$, i.e., for any $y \in \mathrm{Y}$ the payoffs $f_i((1, 1), y)$ cannot be simultaneously smaller than the corresponding payoffs $\bar{f}_i^{\mathrm{B}}$ $(i = 1, 2)$. And this is the highest guarantee (in the sense of Slater) among all the guarantees $f(x^{\mathrm{B}}, y_{\mathrm{S}}) = (x_1^{\mathrm{B}} - \cos\beta, x_2^{\mathrm{B}} - \sin\beta)$ for all $\beta \in [0, \pi/2]$ and any other Berge equilibria x^{B} in game (3.3.15).

3.3.4 Existence in Mixed Strategies and One Remark

> Not the existence theorem is the valuable thing,
> but the construction carried out in the proof.
> Mathematics is, as Brouwer sometimes says,
> more action than theory.
> —Weyl[18]

The existence of a Slater-guaranteed balanced Berge equilibrium in mixed strategies is established for the noncooperative N-player game under uncertainty.

Remark 3.3.1 The auxiliary noncooperative game without uncertainty (3.3.9), (3.3.8) allows us to establish the existence of a Slater-guaranteed balanced Berge equilibrium in mixed strategies in game (3.3.1) under uncertainty. Let us associate with game (3.3.1) its mixed extension

$$\tilde{\Gamma} = \langle\, \mathbb{N}, \{\nu_i\}_{i \in \mathbb{N}}, \{\mu\}, \{f_i(\nu, \mu)\}_{i \in \mathbb{N}} \,\rangle, \tag{3.3.17}$$

where, like in (3.3.1), $\mathbb{N} = \{1, \dots, N\}$ denotes the set of players. Assuming that the sets X_i $(i \in \mathbb{N})$ and Y are compact and the payoff functions $f_i(x, y)$ are continuous on $\mathrm{X} \times \mathrm{Y}$, we will construct the sets $\{\nu_i\}$ of mixed strategies $\nu_i(\cdot)$ of player i. Specifically, $\nu_i(\cdot)$ is a probability measure on the Borel σ-algebra of all subsets of the compact set X_i.

The mixed uncertainties $\mu(\cdot)$ represent probability measures on the compact set Y. Let $\{\mu\}$ denote the set of such uncertainties. The mixed strategy profiles $\nu(\cdot)$ are the product measures $\nu(dx) = \nu_1(dx_1) \cdots \nu_N(dx_N)$. Denote by $\{\nu\}$ the set of such mixed strategy profiles. In a similar fashion, define the product measures $\eta(dxdy) = \nu(dx)\mu(dy)$; then the payoff function of player i in game (3.3.17) is the expectation

[18]Hermann Weyl, (1885–1955), was a German American mathematician with widely varied contributions in pure mathematics and theoretical physics.

$$f_i(\nu, \mu) = \int\limits_X \int\limits_Y f_i(x, y)\mu(dy)\nu(dx) = \int\limits_Y \int\limits_X f_i(x, y)\nu(dx)\mu(dy).$$

Recall that $f = (f_1, \ldots, f_N)$. The following concept is an analog of Definition 3.3.1 for game (3.3.17).

Definition 3.3.2 A pair $(\tilde{\nu}^{\mathrm{B}}(\cdot), \tilde{f}^{\mathrm{S}}) \in \{\nu\} \times \mathbb{R}^N$ is called a Slater-guaranteed balanced Berge equilibrium (*SGBBE*) in the mixed extension (3.3.17) (or an *SGBBE* in mixed strategies in game (3.3.1) under uncertainty) if there exists a mixed uncertainty $\mu_{\mathrm{S}}(\cdot) \in \{\mu\}$ such that

(1°) the mixed strategy profile $\nu^{\mathrm{B}}(\cdot) \in \{\nu\}$ of game (3.3.17) is a Berge equilibrium in game

$$\langle\, \mathbb{N}, \{\nu_i\}_{i \in \mathbb{N}}, \{f_i(\nu, \mu_{\mathrm{S}})\}_{i \in \mathbb{N}} \,\rangle$$

(which is obtained from (3.3.17) by setting $\mu(\cdot) = \mu_{\mathrm{S}}(\cdot)$), i.e.,

$$\max_{\nu(\cdot) \in \{\nu\}} f_i(\nu || \nu_i^{\mathrm{B}}, \mu_{\mathrm{S}}) = f_i(\nu^{\mathrm{B}}, \mu_{\mathrm{S}}) \quad (i \in \mathbb{N}); \tag{3.3.18}$$

(2°) the mixed uncertainty $\mu_{\mathrm{S}}(\cdot) \in \{\mu\}$ is a Slater-minimal alternative in the N-criteria choice problem

$$\langle\, \{\mu\}, \{f_i(\nu^{\mathrm{B}}, \mu)\}_{i \in \mathbb{N}} \,\rangle$$

(which is obtained from (3.3.17) by setting $\nu(\cdot) = \nu^{\mathrm{B}}(\cdot)$), i.e.,

$$f(\nu^{\mathrm{B}}, \mu) \not< f(\nu^{\mathrm{B}}, \mu_{\mathrm{S}}) \quad \forall \mu(\cdot) \in \{\mu\}; \tag{3.3.19}$$

denote by $\{\nu^{\mathrm{B}}, \mu_{\mathrm{S}}\}$ the set of all product measures that satisfy (3.3.18) and (3.3.19) simultaneously;

(3°) the pair $(\tilde{\nu}^{\mathrm{B}}(\cdot), \tilde{\mu}_{\mathrm{S}}(\cdot))$ is a Slater-maximal alternative in the N-criteria choice problem

$$\left\langle \left\{\nu^{\mathrm{B}}, \mu_{\mathrm{S}}\right\}, \{f_i\left(\nu^{\mathrm{B}}, \mu_{\mathrm{S}}\right)\}_{i \in \mathbb{N}} \right\rangle,$$

i.e.,

$$\tilde{f}^{\mathrm{S}} = f(\tilde{\nu}^{\mathrm{B}}, \tilde{\mu}_{\mathrm{S}}) \not< f(\nu, \mu) \quad \forall (\nu, \mu) \in \left\{\nu^{\mathrm{B}}, \mu_{\mathrm{S}}\right\}.$$

Theorem 3.3.2 *Consider game* (3.3.1) *under the assumptions that the sets* X_i *and* Y *are compact and the payoff functions* $f_i(x, y)$ *are continuous on* $\mathrm{X} \times \mathrm{Y}$ $(i \in \mathbb{N})$. *Then there exists a Slater-guaranteed balanced Berge equilibrium in mixed strategies in this game.*

The Advantages of Balanced Berge Equilibrium: Further Clarification The Slater-guaranteed balanced Berge equilibrium $(\bar{x}^{\mathrm{B}}, \bar{f}^{\mathrm{S}})$ introduced by Definition 3.3.1 has the following obvious pleasant features.

First, using their strategies x_i^{B} from a profile x^{B}, the players surely obtain a guaranteed vector payoff f_i^{B}, which is often larger (not smaller) than the vector payoff yielded by the strongly-guaranteed equilibrium; see the next section. Our aim is to increase guarantees as much as possible!

Second, this equilibrium is based on the hypothesis of "the worst-case uncertainty" for the players, i.e., on the generally accepted principle of guaranteed result under "strong uncertainty."

Third, for calculating a Slater-guaranteed balanced Berge equilibrium, it is necessary to construct a Berge equilibrium in an auxiliary game obtained from the original game. This feature has allowed us to prove existence (see Theorem 3.3.2) under the standard assumptions of game theory.

Fourth, condition 3° of Definition 3.3.1 eliminates *the internal instability* of the set of all Berge equilibria, since by Slater maximality it is impossible to find two balanced equilibria $(\bar{x}^{(1)}, \bar{f}^{(1)})$ and $(\bar{x}^{(2)}, \bar{f}^{(2)})$ such that $\bar{f}_i^{(1)} > \bar{f}_i^{(2)}$ $(i \in \mathbb{N})$, where $\bar{f}_i^{(j)} = f_i(\bar{x}^{(j)}, \bar{y}^{(j)})$ $(j = 1, 2)$.

Fifth, in the special case (3.3.10) of noncooperative games, such guaranteed equilibria are interchangeable, in the sense that a pair $(\bar{x}, \bar{y})$ satisfies conditions 1° and 2° of Definition 3.3.1 if and only if $\bar{x} \in \mathrm{X}^{\mathrm{B}}$ and $\bar{y} \in \mathrm{Y}_{\mathrm{S}}$ (see Steps 1 and 2 in Sect. 3.3.3).

In conclusion, yet note that the concept of balanced equilibrium suffers from several drawbacks: no garden without its weeds. Their detailed description as well as some "recipes" will be given in Sects. 3.4 and 3.5.

3.4 Strongly-Guaranteed Berge Equilibrium

> In the final analysis, people are equal but not always,
> not everywhere and not in all respects.
> —Grzegorczyk[19]

[19]Wladislaw Grzegorczyk, a Polish aphorist.

3.4.1 Introduction

The last thing we decide in writing a book
is what to put first.
—Pascal[20]

In the previous section, we have considered a solution concept for noncooperative games under uncertainty (NGUs) known as balanced Berge equilibrium, which was suggested by Zhukovskiy in [93, p. 233] back in 1994, using an appropriate modification of the concept of saddle point. The saddle point-based approach was also used for different types of equilibria in his later publications [51] and [52], the latter devoted to cooperative games. Section 3.4 presents a novel formalization for the guaranteed solutions of NGUs that relies on maximin.

3.4.2 Maximin and Its Interpretation Using Two-Level Game

Some man married a very skinny woman.
Being asked why, he said,
"I have chosen the least evil."
—Bar Hebraeus[21]

A hierarchical interpretation of the maximin as a two-level game is suggested.

As mentioned earlier, a single-criterion choice problem under uncertainty (SCCPU) is described by a triplet

$$\langle \mathrm{X}_1, \mathrm{Y}, f_1(x_1, y) \rangle, \tag{3.4.1}$$

where $\mathrm{X}_1 \subseteq \mathbb{R}^{n_1}$ denotes the set of admissible alternatives of a decision maker (DM); $\mathrm{Y} \subseteq \mathbb{R}^m$ is the set of uncertain factors y; $f_1(x_1, y)$ is a DM's objective function defined on the set $\mathrm{X}_1 \times \mathrm{Y}$. He seeks to maximize this function by choosing an appropriate alternative $x_1 \in \mathrm{X}_1$, under any realization of the uncertain factor $y \in \mathrm{Y}$.

In operations research, a solution of problem (3.4.1) is a pair $(x_1^{\mathrm{g}}, f_1^{\mathrm{g}}) \in \mathrm{X}_1 \times \mathbf{R}$ such that

$$f_1^{\mathrm{g}} = \max_{x_1 \in \mathrm{X}_1} \min_{y \in \mathrm{Y}} f_1(x_1, y) = \min_{y \in \mathrm{Y}} f_1(x_1^{\mathrm{g}}, y). \tag{3.4.2}$$

[20]Blaise Pascal, (1623–1662), was a French mathematician, physicist, religious philosopher, and master of prose.

[21]Bar Hebraeus, Arabic Ibn Al-'Ibri ("Son of the Hebrew"), or Abu al-Faraj, Latin name Gregorius, (1226–1286), was a medieval Syrian scholar noted for his encyclopaedic learning in science and philosophy. An English translation of a quote from [119, p. 21].

It was introduced by Wald [282] in 1939. More specifically, using the alternative x_1^g the DM achieves the highest guarantee $f_1^g \leqslant f_1(x_1^g, y)$ for all $y \in Y$ (also see Remark 3.4.1).

Let us again consider problem (3.4.1), this time as the following hierarchical two-player game. Player 1 (the DM) chooses $x_1 \in X_1$, while player 2 chooses $y \in Y$. Assume this game has a fixed sequence of moves [134, p. 79], i.e., player 1 is given priority in actions over player 2. Such a setup with the first move of player 1 describes well, e.g., an interaction of conflicting parties in a two-level hierarchical system with a single player at each level. We will also accept the hypothesis that, whenever the outcome depends on the choice of player 2 only, he always minimizes the function $f_1(x_1, y)$. Player 1 is informed about this behavior.

Then player 1 takes advantage of the *first move*, reporting his strategy $x_1 \in X_1$ to player 2. Making the second move in this game, player 2 responds with a counter strategy $y(x_1) : X_1 \to Y$ that minimizes the function $f_1(x_1, y)$ in y for each $x_1 \in X_1$. If for each x_1 this minimum is achieved at a unique point $y(x_1)$, then the best (guaranteed) result of player 1 gives

$$f_1^g = \max_{x_1 \in X_1} \min_{y \in Y} f_1(x_1, y) = \max_{x_1 \in X_1} f_1(x_1, y(x_1))$$
$$= f_1\left(x_1^g, y\left(x_1^g\right)\right) = \min_{y \in Y} f_1\left(x_1^g, y\right).$$

The sequence of moves of the DM and of player 2 is illustrated in Fig. 3.6.

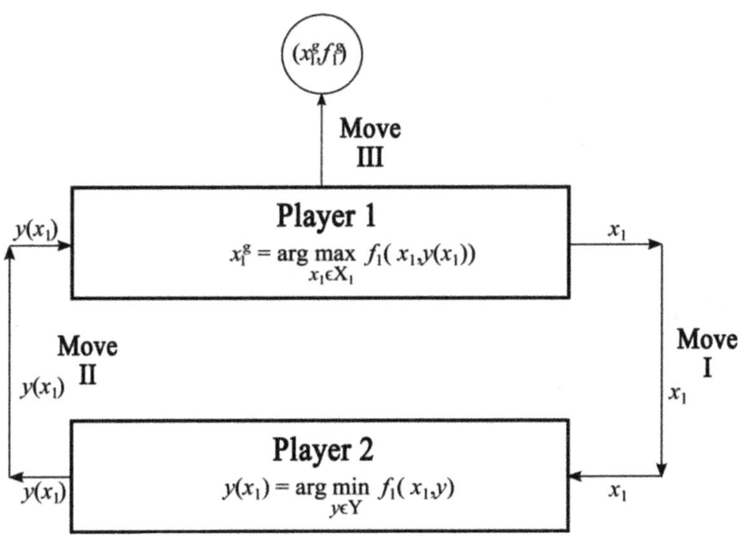

Fig. 3.6 Hierarchy in maximin setup

As a result, the DM prefers the maximin strategy x_1^g, which yields the best guaranteed payoff

$$f_1^g \leqslant f(x_1^g, y) \quad \forall y \in Y.$$

Note that, for all $x_1 \in X_1$, this payoff exceeds all other guaranteed payoffs

$$f_1(x_1, y(x_1)) = \min_{y \in Y} f_1(x_1, y) \leqslant f_1(x_1, y) \quad \forall x_1 \in X_1.$$

Remark 3.4.1 The design operation $y(x_1) : X_1 \to Y$ corresponds to the calculation of the *inner minimum*

$$f_1(x_1, y(x_1)) = \min_{y \in Y} f_1(x_1, y) \quad \forall x_1 \in X_1$$

in the maximin formula (3.4.2). On the other hand, the definition of x_1^g using

$$f_1(x^g, y\left(x_1^g\right) = \max_{x_1 \in X_1} f_1(x_1, y(x_1))$$

matches the *outer maximum* in (3.4.2). Actually, the application of these operations (inner minimum and outer maximum) to the NGUs underlies the concepts of guaranteed equilibria formalized below.

3.4.3 Drawback of Balanced Equilibrium as Solution of Noncooperative Game Under Uncertainty

Nobody can be perfect unless he admits his faults,
but if he has faults how can he be perfect?
—Peter[22]

A major drawback of the balanced equilibrium is identified and two alternative types of guaranteed equilibria for the NGU are suggested.

In Sect. 3.3, we have considered the NGU

$$\langle \mathbb{N}, \{X_i\}_{i \in \mathbb{N}}, Y, \{f_i(x, y)\}_{i \in \mathbb{N}} \rangle, \tag{3.4.3}$$

where $\mathbb{N} = \{1, \dots, N\}$ is the set of players; $X_i \subseteq \mathbb{R}^{n_i}$ is the set of pure strategies x_i of player i; $X = \prod_{i \in \mathbb{N}} X_i$ is the set of all pure strategy profiles $x = (x_1, \dots, x_N)$;

[22]Laurence Johnston Peter, (1919–1990), was a Canadian educator and hierarchiologist, author of the Peter principle.

$\mathrm{Y} \subseteq \mathbb{R}^m$ is the set of pure uncertainties y; finally, $f_i(x, y)$ is the payoff function of player i, defined on $\mathrm{X} \times \mathrm{Y}$. Using an appropriate modification of the saddle point, balanced equilibrium has been formalized by Definition 3.3.1 as a first concept of guaranteed solution of game (3.4.3).

At the end of Sect. 3.3.3, we have also pointed to a negative feature of this concept, which stems from the following circumstance. In accordance with condition 1° of Definition 3.3.1, a strategy profile $\bar{x}^{\mathrm{B}} \in \mathrm{X}$ is a Berge equilibrium if

$$\max_{x \in \mathrm{X}} f_i\left(\bar{x}||x_i^{\mathrm{B}}, y_S\right) = f_i\left(\bar{x}^{\mathrm{B}}, y_{\mathrm{S}}\right), \tag{3.4.4}$$

where the uncertain factor y_{S} has a frozen value. However, even the problem statement postulates that the uncertain factor y may take *arbitrary values* from Y, and orientation towards a specific value y_{S} is quite delusive (note that equalities (3.4.4) do not necessarily hold for other $y \neq y_{\mathrm{S}}$). If some value $y \in \mathrm{Y}$, $y \neq y_{\mathrm{S}}$, is realized in game (3.4.3), then generally the strategy profile x^{B} fails to be a Berge equilibrium; moreover, x^{B} yields the vector guarantee $\bar{f}^{\mathrm{S}} = f(\bar{x}^{\mathrm{B}}, \bar{y}_{\mathrm{S}})$ only if all players adhere to their strategies from the profile x^{B} (without any deviations from x^{B} allowed). Nevertheless, a series of considerable advantages in favor of Slater-guaranteed balanced Berge equilibrium have been outlined in Sect. 3.3; in some cases (e.g., for payoff functions with separate components in x and y), this equilibrium becomes rather useful in applications. The negative feature can be eliminated using a strongly-guaranteed equilibrium or Slater-guaranteed equilibrium as the solution concepts of the NGUs; see Sects. 3.4.4 and 3.4.5 for a detailed description.

3.4.4 Formalization

> ...nothing whatsoever takes place in the universe in which some relation of maximum and minimum does not appear.
>
> —L. Euler[23]

A guaranteed solution of a noncooperative game under uncertainty is proposed, which (in our view) is the most obvious concept among the ones analyzed in Sect. 3.3 and below.

Consider the noncooperative game under uncertainty with a possible *information discrimination* of players:

$$\Gamma = \langle \mathbb{N}, \{\mathrm{X}_i\}_{i \in \mathbb{N}}, \mathrm{Y}^{\mathrm{X}}, \{f_i(x, y)\}_{i \in \mathbb{N}} \rangle. \tag{3.4.5}$$

[23]Leonhard Euler, (1707–1783), was a Swiss mathematician and physicist. Recognized as one of the greatest mathematicians of all time. A quote from *Leonhard Euler's Elastic Curves*, by W.A. Oldfather, C.A. Ellis and D.M. Brown, *Isis*, vol. 20, no. 1 (Nov., 1933), pp. 72–160.

In this game, $\mathbb{N} = \{1, 2, \ldots, N\}$ denotes the set of players; $X_i \subseteq \mathbb{R}^{n_i}$ is the set of pure strategies x_i of player i, and a vector $x = (x_1, \ldots, x_N) \in X = \prod X_i$ forms *a pure strategy profile* in the game Γ; $Y \subseteq \mathbb{R}^m$ is the set of uncertain factors y; Y^X is the set of functions $y(x)$ defined on X and taking values from Y; these m-dimensional vector functions $y(x)$ will be called *"aware" uncertainties in game* (3.4.5); finally, $f_i(x, y) = f_i(x, y(x))$ gives the payoff function of player i $(i \in \mathbb{N})$.

This game runs as follows. The players simultaneously choose their individual strategies $x_i \in X_i$ $(i \in \mathbb{N})$ without building any coalitions. As a result, we have a strategy profile in the game Γ, i.e., an ordered collection of strategies $x = (x_1, \ldots, x_N) \in X = X_1 \times \cdots \times X_N$. Let us accept the hypotheses about the *information discrimination* of players and the *additional awareness of uncertainty*. That is, by analogy with the hierarchical games considered in Sect. 3.4.2, *the first move* belongs to the players: they choose and then report their strategies $x_i \in X_i$ to a DM, who is "in charge of" uncertainty design. The *second move* is given to the DM—he generates N uncertain factors in the form of continuous m-dimensional vector functions $y^{(i)}(x)$ $(i \in \mathbb{N})$ defined on the set X and then reports them to all N players. Assume the worst-case uncertainties, which spoil the individual payoff of each player as much as possible. Using this information, the players choose a strategy profile $x^B \in X$ yielding a "good" payoff $f_i(x^B, y(x^B))$ (e.g., a Berge equilibrium) for each player i $(i \in \mathbb{N})$. The Slater-maximal profile $\bar{x}^B$ is selected from the set of all good profiles. The point is that the set of Berge equilibria $\{x^B\}$ has *internal instability* (see Example 3.3.1), i.e., there may exist two profiles $x^{(j)} \in \{x^B\}$ $(j = 1, 2)$ such that $f_i[x^{(1)}] > f_i[x^{(2)}]$ $(i \in \mathbb{N})$. This drawback is eliminated by using the Slater maximality of $\bar{x}^B$. The hierarchical decision-making procedure of NGU (3.4.5) is illustrated in Fig. 3.7.

Note that sometimes it is necessary to adopt mixed strategies instead of the pure ones in order to prove the existence of these good solutions—the strategy profiles in game (3.4.5). In fact, this approach will be used in the current and forthcoming sections.

Recall that the guaranteed solution (x_1^g, f_1^g) of a single-criterion choice problem

$$\langle X_1, Y, f_1(x_1, y)\rangle$$

is described by the chain of equalities

$$f_1^g = \max_{x_1 \in X_1} \min_{y \in Y} f_1(x_1, y) = \min_{y \in Y} f_1\left(x_1^g, y\right).$$

First, we have to calculate *the inner minimum*

$$y(x_1) = \arg\min_{y \in Y} f_1(x_1, y),$$

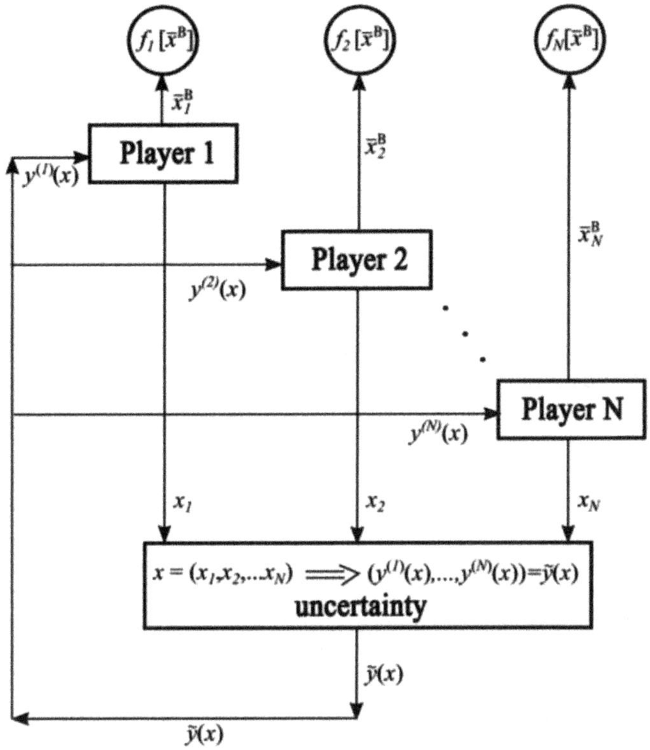

Fig. 3.7 Decision-making in the NGU (3.4.5)

and then *outer maximum*

$$x_1^g = \arg\max_{x_1 \in X_1} f_1(x_1, y(x_1)), \quad f_1^g = f_1\left(x_1^g, y\left(x_1^g\right)\right).$$

Let us clarify the optimal meaning of these concepts.

First, it follows from $f_1^g = \min_{y \in Y} f_1(x_1^g, y)$ that

$$f_1^g \leqslant f_1\left(x_1^g, y\right) \quad \forall y \in Y,$$

i.e., with the strategy x_1^g the DM obtains the guaranteed outcome f_1^g under any realization of the uncertain factor $y \in Y$.

Second, since $f_1[x_1] = \min_{y \in Y} f_1(x_1, y) = f_1(x_1, y(x_1))$, with any strategy $x_1 \in X_1$ the DM obtains a guaranteed outcome

$$f_1[x_1] \leqslant f_1(x_1, y) \quad \forall y \in Y,$$

and the guarantee f_1^{g} is *highest* because

$$f_1^{\mathrm{g}} = f_1\left(x_1^{\mathrm{g}}, y\left(x_1^{\mathrm{g}}\right)\right) \geqslant f_1[x_1] = f_1(x_1, y(x_1)) \quad \forall x_1 \in \mathrm{X}_1.$$

The concept of strongly-guaranteed solution of game (3.4.5) that is introduced below relies on a modification of these two properties of maximin. The modification itself consists in replacing the inner minimum by N scalar minima, i.e.,

$$\min_{y \in \mathrm{Y}} f_i(x, y) = f_i(x, y^{(i)}(x)) = f_i[x] \;\; \forall x \in \mathrm{X} \;\; (i \in \mathbb{N}),$$

and also in replacing the outer maximum by the concept of Berge equilibrium, i.e.,

$$\max_{x \in \mathrm{X}} f_i\left[x||x_i^{\mathrm{B}}\right] = f_i\left[x^{\mathrm{B}}\right] \quad (i \in \mathbb{N}),$$

where $[x||x_i^{\mathrm{B}}] = [x_1, \ldots, x_{i-1}, x_i^{\mathrm{B}}, x_{i+1}, \ldots, x_N]$.

We will formalize the concept of Slater-strongly-guaranteed Berge equilibrium in three steps as follows.

Step 1. Associate with each strategy profile $x \in \mathrm{X}$ and each player $i \in \mathbb{N}$ a unique continuous vector function $y^{(i)}(x)$ on X such that

$$f_i\left(x, y^{(i)}(x)\right) = \min_{y \in \mathrm{Y}} f_i(x, y) = f_i[x] \quad (i \in \mathbb{N}). \tag{3.4.6}$$

Step 2. Associate with game (3.4.5) the noncooperative N-player game (without uncertainty)

$$\langle \mathbb{N}, \{\mathrm{X}_i\}_{i \in \mathbb{N}}, \{f_i[x]\}_{i \in \mathbb{N}} \rangle, \tag{3.4.7}$$

further referred to as the game of guarantees. For this game, find a Berge equilibrium $x^{\mathrm{B}} \in \mathrm{X}$ from the equalities

$$\max_{x \in \mathrm{X}} f_i\left[x||\, x_i^{\mathrm{B}}\right] = f_i\left[x^{\mathrm{B}}\right] \quad (i \in \mathbb{N}). \tag{3.4.8}$$

Step 3. From the set of all Berge equilibria $\{x^{\mathrm{B}}\}$, choose the maximal one $\bar{x}^{\mathrm{B}}$ in the vector sense, e.g., find a Slater-maximal alternative $\bar{x}^{\mathrm{B}}$ in the N-criteria choice problem

$$\left\langle \left\{x^{\mathrm{B}}\right\}, \{f_i[x]\}_{i \in \mathbb{N}} \right\rangle.$$

In the case of Slater maximum, it suffices to calculate $\bar{x}^{\mathrm{B}}$ using the condition

$$\max_{x\in\{x^B\}} \sum_{i\in\mathbb{N}} \alpha_i f_i[x] = \sum_{i\in\mathbb{N}} \alpha_i \bar{f}_i\left[\bar{x}^{\mathrm{B}}\right],$$

where all the constants $\alpha_i \geqslant 0 (i \in \mathbb{N}) \wedge \sum_{i\in\mathbb{N}} \alpha_i > 0$, see [152, pp. 68–69].

Finally, construct the N-dimensional vector

$$\bar{f}\left[\bar{x}^{\mathrm{B}}\right] = \left(\bar{f}_1\left[\bar{x}^{\mathrm{B}}\right], \ldots, \bar{f}_N\left[\bar{x}^{\mathrm{B}}\right]\right).$$

The resulting pair $(\bar{x}^{\mathrm{B}}, \bar{f}[\bar{x}^{\mathrm{B}}]) \in \mathrm{X} \times \mathbb{R}^N$, where $f = (f_1, \ldots, f_N)$, will be called *the Slater-strongly-guaranteed Berge equilibrium* in game (3.4.5); in addition, $\bar{x}^{\mathrm{B}}$ is *the strongly-guaranteeing strategy profile* in game (3.4.5) while $\bar{f}_i[\bar{x}^{\mathrm{B}}]$ is *the strongly-guaranteed payoff of player* $i \in \mathbb{N}$.

The game-theoretic meaning of the suggested solution consists in the following. If the players have chosen the strategies $x_i \in \mathrm{X}_i$ $(i \in \mathbb{N})$, thereby forming the profile $x = (x_1, \ldots, x_N)$, then each player i obtains a payoff $f_i(x, y)$ not smaller than $f_i[x]$ (3.4.6) under any realization of the uncertain factor $y \in \mathrm{Y}$. (This fact follows from the last equality of (3.4.6), written in the form $f_i[x] \leqslant f_i(x, y) \quad \forall y \in \mathrm{Y}$). In other words, the value $f_i[x]$ is the guarantee for player i under the players' strategies from the profile $x \in \mathrm{X}$ and any realization of the uncertain factor $y \in \mathrm{Y}$, regardless of their choice.

Next, in accordance with Step 2 (see the definition), instead of the noncooperative game under uncertainty (3.4.5) one has to consider the game of guarantees (3.4.7), (3.4.6) without uncertainty. In this game, the payoff functions of the players are their guarantees $f_i[x]$ $(i \in \mathbb{N})$, while the Berge equilibrium is defined by the same principle, now applied to the new payoff functions—the guarantees $f_i[x]$ $(i \in \mathbb{N})$ of the original payoff functions $f_i(x, y)$.

The strongly-guaranteed equilibrium is stable in the sense that, if the players choose their strategies from the profile $x^{\mathrm{B}} = (x_1^{\mathrm{B}}, \ldots, x_N^{\mathrm{B}})$, then

First, under any realization of the uncertain factor $y \in \mathrm{Y}$ the conflicting parties obtain guaranteed payoffs $f_i(x^{\mathrm{B}}, y) \geqslant f_i[x^{\mathrm{B}}] = f_i^{\mathrm{B}}$ $(i \in \mathbb{N})$ that are not smaller than their guarantees;

Second, any deviation, e.g., of player 1 from the strategy x_1^{B} (i.e., the choice of another strategy $\tilde{x}_1 \in \mathrm{X}$ such that $\tilde{x}_1 \neq x_1^{\mathrm{B}}$) gives, e.g., to player 2 a payoff $f_2(x^{\mathrm{B}}||\tilde{x}_1, y)$ with a guarantee $f_2[x^{\mathrm{B}}||\tilde{x}_1]$ not higher than the guarantee $f_2[x^{\mathrm{B}}]$ in the equilibrium x^{B} (the noncooperative game under uncertainty (3.4.5) is assessed using the game of guarantees (3.4.7)).

3.4.5 Existence in Mixed Strategies

The existence of a strongly-guaranteed Berge equilibrium in mixed strategies is established for the noncooperative two-player game under uncertainty with continuous payoff functions that are strictly convex in the uncertain factors, and also with compact sets of strategies and uncertain factors.

To simplify notation, our analysis below will be confined to game (3.4.5) with two players, i.e., $\mathbb{N} = \{1, 2\}$.

Let the sets X_i $(i=1,2)$ be convex and compact and consider the Borel σ-algebra of all subsets of the set X_i (the details can be found in Remark 3.3.1); as an extension of the set of (pure) strategies $x_i \in X_i$ of player i, consider his mixed strategies $\mu_i(\cdot)$—probability measures on the compact set X_i, i.e., on the Borel σ-algebra of the set X_i. Denote by $\{\mu_i\}$ $(i = 1, 2)$ the set of mixed strategies of player i. Note that a measure of the form $\delta(x_i - x_i^*)(dx_i)$, where $\delta(\cdot)$ is the Dirac function, is also a mixed strategy of player i. The product measures $\mu(dx_1, dx_2)$ introduced by the definitions in [122, p. 271] with the notations [108, p. 284],

$$\mu(dx_1, dx_2) = \mu_1(dx_1)\mu_2(dx_2),$$

are probability measures on the product $X = X_1 \times X_2$ of the compact sets X_1 and X_2. To construct the product measure $\mu(dx_1, dx_2)$, as the σ-algebra of all subsets $X_1 \times X_2$ one takes the smallest Borel σ-algebra containing all the products $Q_1 \times Q_2$, where Q_i is an element of the Borel σ-algebra of the compact set X_i $(i = 1, 2)$.

If the payoff functions $f_i[x_1, x_2]$ are continuous on $X_1 \times X_2$, we define the following integrals in terms of expectation:

$$f_i[\mu_1, x_2] = \int\limits_{X_1} f_i[x_1, x_2]\mu_1(dx_1), \qquad f_i[x_1, \mu_2] = \int\limits_{X_2} f_i[x_1, x_2]\mu_2(dx_2).$$

Since the functions $f_i[x_1, x_2]$ are continuous on $X_1 \times X_2$, the integrals $f_i[\mu_1, x_2]$ and $f_i[x_1, \mu_2]$ are continuous functionals on X_2 and X_1, respectively; see [24, p. 113]. Then there exist the double integrals

$$f_i[\mu_1, \mu_2] = \int\limits_{X_2} f_i[\mu_1, x_2]\mu_2(dx_2) = \int\limits_{X_2}\int\limits_{X_1} f_i[x_1, x_2]\mu_1(dx_1)\mu_2(dx_2),$$

$$\int\limits_{X_1} f_i[x_1, \mu_2]\mu_1(dx_1) = \int\limits_{X_1}\int\limits_{X_2} f_i[x_1, x_2]\mu_2(dx_2)\mu_1(dx_1),$$

which take the same value by Fubini's theorem.

Now let us pass to the mixed extension of game (3.4.7) with $\mathbb{N} = \{1, 2\}$, i.e., to the noncooperative game

$$\tilde{\Gamma}_2 = \langle \{1, 2\}, \{\mu_i\}_{i=1,2}, \{f_i[\mu_1, \mu_2]\}_{i=1,2} \rangle,$$

where $\{\mu_i\}$ is the set of mixed strategies $\mu_i(\cdot)$ of player i, which are probability measures on the compact set X_i; the expectation

$$f_i[\mu_1, \mu_2] = \int\limits_{X_1 \times X_2} f_i[x_1, x_2]\mu_1(dx_1)\mu_2(dx_2)$$

gives *the mixed extension of the payoff function* $f_i[x_1, x_2]$ $(i = 1, 2)$.

A pair of mixed strategies $(\mu_1^B(\cdot), \mu_2^B(\cdot)) \in \{\mu_1\} \times \{\mu_2\}$ is *called a Berge equilibrium in game* $\tilde{\Gamma}_2$ if

$$\begin{aligned} f_1[\mu_1^B, \mu_2] &\leqslant f_1[\mu_1^B, \mu_2^B] \quad \forall \mu_2(\cdot) \in \{\mu_2\}, \\ f_2[\mu_1, \mu_2^B] &\leqslant f_2[\mu_1^B, \mu_2^B] \quad \forall \mu_1(\cdot) \in \{\mu_1\}. \end{aligned} \tag{3.4.9}$$

Interestingly, the set of all payoffs $f[\mu^B] = (f_1[\mu^B], f_2[\mu^B])$ on the set of all Berge equilibria $\{\mu^B(\cdot) = \mu_1^B(\cdot)\mu_2^B(\cdot)\}$ is compact in $\mathbb{R}^2$ (this follows from Proposition 3.4.1 below).

In accordance with [22, pp. 117–119], *if in the game* $\tilde{\Gamma}_2$ *the payoff functions* $f_i[x_1, x_2]$ *are continuous on* $X_1 \times X_2$ *and the sets* X_i *are compact* $(i = 1, 2)$, *then the game* $\tilde{\Gamma}_2$ *possesses a Berge equilibrium*

$$\mu^B(\cdot) = \left(\mu_1^B(\cdot), \mu_2^B(\cdot)\right) \in \{\mu_1\} \times \{\mu_2\}.$$

Sometimes, this profile is called *a mixed strategy Berge equilibrium in game* (3.4.7) *with* $\mathbb{N} = \{1, 2\}$.

Proposition 3.4.1 *Assume that in game* (3.4.7) *with* $\mathbb{N} = \{1, 2\}$ *the sets* X_i $(i = 1, 2)$ *are convex and compact and the payoff functions* $f_i[x_1, x_2]$ *are continuous on* $X_1 \times X_2$. *Then the set* $\mathcal{F}^B = \{f_1[\mu^B], f_2[\mu^B]\}$ *of all Berge equilibrium payoffs in the game* $\tilde{\Gamma}_2$ *is a non-empty and compact set, i.e., a closed bounded subset of* $\mathbb{R}^2$.

Proof In view of the well-known properties of probability measures [41, p. 288]; [122, p. 254], the set of all possible product measures $\mu(dx_1, dx_2) = \mu_1(dx_1)\mu_2(dx_2)$ is *weakly closed and weakly compact* [122, pp. 212, 254]; [180, pp. 48, 49]. Hence, from each sequence

$$\left\{\mu^{(k)}(dx) = \mu_1^{(k)}(dx_1)\mu_2^{(k)}(dx_2)\right\} \quad (k = 1, 2, \ldots)$$

one can extract a subsequence

$$\left\{\mu^{(k_j)}(dx) = \mu_1^{(k_j)}(dx_1)\mu_2^{(k_j)}(dx_2)\right\} \quad (j = 1, 2, \ldots)$$

that *weakly converges* [122, p. 212, 254]; [105, p. 199] to a function $\mu(\cdot) \in \in \{\mu\}$, i.e., for any choice of a continuous scalar function $\varphi[x_1, x_2]$ defined on X, it holds that

$$\lim_{j\to\infty} \int_{\mathrm{X}} \varphi[x_1, x_2]\mu^{(k_j)}(dx) = \int_{\mathrm{X}} \varphi[x_1, x_2]\mu(dx).$$

Denote by $\mathfrak{M}^B$ the set of all Berge equilibria $\mu^{\mathrm{B}}(dx) = \mu_1^{\mathrm{B}}(dx_1)\mu_2^{\mathrm{B}}(dx_2)$ described by formulas (3.4.9). Then $\mathfrak{M}^B \neq \varnothing$, as shown in [22, pp. 117–119]. Now, take an arbitrary infinite sequence of such equilibria $\mu^{(k)}(\cdot) \in \mathfrak{M}^B$ $(k = 1, 2, \ldots)$. Owing to the weak compactness of the set of probability measures, there exist a subsequence of measures $\mu^{(k_j)}(\cdot) \in \mathfrak{M}^B$ $(j = 1, 2, \ldots)$ and a probability measure $\mu^{(o)}(\cdot) \in \{\mu\}$ such that, for a continuous function $f_i[x] = f_i[x_1, x_2]$ on X,

$$\lim_{j\to\infty} f_i\left[\mu^{(k_j)}\right] = \lim_{j\to\infty} \int_{\mathrm{X}} f_i[x]\mu^{(k_j)}(dx) = \int_{\mathrm{X}} f_i[x]\mu^{(\mathrm{o})}(dx) = f_i\left[\mu^{(\mathrm{o})}\right].$$

Let us show that the limiting measure $\mu^{(\mathrm{o})}(\cdot) = \mu_1^{(\mathrm{o})}(\cdot)\mu_2^{(\mathrm{o})}(\cdot)$ is also a Berge equilibrium, i.e.,

$$\begin{aligned} f_1[\mu_1^{(o)}, \mu_2] &\leqslant f_1[\mu^{(\mathrm{o})}] \quad \forall \mu_2(\cdot) \in \{\mu_2\}, \\ f_2[\mu_1, \mu_2^{(o)}] &\leqslant f_2[\mu^{(\mathrm{o})}] \quad \forall \mu_1(\cdot) \in \{\mu_1\}. \end{aligned}$$

Assume on the contrary that there exists a measure $\bar{\mu}_1(\cdot) \in \{\mu_1\}$ or a measure $\bar{\mu}_2(\cdot) \in \{\mu_2\}$ such that

$$f_1\left[\mu_1^{(\mathrm{o})}, \bar{\mu}_2\right] > f_1\left[\mu^{(\mathrm{o})}\right] \quad \vee \quad f_2[\bar{\mu}_1, \mu_2^{(\mathrm{o})}] > f_2\left[\mu^{(\mathrm{o})}\right].$$

For example, let

$$f_1\left[\mu_1^{(\mathrm{o})}, \bar{\mu}_2\right] > f_1\left[\mu^{(\mathrm{o})}\right],$$

which is equivalently written as

$$\int_{\mathrm{X}} f_1[x]\mu_1^{(\mathrm{o})}(dx_1)\bar{\mu}_2(dx_2) > \int_{\mathrm{X}} f_1[x]\mu^{(\mathrm{o})}(dx).$$

Then, for sufficiently large j,

$$\int_{X} f_1[x]\mu_1^{(k_j)}(dx_1)\bar{\mu}_2(dx_2) > \int_{X} f_1[x]\mu^{(k_j)}(dx),$$

which contradicts the inclusion $\mu^{(k_j)}(\cdot) \in \mathfrak{M}^B$, i.e., the Berge equilibrium condition of each mixed strategy profile $\mu^{(k_j)}(\cdot) \in \mathfrak{M}^B$ in the game $\tilde{\Gamma}_2$. Hence, the set $\mathcal{F}^{\mathrm{B}} =$ $= \{f_1[\mu^B], f_2[\mu^B] \mid \forall \mu^B(\cdot) \in \mathfrak{M}^B\}$ is compact in $\mathbb{R}^2$. ∎

Our next task is to construct a strongly-guaranteed Berge equilibrium in mixed strategies for this game using Steps 1–3 above.

Consider game (3.4.5) with $N = 2$ in which the sets X_i $(i = 1, 2)$ and Y are compact and the payoff functions $f_i(x_1, x_2, y)$ $(i = 1, 2)$ are continuous on $\mathrm{X}_1 \times \mathrm{X}_2 \times \mathrm{Y}$.

A quadruple $(\bar{\mu}_1^{\mathrm{B}}(\cdot), \bar{\mu}_2^{\mathrm{B}}(\cdot), \bar{f}_1^{\mathrm{B}}, \bar{f}_2^{\mathrm{B}}) \in \{\mu_1\} \times \{\mu_2\} \times \mathbb{R}^2$ is called *a strongly-guaranteed Berge equilibrium in mixed strategies in game* (3.4.5) *with* $N = 2$ if for each i there exists a unique continuous m-dimensional vector functions $y^{(i)}(x) : \mathrm{X}_1 \times \mathrm{X}_2 \to \mathrm{Y}$ $(i = 1, 2)$ such that inequalities (3.4.9) hold for the function $f_i[\mu_1, \mu_2]$ $(i = 1, 2)$ and the product measure $\bar{\mu}^{\mathrm{B}}(\cdot) = \bar{\mu}_1^{\mathrm{B}}(\cdot)\bar{\mu}_2^{\mathrm{B}}(\cdot)$ yields a Slater-maximal alternative in the bicriteria choice problem

$$\langle \{\mu^{\mathrm{B}}\}, \{f_i[\mu]\}_{i=1,2} \rangle.$$

Here $f_i[\mu] = f_i[\mu_1, \mu_2] = \int\limits_{X} f_i[x]\mu_1(dx_1)\mu_2(dx_2)$, $f_i[x] = f_i(x, y^{(i)}(x)) = \min_{y \in \mathrm{Y}} f_i(x, y)$, $\bar{\mu}_i^{\mathrm{B}}(\cdot) \in \{\mu_i\}$ indicates the mixed strategy of player i, and $\bar{f}_i^{\mathrm{B}} = \bar{f}_i[\mu_1^{\mathrm{B}}, \mu_2^{\mathrm{B}}]$ $(i = 1, 2)$ is his guaranteed payoff.

Theorem 3.4.1 *Consider the noncooperative two-player game under uncertainty*

$$\Gamma_2 = \langle \{1, 2\}, \{\mathrm{X}_i\}_{i=1,2}, \mathrm{Y}^{\mathrm{X}}, \{f_i(x_1, x_2, y)\}_{i=1,2} \rangle$$

under the assumptions that

(1^0) *the set* $\mathrm{X}_i \subset \mathbb{R}^{n_i}$ *of all pure strategies* x_i *of player* i *is convex and compact* $(i = 1, 2)$ *and the set* $\mathrm{Y} \subset \mathbb{R}^m$ *of uncertain factors* y *is convex and compact;*

(2^0) *the payoff function* $f_i(x, y)$ *of player* i $(i = 1, 2)$ *is continuous on* $\mathrm{X}_1 \times \mathrm{X}_2 \times \mathrm{Y}$ *and strictly convex in* $y \in \mathrm{Y}$ *for each* $(x_1, x_2) \in \mathrm{X}_1 \times \mathrm{X}_2$.

Then there exists a strongly-guaranteed Berge equilibrium in mixed strategies in this game.

Proof Using the compactness of the sets X_i $(i = 1, 2)$ and Y, the concavity of Y and also the continuity of the payoff functions $f_i(x_1, x_2, y)$ on $\mathrm{X}_1 \times \mathrm{X}_2 \times \mathrm{Y}$ and their strict convexity in $y \in \mathrm{Y}$ for each $x = (x_1, x_2) \in \mathrm{X}_1 \times \mathrm{X}_2$, we conclude (see [14, p. 54]) that there exist two continuous m-dimensional vector functions $y^{(i)}(x_1, x_2)$

defined on $X_1 \times X_2$ such that

$$\min_{y \in Y} f_i(x_1, x_2, y) = f_i\left(x_1, x_2, y^{(i)}(x_1, x_2)\right) = f_i[x_1, x_2] \quad (i = 1, 2)$$

for any $(x_1, x_2) \in X_1 \times X_2$. The functions

$$f_i\left(x_1, x_2, y^{(i)}(x_1, x_2)\right) = f_i[x_1, x_2] \quad (i = 1, 2)$$

are continuous on $X_1 \times X_2$ as superpositions of the continuous functions $f_i(x_1, x_2, y)$ and $y = y^{(i)}(x_1, x_2)$.

Now, design a noncooperative two-player game—the game of guarantees

$$\langle \{1, 2\}, \{X_i\}_{i=1,2}, \{f_i[x_1, x_2]\}_{i=1,2} \rangle. \tag{3.4.10}$$

As established earlier, in this game the payoff function $f_i[x_1, x_2]$ of player i $(i = 1, 2)$ is continuous on the product $X_1 \times X_2$ of compact sets. Consequently, by [22, pp. 117–119], there exists a mixed strategy Berge equilibrium $(\mu_1^B(\cdot), \mu_2^B(\cdot)) \in \{\mu_1\} \times \{\mu_2\}$, which satisfies inequalities (3.4.9). Then construct the pair

$$f_i\left[\mu^B\right] = f_i\left[\mu_1^B, \mu_2^B\right] = \int\limits_{X_1 \times X_2} f_i[x_1, x_2]\mu_1^B(dx_1)\mu_2^B(dx_2) \quad (i = 1, 2),$$

in which the set $\{f_i[\mu^B] = f_i[\mu_1^B, \mu_2^B]\}$ is compact in $\mathbb{R}^2$ (see Proposition 3.4.1); $\mathfrak{M}^B$ forms the set of all Berge equilibria $\mu^B(\cdot) = \mu_1^B(\cdot)\mu_2^B(\cdot)$ (each of them satisfies inequalities (3.4.9)). This compact set is nonempty [22, pp. 117–119]; denote it by $\mathcal{F}^B$. Consider a continuous function $\sum_{i=1}^2 \alpha_i f_i$, where $\alpha_i = \text{const} > 0$ and $i \in \mathbb{N} = \{1, 2\}$, on the compact set $\mathcal{F}^B$. By the Weierstrass theorem, there exists a vector $\bar{f}^B = (\bar{f}_1^B, \bar{f}_2^B) \in \mathcal{F}^B$ such that

$$\max_{f \in \mathcal{F}^B} \sum_{i=1}^{2} \alpha_i f_i = \sum_{i=1}^{2} \alpha_i \bar{f}_i^B.$$

Finally, find the product measure $\bar{\mu}^B(\cdot) = \bar{\mu}_1^B(\cdot)\bar{\mu}_2^B(\cdot)$ from the equalities $\bar{f}_i^B = f_i[\bar{\mu}^B]$ $(i = 1, 2)$.

By definition, the resulting triplet $(\bar{\mu}^B(\cdot), \bar{f}_1^B, \bar{f}_2^B)$ is a strongly-guaranteed Berge equilibrium in mixed strategies in game (3.4.5) with $\mathbb{N} = \{1, 2\}$. ■

Remark 3.4.2 *First*, the assumptions of Theorem 3.4.1 can be relaxed by requiring only the compactness of the sets X_i $(i = 1, 2)$ and Y and the continuity of the payoff functions $f_i(x, y)$ on the set $X_1 \times X_2 \times Y$ (see Theorem 3.5.1 below). Theorem 3.4.1 itself is placed here to illustrate an original method for establishing the existence of guaranteed equilibria.

Second, Theorem 3.4.1 generalizes directly to the games with $N > 2$ players. In this case, the definition of a strongly-guaranteed equilibrium involves a vector guarantee $f[x] = (f_1[x], \dots, f_N[x])$, since for each $x \in \mathrm{X}$ and for all $y \in \mathrm{Y}$ the value $f_i(x, y)$ cannot be smaller than $f_i[x]$ $(i \in \mathbb{N})$ (see (3.4.6)). This vector guarantee is lowest among all other vector guarantees $f^{\mathrm{S}}[x]$ (Slater guarantees, see Sect. 3.5) because $f_i^{\mathrm{S}}[x] \geqslant f_i[x]$ $\forall x \in \mathrm{X}$, $i \in \mathbb{N}$ (here $f^{\mathrm{S}}[x] = f(x, y_{\mathrm{S}}(x))$ and $y_{\mathrm{S}}(x)$ yields the Slater-minimal alternative in the N-criteria choice problem $\langle \mathrm{Y}, f(x, y)\rangle$ for each frozen $x \in \mathrm{X}$). This fact explains the term "strongly-guaranteed equilibrium." However, keep in mind that the players seek for *as high guarantees as possible*.

Remark 3.4.3 Once again we will stress the game-theoretic meaning and advantages of strongly-guaranteed equilibrium.

First, in accordance with (3.4.6), each strategy profile $x \in \mathrm{X}$ is associated with a vector guarantee $f[x] = (f_1[x], \dots, f_N[x])$: by the inequality $f_i(x, y) \geqslant f_i[x]$ $\forall y \in \mathrm{Y}$ $(i \in \mathbb{N})$, the payoffs $f_i(x, y)$ cannot be smaller than $f_i[x]$ $(i \in \mathbb{N})$ for all $y \in \mathrm{Y}$. Indeed, with his strategy $x_i \in \mathrm{X}_i$ player i obtains a payoff $f_i(x, y)$ that is surely not less than $f_i[x]$ under any realization of the uncertain factors $y \in \mathrm{Y}$. Therefore, transition to the same game of guarantees (3.4.7) for all $y \in \mathrm{Y}$ allows the players to forget about the existing uncertainty and to be guided by an increase of their guarantees only (which depend on the strategy profile x formed by their choice).

Second, the aspiration of player $i \in \mathbb{N}$ to increase his guarantee $f_i[x]$ also results in a Berge equilibrium (an analog of the outer maximum in the noncooperative game of guarantees (3.4.7)). Being a Berge equilibrium, the strategy profile $x^{\mathrm{B}} = (x_1^{\mathrm{B}}, \dots, x_N^{\mathrm{B}})$ is stable against the deviation of any coalition of $N - 1$ players. For example, if player 1 is deviating from x_1^{B} with a choice $x_1 \neq x_1^{\mathrm{B}}$, then say the guarantee $f_2[x^{\mathrm{B}} \| x_1]$ of player 2 in the strategy profile $[x^{\mathrm{B}} \| x_1] = [x_1, x_2^{\mathrm{B}}, \dots, x_N^{\mathrm{B}}]$ cannot exceed $f_2[x^{\mathrm{B}}]$ (which follows from (3.4.8)), yet may decrease. (Each player seeks to maximize his guarantee!) Therefore, in contrast to the balanced equilibria considered in Sect. 3.3, the strategy profile x^{B} still satisfies the Berge equilibrium conditions for all uncertain factors $y \in \mathrm{Y}$ (we again emphasize that the guarantees $f_i[x]$ are independent of y).

Third, the set of all Berge equilibria $\{x^{\mathrm{B}}\} = \mathrm{X}^{\mathrm{B}}$ in the game (3.4.7) is *internally unstable* (see Example 3.3.1). This nuisance is eliminated using the Slater maximality of the suggested solution $\bar{x}^{\mathrm{B}}$.

Then a strongly-guaranteed Berge equilibrium $(\bar{x}^{\mathrm{B}}, \bar{f}^{\mathrm{B}})$ in the NGU (3.4.5) is a pair $(\bar{x}^{\mathrm{B}}, f[\bar{x}^{\mathrm{B}}])$ composed of a Berge equilibrium $\bar{x}^{\mathrm{B}}$ in the game of guarantees to be used by the players and a vector guarantee $f[\bar{x}^{\mathrm{B}}] = \bar{f}^{\mathrm{B}}$ yielded by them in this equilibrium.

Remark 3.4.4 As follows from Remark 3.4.3, an analog of the inner minimum (in the maximin definition) is Step 1 of strongly-guaranteed equilibrium design. In turn, Steps 2 and 3 correspond to the outer maximum in the maximin definition. Let us show that each vector guarantee in pure strategies

$$f[x] = \left(f_1(x, y^{(1)}(x)\right) = f_1[x], \ldots, f_N\left(x, y^{(N)}(x)\right) = f_N[x])$$

induces a vector guarantee in mixed strategies

$$f[\mu] = (f_1[\mu], \ldots, f_N[\mu]),$$

where

$$f_i[\mu] = \int\limits_{\mathrm{X}} f_i\left(x, y^{(i)}(x)\right) \mu(dx), \quad i \in \mathbb{N}.$$

Indeed, from (3.4.6) for each $x \in \mathrm{X}$ we have N inequalities of the form

$$f_i[x] \leqslant f_i(x, y) \quad \forall y \in \mathrm{Y}.$$

Integrating both sides of these inequalities with an arbitrary mixed strategy profile $\mu(\cdot)$ as the integration measure gives

$$f_i[\mu]=\int\limits_{\mathrm{X}} f_i\left(x, y^{(i)}(x)\right) \mu(dx)\leqslant \int\limits_{\mathrm{X}} f_i(x, y)\mu(dx)=f_i[\mu, y] \quad \forall y \in \mathrm{Y}\ (i \in \mathbb{N}).$$

Equivalently, every mixed strategy profile $\mu(\cdot) \in \{\mu\}$ in the game

$$\langle\, \mathbb{N}, \{\mu_i\}_{i\in\mathbb{N}}, \mathrm{Y}, \{f_i[\mu, y]\}_{i\in\mathbb{N}}\, \rangle$$

induces a vector guarantee $f[\mu] = (f_1[\mu], \ldots, f_N[\mu])$: for any $y \in \mathrm{Y}$, the payoffs $f_i[\mu, y]$ cannot be smaller than $f_i[\mu]$.

Then, in accordance with Steps 2 and 3 of strongly-guaranteed Berge equilibrium design in game (3.4.5) with mixed strategies, it is necessary to build the vector guarantees $f[\mu^{\mathrm{B}}]$ achieved on all mixed strategy Berge equilibria $\mu^{\mathrm{B}}(\cdot) \in \{\mu\}$. Finally, among them we have to choose the Slater-maximal strategy profile $\bar{\mu}^{\mathrm{B}}(\cdot)$.

3.4.6 Linear-Quadratic Setup of Game

A good example is the best sermon.
—English proverb

An explicit form of a strongly-guaranteed Berge equilibrium in mixed strategies is obtained for the noncooperative linear-quadratic two-player game under uncertainty.

This section considers game (3.4.5) with $\mathbb{N} = \{1, 2\}$, the sets $\mathrm{X}_i = \mathbb{R}^n$ and $\mathrm{Y} = \mathbb{R}^m$ (no constraints), and the linear-quadratic payoff functions in x_i and y given by

$$\begin{aligned} f_1(x, y) &= x_2' A_1 x_2 + 2x_1' x_2 + 2x_2' C_1 y + y' D_1 y + 2a_1' x_2 + \varphi_1(x_1), \\ f_2(x, y) &= x_1' A_2 x_1 - 2x_2' x_1 + 2x_1' C_2 y + y' D_2 y + 2a_2' x_1 + \varphi_2(x_2). \end{aligned} \tag{3.4.11}$$

In this game, x_1 and x_2 are n-dimensional column vectors, y is an m-dimensional column vector, prime denotes transposition, constant vectors a_i and matrices A_i, C_i, D_i have compatible dimensions, and the matrices A_i and D_i are symmetric $(i = 1, 2)$. Recall that the notation $A_i < 0$ $(D_i > 0)$ means the negative (positive) definiteness of the quadratic form $x' A_i x$ for all $x \in \mathbb{R}^n$ $(y' D_i y$ for all $y \in \mathbb{R}^m$, respectively), while the notation $K \leqslant 0$ the negative semidefiniteness of the quadratic form $x' K x$ for all $x \in \mathbb{R}^n$. Also, 0_n stands for an n-dimensional zero vector, $\varphi_i(x_i)$ $(i = 1, 2)$ are scalar continuous functions.

Thus, we are studying the noncooperative two-player game under uncertainty

$$\langle \{1, 2\}, \{\mathrm{X}_i = \mathbb{R}^n\}_{i=1,2}, \mathrm{Y} = \mathbb{R}^m, \{f_i(x_1, x_2, y)\}_{i=1,2} \rangle, \tag{3.4.12}$$

in which the payoff functions $f_i(x_1, x_2, y)$ are defined by (3.4.11), player i chooses the n-dimensional column vector $x_i \in \mathbb{R}^n$ as his strategy, and the uncertain factors are $y \in \mathbb{R}^m$. The special form (3.4.11) of the payoff functions $f_i(x_1, x_2, y)$ covers all linear and quadratic terms in x_j $(i, j = 1, 2; i \neq j)$. An attempt to consider other possible terms would run into cumbersome calculations, still remaining the same in principle.

Proposition 3.4.2 *Consider game* (3.4.12) *with*

$$A_i < 0, \quad D_i > 0 \quad (i = 1, 2). \tag{3.4.13}$$

For any continuous scalar functions $\varphi_i(x_i)$ $(i = 1, 2)$, the strongly-guaranteed Berge equilibrium $(x_1^{\mathrm{B}}, x_2^{\mathrm{B}}, f_1^{\mathrm{B}}, f_2^{\mathrm{B}})$ has the form

$$\begin{aligned} x_1^{\mathrm{B}} &= -\left[\left(A_1 - C_1 D_1^{-1} C_1'\right)^{-1} + \left(A_2 - C_2 D_2^{-1} C_2'\right)\right]^{-1} \times \\ &\quad \times \left[\left(A_1 - C_1 D_1^{-1} C_1'\right)^{-1} a_1 + a_2\right], \\ x_2^{\mathrm{B}} &= \left[\left(A_2 - C_2 D_2^{-1} C_2'\right)^{-1} + \left(A_1 - C_1 D_1^{-1} C_1'\right)\right]^{-1} \times \\ &\quad \times \left[\left(A_2 - C_2 D_2^{-1} C_2'\right)^{-1} a_2 - a_1\right], \\ f_1^{\mathrm{B}} &= -\left[x_2^{\mathrm{B}}\right]' \left[A_1 - C_1 D_1^{-1} C_1'\right] x_2^{\mathrm{B}} + \varphi_1\left(x_1^{\mathrm{B}}\right), \\ f_2^{\mathrm{B}} &= -\left[x_1^{\mathrm{B}}\right]' \left[A_2 - C_2 D_2^{-1} C_2'\right] x_1^{\mathrm{B}} + \varphi_2\left(x_2^{\mathrm{B}}\right). \end{aligned} \tag{3.4.14}$$

Proof The following chain of implications is immediate from (3.4.13) and [93]:

$$
\begin{gathered}
[D_i > 0] \Rightarrow [\det D_i \neq 0] \Rightarrow \left[\exists D_i^{-1}\right], \\
[D_i > 0] \Rightarrow \left[D_i^{-1} > 0\right] \Rightarrow \left[C_i D_i^{-1} C_i' \geqslant 0\right] \Rightarrow \left[-C_i D_i^{-1} C_i' \leqslant 0\right], \\
\left[A_i < 0 \wedge -C_i D_i^{-1} C_i' \leqslant 0\right] \Rightarrow \left[A_i - C_i D_i^{-1} C_i' < 0\right].
\end{gathered}
\tag{3.4.15}
$$

Next, the proof will proceed along Steps 1 and 2 of strongly-guaranteed Berge equilibrium design for game (3.4.5) with $\mathbb{N} = \{1, 2\}$.

Step 1. Find $y^{(i)}(x_1, x_2)$ from the condition

$$
f_i(x_1, x_2, y^{(i)}(x_1, x_2)) = \min_y f_i(x_1, x_2, y). \tag{3.4.16}
$$

Without any constraints imposed on the strategy profiles $x = (x_1, x_2) \in \mathbb{R}^{2n}$ ($x_i \in \mathbb{R}^n$ and $y \in \mathbb{R}^m$), in expression (3.4.16) the sufficient conditions of minimum over all m-dimensional vector functions $y^{(i)}(x)$ reduce to

$$
\begin{gathered}
\operatorname{grad}_y f_i(x, y^{(i)}(x)) = \left.\frac{\partial f_i(x, y)}{\partial y}\right|_{y^{(i)}(x)} = 2D_i y^{(i)}(x) + 2C_i' x_j = 0_m, \\
(i, j = 1, 2;\ i \neq j) \\
\left.\frac{\partial^2 f_i(x, y)}{\partial y^2}\right|_{y^{(i)}(x)} = 2D_i > 0 \quad (i = 1, 2),
\end{gathered}
\tag{3.4.17}
$$

where $\frac{\partial^2 f_i}{\partial y^2}$ denotes the Hessian of $f_i(x, y)$ with respect to the components of the m-dimensional vector y; here we have used the inequalities $D_i > 0$ from (3.4.13) and also the gradient calculation formulas

$$
\frac{\partial}{\partial y}(y' L x) = Lx, \quad \frac{\partial}{\partial y}(x' K y) = K' x, \quad \frac{\partial}{\partial y}(y' D y) = 2Dy
$$

from [93, pp. 13–16]. In accordance with (3.4.17),

$$
y^{(i)}(x) = -D_i^{-1} C_i' x_j \quad (i, j = 1, 2;\ i \neq j). \tag{3.4.18}
$$

For all $x \in \mathbb{R}^{2n}$ $(i, j = 1, 2;\ i \neq j)$, from (3.4.18) we also have the identity

$$
[y^{(i)}(x)]' D_i y^{(i)}(x) + 2x_j' C_i y^{(i)}(x) = -[y^{(i)}(x)]' D_i y^{(i)}(x). \tag{3.4.19}
$$

Using (3.4.18) and (3.4.19), find

$$\begin{aligned}
f_1[x_1, x_2] &= f_1\left(x_1, x_2, y^{(1)}(x)\right) = x_2' A_1 x_2 + 2x_1' x_2 \\
&- \left[y^{(1)}(x)\right]' D_1 y^{(1)}(x) + 2a_1' x_2 + \varphi_1(x_1) \\
&= x_2' \left[A_1 - C_1 D_1^{-1} C_1'\right] x_2 + 2x_1' x_2 \\
&+ 2a_1' x_2 + \varphi_1(x_1),\ f_2[x_1, x_2] = f_2\left(x_1, x_2, y^{(2)}(x)\right) \\
&= x_1' \left[A_2 - C_2 D_2^{-1} C_2'\right] x_1 - 2x_2' x_1 + 2a_2' x_1 + \varphi_2(x_2),
\end{aligned} \tag{3.4.20}$$

where, by (3.4.15),

$$A_i - C_i D_i^{-1} C_i' < 0 \quad (i = 1, 2). \tag{3.4.21}$$

Step 2. To construct the strategy profile $(x_1^{\mathrm{B}}, x_2^{\mathrm{B}})$ that yields maximum in (3.4.8), one again employs the sufficient conditions

$$\begin{aligned}
\left.\frac{\partial f_1\left[x_1^{\mathrm{B}}, x_2\right]}{\partial x_2}\right|_{x_2^{\mathrm{B}}} &= 2\left[A_1 - C_1 D_1^{-1} C_1'\right] x_2^{\mathrm{B}} + 2x_1^{\mathrm{B}} + 2a_1 = 0_n, \\
\left.\frac{\partial f_2\left[x_1, x_2^{\mathrm{B}}\right]}{\partial x_1}\right|_{x_1^{\mathrm{B}}} &= -2x_2^{\mathrm{B}} + 2\left[A_2 - C_2 D_2^{-1} C_2'\right] x_1^{\mathrm{B}} + 2a_2 = 0_n, \\
\left.\frac{\partial^2 f_1\left[x_1^{\mathrm{B}}, x_2\right]}{\partial x_2^2}\right|_{x_2^{\mathrm{B}}} &= 2\left[A_1 - C_1 D_1^{-1} C_1'\right] < 0, \\
\left.\frac{\partial^2 f_2\left[x_1, x_2^{\mathrm{B}}\right]}{\partial x_1^2}\right|_{x_1^{\mathrm{B}}} &= 2\left[A_2 - C_2 D_2^{-1} C_2'\right] < 0.
\end{aligned} \tag{3.4.22}$$

A special *remark* is in order. When $A_i < 0$ and $D_i > 0$, the first two equalities in (3.4.22) are necessary conditions for the existence of a Berge equilibrium $(x_1^{\mathrm{B}}, x_2^{\mathrm{B}})$; this system of equations has a unique solution and hence the resulting equilibrium is also unique.

The last two inequalities in (3.4.22) follow directly from (3.4.21). Using the first two equalities, we arrive at the following system of two linear algebraic equations in the two unknown vectors x_1^{B} and x_2^{B}:

$$\begin{cases} (A_1 - C_1 D_1^{-1} C_1')x_2^{\mathrm{B}} + x_1^{\mathrm{B}} = -a_1, \\ -x_2^{\mathrm{B}} + (A_2 - C_2 D_2^{-1} C_2')x_1^{\mathrm{B}} = -a_2. \end{cases} \tag{3.4.23}$$

Multiplication of the first equation on the right by the inverse of the nondegenerate matrix $A_1 - C_1 D_1^{-1} C_1'$ (see (3.4.21)) and summation by columns yields

$$\left[(A_1 - C_1 D_1^{-1} C_1')^{-1} + (A_2 - C_2 D_2^{-1} C_2')\right] x_1^{\mathrm{B}} = -\left[(A_1 - C_1 D_1^{-1} C_1')^{-1} a_1 + a_2\right]. \tag{3.4.24}$$

In the same way, multiplication of the second equation on the right by the inverse of the nondegenerate matrix $A_2 - C_2 D_2^{-1} C_2'$ with minus sign and summation by columns yields

$$\left[(A_2 - C_2 D_2^{-1} C_2')^{-1} + (A_1 - C_1 D_1^{-1} C_1')\right] x_2^{\mathrm{B}} = = (A_2 - C_2 D_2^{-1} C_2')^{-1} a_2 - a_1. \tag{3.4.25}$$

From (3.4.21) we have

$$(A_i - C_i D_i^{-1} C_i')^{-1} < 0 \quad (i = 1, 2),$$

and, by (3.4.21),

$$\begin{aligned} (A_1 - C_1 D_1^{-1} C_1')^{-1} + (A_2 - C_2 D_2^{-1} C_2') < 0, \\ (A_2 - C_2 D_2^{-1} C_2')^{-1} + (A_1 - C_1 D_1^{-1} C_1') < 0. \end{aligned}$$

Hence, these matrices are invertible.

Then the first two formulas of (3.4.14) follow from (3.4.24) and (3.4.25). To construct $f_i^{\mathrm{B}} = f_i[x_1^{\mathrm{B}}, x_2^{\mathrm{B}}]$ $(i = 1, 2)$, we will again utilize the first two equalities in (3.4.22). In particular,

$$\begin{aligned} f_1^{\mathrm{B}} = f_1\left[x_1^{\mathrm{B}}, x_2^{\mathrm{B}}\right] = \left[x_2^{\mathrm{B}}\right]' \left[A_1 - C_1 D_1^{-1} C_1'\right] x_2^{\mathrm{B}} + 2\left[x_1^{\mathrm{B}}\right]' x_2^{\mathrm{B}} \\ + 2a_1' x_2^{\mathrm{B}} + \varphi_1\left(x_1^{\mathrm{B}}\right) = -\left[x_2^{\mathrm{B}}\right]' \left[A_1 - C_1 D_1^{-1} C_1'\right] x_2^{\mathrm{B}} + \varphi_1\left(x_1^{\mathrm{B}}\right), \end{aligned}$$

and similarly

$$f_2^{\mathrm{B}} = -\left[x_1^{\mathrm{B}}\right]' \left[A_2 - C_2 D_2^{-1} C_2'\right] x_1^{\mathrm{B}} + \varphi_2\left(x_2^{\mathrm{B}}\right).$$

Step 3. The quadruple $(x_1^B, x_2^B, f_1^B, f_2^B)$ is unique due to the strict convexity of $f_i(x, y)$ (3.4.21) in y for each $x \in \mathbb{R}^{2n}$, conditions (3.4.13), and the special remark for Step 2. ■

Example 3.4.1 To apply Proposition 3.4.2, we first have to verify requirements (3.4.13) and then construct the strongly-guaranteed Berge equilibrium $(x_1^B, x_2^B, f_1^B, f_2^B)$ by formulas (3.4.14). Let the variables x_1, x_2, and y in (3.4.14) as well as the constants a_i $(i = 1, 2)$ be scalar and choose the matrices $A_i = -\frac{1}{2}$, $C_i = 1$, and $D_i = 2$. In this case, $A_i - C_i D_i^{-1} C_i' = -\frac{1}{2} - \frac{1}{2} = -1$ $(i = 1, 2)$, and formulas (3.4.14) yield

$$x_1^B = [-1 - 1]^{-1}[-1a_2 - a_1] = \frac{1}{2}(a_1 + a_2), x_2^B = -[-1 - 1]^{-1}[-1a_1 + a_2]$$
$$= \frac{1}{2}(a_1 - a_2), f_1^B = -\frac{1}{4}(a_1 + a_2)^2, f_2^B = -\frac{1}{4}(a_1 - a_2)^2.$$

The dependence of the strong guarantees f_i^B on a_1 and a_2 is illustrated in Figs. 3.8 and 3.9 by the parabolic cylinders $f_1^B = -\frac{1}{4}(a_1 + a_2)^2$ and $f_2^B = -\frac{1}{4}(a_1 - a_2)^2$.

The vertex of the parabola $f_1^B = -\frac{1}{4}v^2$ in Fig. 3.8 is "sliding" along the line $a_1 = a_2$. By analogy, the vertex of the parabola $f_2^B = -\frac{1}{4}u^2$ in Fig. 3.9 is "sliding" along the line $a_1 = -a_2$, also forming a parabolic cylinder. Here $a_1 + a_2 = v$ and $a_1 - a_2 = u$.

Fig. 3.8 Strong guarantee f_1^B as function of a_1 and a_2

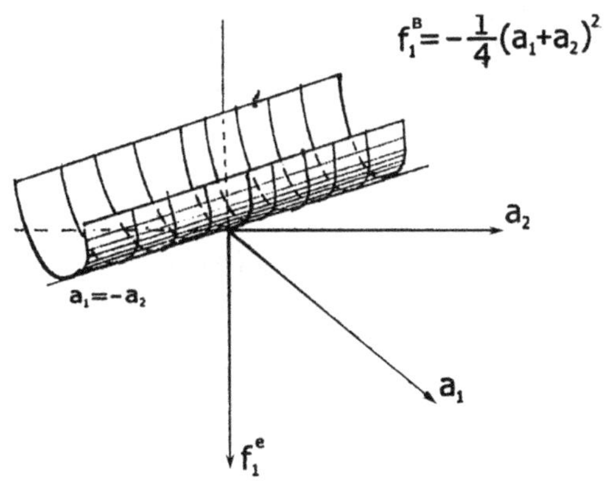

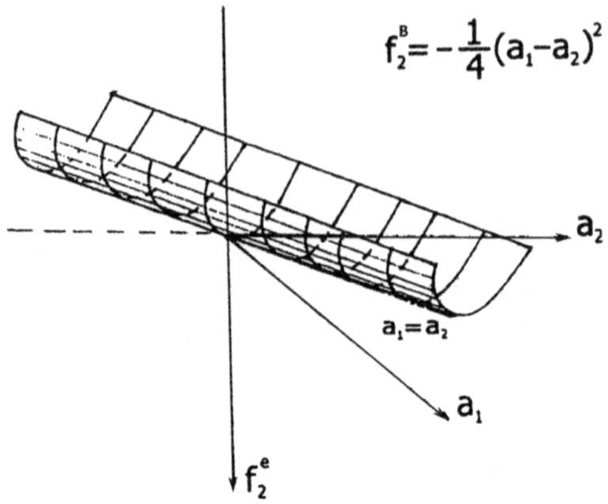

Fig. 3.9 Strong guarantee f_2^{B} as function of a_1 and a_2

3.5 Slater-Guaranteed Equilibria

> The mathematicians and physics men
> Have their mythology; they work alongside the truth,
> Never touching it; their equations are false
> But the things work. Or, when gross error appears,
> They invent new ones; they drop the theory of waves
> In universal ether and imagine curved space.
> Nevertheless their equations bombed Hiroshima.
> The terrible things worked.
>
> —Jeffers[24]

In this section, the third type of guaranteed solutions of a conflict (noncooperative N-player game under uncertainty) is suggested, the central concept for Chap. 3, based on an appropriate modification of maximin. The properties of this solution as well as its existence in the class of mixed strategies are established.

3.5.1 Definition and Properties

> Hier liegt der Hund begraben.
>
> —German proverb[25]

To formalize another guaranteed solution of the game

$$\langle\, \mathbb{N}, \{\mathrm{X}_i\}_{i\in\mathbb{N}}, \mathrm{Y}^{\mathrm{X}}, \{f_i(x,y)\}_{i\in\mathbb{N}}\,\rangle \tag{3.5.1}$$

[24]John Robinson Jeffers, (1887–1962), was an American poet. A fragment from his poem *The Great Wound.*

[25]German "That's where the dog lies buried." Close to the English proverb "That's where the shoe pinches!" Used to emphasize the essence of something.

using the maximin-type approach, we will again consider the following zero-sum game with a scalar payoff function $f(x, y)$:

$$\langle \{1, 2\}, \mathrm{X}, \mathrm{Y}^{\mathrm{X}}, f(x, y) \rangle, \tag{3.5.2}$$

where $\mathrm{X} \subseteq \mathbb{R}^n$ denotes the set of all strategies x of player 1 and $\mathrm{Y}^{\mathrm{X}} = \{y(x) | \mathrm{X} \to \mathrm{Y}\}$ is the set of counter-strategies $y(x)$ of player 2. In game (3.5.2), player 1 seeks *to maximize the scalar payoff function* $f(x, y)$ with an appropriate choice of his strategy $x \in \mathrm{X}$ under *information discrimination*, as follows. Making *the first move* in game (3.5.2), player 1 informs the opponent about his intended strategies $x \in \mathrm{X}$. Using this information, player 2 forms a counter-strategy $y(x) : \mathrm{X} \to \mathrm{Y}$ in order *to minimize* $f(x, y)$ with $y = y(x)$. Next, player 2 makes *the second move*, reporting the chosen strategy $y(\cdot) \in \mathrm{Y}^{\mathrm{X}}$ to player 1. The final decision is left to player 1: he designs a strategy $x^{\mathrm{g}} \in \mathrm{X}$ with maximization of $f(x, y(x))$, i.e., calculates

$$x^{\mathrm{g}} = \arg\max_{x \in \mathrm{X}} f(x, y(x)).$$

As a result, player 1 obtains the guaranteed payoff $f^{\mathrm{g}} = f(x^{\mathrm{g}}, y(x^{\mathrm{g}}))$ because

$$f(x^{\mathrm{g}}, y(x^{\mathrm{g}})) \leqslant f(x^{\mathrm{g}}, y) \quad \forall y \in \mathrm{Y}, \tag{3.5.3}$$

which follows from the design rule of the counter-strategy $y(x^{\mathrm{g}}) = \arg\min_{y \in \mathrm{Y}} f(x^{\mathrm{g}}, y)$.

Recall that the formalization procedure of the maximin f^{g} and maximin strategy x^{g} consists of two sequential operations:

- *first, the inner minimum*, i.e., for all $x \in \mathrm{X}$ it is necessary to find a counter-strategy $y(x) : \mathrm{X} \to \mathrm{Y}$ such that

$$\min_{y \in \mathrm{Y}} f(x, y) = f(x, y(x)) \quad \forall x \in \mathrm{X}; \tag{3.5.4}$$

- *second, the outer maximum*

$$\max_{x \in \mathrm{X}} f(x, y(x)) = f(x^{\mathrm{g}}, y(x^{\mathrm{g}})) = f^{\mathrm{g}}. \tag{3.5.5}$$

In accordance with (3.5.4), for $x = x^{\mathrm{g}}$ we have inequality (3.5.3), i.e., the strategy $x = x^{\mathrm{g}}$ gives player 1 the guaranteed payoff $f^{\mathrm{g}} \leqslant f(x^{\mathrm{g}}, y)$ $\forall y \in \mathrm{Y}$. Moreover, by (3.5.5) this guarantee f^{g} is highest among all guarantees $f(x, y(x))$ (for any strategies $x \in \mathrm{X}$ of player 1), since

$$f(x, y(x)) \leqslant f(x^{\mathrm{g}}, y(x^{\mathrm{g}})) = f^{\mathrm{g}} \quad \forall x \in \mathrm{X}.$$

Now, introduce the concept of *Slater-guaranteed Berge equilibrium* (SGBE) for the noncooperative game (3.5.1) using an appropriate modification of maximin, i.e., *first*, replacing the scalar inner minimum with a *vector minimum* (here the *Slater minimum*)

and *second*, replacing *the outer maximum initially* with Berge equilibria design and *then* with the vector maximum on the set of all Berge equilibria (here *the Slater maximum on the set of all Berge equilibria*).

Definition 3.5.1 A pair $(\bar{x}^{\mathrm{B}}, \bar{f}^{\mathrm{S}}) \in \mathrm{X} \times \mathbb{R}^N$ is called a Slater-guaranteed Berge equilibrium in game (3.5.1) if there exists an uncertain factor $y_{\mathrm{S}}(x) : \mathrm{X} \to \mathrm{Y}$ such that

(1) $\bar{f}^{\mathrm{S}} = (\bar{f}_1^{\mathrm{S}}, \ldots, \bar{f}_N^{\mathrm{S}}) = f\,(\bar{x}^{\mathrm{B}}, y_S\,(\bar{x}^{\mathrm{B}}))$, i.e.,

$$\bar{f}_i^{\mathrm{S}} = f_i\left(\bar{x}^{\mathrm{B}}, y_{\mathrm{S}}\left(\bar{x}^{\mathrm{B}}\right)\right) \quad (i \in \mathbb{N});$$

(2) for each $x \in \mathrm{X}$, the uncertain factor $y_{\mathrm{S}}(x)$ is a Slater-minimal alternative in the N-criteria choice problem $\langle \mathrm{Y}, f(x, y)\rangle$, i.e., for any alternative $x = (x_1, \ldots, x_N) \in \mathrm{X}_1 \times \cdots \times \mathrm{X}_N = \mathrm{X}$ the system of N strict inequalities

$$f_i[x] = f_i(x, y_{\mathrm{S}}(x)) > f_i(x, y) \quad \forall y \in \mathrm{Y} \quad (i \in \mathbb{N}) \tag{3.5.6}$$

is inconsistent;

(3) the strategy profiles $x^{\mathrm{B}} \in \mathrm{X}$ are Berge equilibria in the noncooperative game

$$\langle\, \mathbb{N}, \{\mathrm{X}_i\}_{i\in\mathbb{N}}, \{f_i(x, y_{\mathrm{S}}(x))\}_{i\in\mathbb{N}} \,\rangle, \tag{3.5.7}$$

i.e.,

$$\max_{x\in\mathrm{X}} f_i\left(x||x_i^{\mathrm{B}}, y_{\mathrm{S}}\left(x||x_i^{\mathrm{B}}\right)\right) = f_i\left[x^{\mathrm{B}}\right], \quad i \in \mathbb{N}, \tag{3.5.8}$$

where $(x||x_i^{\mathrm{B}}) = (x_1, \ldots, x_{i-1}, x_i^{\mathrm{B}}, x_{i+1}, \ldots, x_N)$; denote by X^{B} the set of all Berge equilibria;

4. the strategy profile $\bar{x}^{\mathrm{B}} \in \mathrm{X}^{\mathrm{B}}$ is a Slater-maximal alternative [81] in the N-criteria choice problem

$$\left\langle \mathrm{X}^{\mathrm{B}}, \{f_i(x, y_S(x))\}_{i\in\mathbb{N}} \right\rangle,$$

i.e., for all $x \in \mathrm{X}^{\mathrm{B}}$ the system of strict inequalities

$$\bar{f}_i^{\mathrm{S}} = f_i\left(\bar{x}^{\mathrm{B}}, y_{\mathrm{S}}\left(\bar{x}^{\mathrm{B}}\right)\right) < f_i(x, y_{\mathrm{S}}(x)), \quad i \in \mathbb{N}, \tag{3.5.9}$$

is inconsistent.

Remark 3.5.1

(a) As inequalities (3.5.6) are inconsistent for $x = \bar{x}^{\mathrm{B}}$, the N-dimensional vector $\bar{f}^{\mathrm{S}}$ forms the Slater guarantee: if the players choose their strategies from the profile $\bar{x}^{\mathrm{B}}$, then it is impossible to reduce all payoffs $\bar{f}_i^{\mathrm{S}}$ $(i \in \mathbb{N})$ simultaneously with any choice $y \in \mathrm{Y}$, because for all $y \in \mathrm{Y}$ the inequalities $f_i(\bar{x}^{\mathrm{B}}, y) < f_i(\bar{x}^{\mathrm{B}}, y_S(\bar{x}^{\mathrm{B}})) = \bar{f}_i^{\mathrm{S}}$, $i \in \mathbb{N}$, fail.
(b) Condition (3.5.8) implies that each strategy profile $x^{\mathrm{B}} \in \mathrm{X}^{\mathrm{B}}$ is a Berge equilibrium in the noncooperative game (3.5.7) and hence is stable against the deviations of any coalitions of size $N - 1$.
(c) Due to the inconsistency of inequalities (3.5.9), the vector guarantee $\bar{f}^{\mathrm{S}} = (\bar{f}_1^{\mathrm{S}}, \ldots, \bar{f}_N^{\mathrm{S}})$ is highest in the vector sense among all guarantees $f(x^{\mathrm{B}}, y_{\mathrm{S}}(x^{\mathrm{B}}))$ $\forall x^{\mathrm{B}} \in \mathrm{X}^{\mathrm{B}}$.

Therefore, following their strategies $\bar{x}_i^{\mathrm{B}}$ $(i \in \mathbb{N})$ from the Berge equilibrium $\bar{x}^{\mathrm{B}} = (\bar{x}_1^{\mathrm{B}}, \ldots, \bar{x}_N^{\mathrm{B}})$, the players obtain the vector guarantee $\bar{f}^{\mathrm{S}}$ for all $y \in \mathrm{Y}$; furthermore, this guarantee is highest (Slater-maximal, see (3.5.9)) among all guarantees yielded by the strategies x_i^{B} $(i \in \mathbb{N})$ from the other Berge equilibria $x^{\mathrm{B}} \in \mathrm{X}^{\mathrm{B}}$. (Note that in Example 2.4.1 the set of Slater-guaranteed Berge equilibria is $(\bar{x}^{\mathrm{B}}, \bar{f}^{\mathrm{B}}) = ((1; 1), (1 - \cos\beta); (1 - \sin\beta) | \beta \in [0, \frac{\pi}{2}]))$.

3.5.2 Existence of Guaranteed Equilibrium in Mixed Strategies

Grau, teurer Freund, ist alle Theorie,
Und grün des Lebens goldner Baum.[26]

The existence of a Slater-guaranteed Berge equilibrium is established in the noncooperative game under uncertainty in the class of mixed strategies, under standard assumptions of game theory.

Problem Statement and Auxiliary Results Consider the noncooperative N-player game under uncertainty defined by an ordered quadruple

$$\Gamma = \langle\, \mathbb{N}, \{\mathrm{X}_i\}_{i\in\mathbb{N}}, \mathrm{Y}^{\mathrm{X}}, \{f_i(x, y)\}_{i\in\mathbb{N}}\,\rangle. \tag{3.5.10}$$

Recall that in the game Γ,

$\mathbb{N} = \{1, \ldots, N\}$ denotes the set of players, with an integer $N \geqslant 2$;
$\mathrm{X}_i \subseteq \mathbb{R}^{n_i}$ is the set of *pure strategies* x_i of player i $(i \in \mathbb{N})$;
$\mathrm{Y} \subseteq \mathbb{R}^m$ is the set of uncertain factors y.

[26]German "My worthy friend, gray are all theories,
And green alone Life's golden tree." A quote from *Faust*, The Study (Mephistopheles), by J.W. von Goethe.

In this game, the players do not build any coalitions and each player i ($i \in \mathbb{N}$) chooses his pure strategy x_i, which yields *a pure strategy profile* $x = (x_1, \ldots, x_N)$ of the game Γ, and $x \in \mathrm{X} = \prod_{i \in \mathbf{N}} \mathrm{X}_i$.

By analogy with the inner minimum in maximin, we will assume the *information discrimination* of the players: they report their chosen strategies x_i (more precisely, the strategy profile $x = (x_1, \ldots, x_N) \in \mathrm{X}$) to a DM, who is responsible for uncertainty generation. This DM generates the uncertain factors in the form of a counter-strategy profile $y(x) : \mathrm{X} \to \mathrm{Y}$, $y(\cdot) \in \mathrm{Y}^{\mathrm{X}}$. Thus, the uncertainty in the game Γ will be identified with the m-dimensional vector function $y(x) : \mathrm{X} \to \mathrm{Y}$. Note that the DM chooses $y(x) = y_{\mathrm{S}}(x)$ in order to achieve the Slater minimum of $f(x, y_{\mathrm{S}}(x))$ in the N-criteria choice problem

$$\Gamma(x) = \langle\, \mathrm{Y}, \{f(x, y) = (f_1(x, y), \ldots, f_N(x, y))\}\,\rangle \tag{3.5.11}$$

for each $x \in \mathrm{X}$. In other words, for each $x \in \mathrm{X}$ the system of strict inequalities

$$f_i(x, y) \;<\; f_i(x, y_{\mathrm{S}}(x)) \quad \forall y \in \mathrm{Y},\ i \in \mathbb{N},$$

is inconsistent. Then the following result holds.

Proposition 3.5.1 *Consider the game Γ under the assumptions that*

(a) the sets X_i ($i \in \mathbb{N}$) and Y are nonempty, convex and compact;
(b) the scalar functions $f_i(x, y)$ ($i \in \mathbb{N}$) are continuous on $\mathrm{X} \times \mathrm{Y}$ and there exists at least one $j \in \mathbb{N}$ such that for each $x \in \mathrm{X}$ the function $f_j(x, y)$ is strictly convex in $y \in \mathrm{Y}$, i.e. for any $y^{(1)}, y^{(2)} \in \mathrm{Y}$ and any $\lambda \in (0, 1)$,

$$f_j\left(x, \lambda y^{(1)} + (1-\lambda) y^{(2)}\right) \;<\; \lambda f_j\left(x, y^{(1)}\right) + (1-\lambda) f_j\left(x, y^{(2)}\right).$$

Then there exists a unique Slater-minimal aware uncertainty $y_{\mathrm{S}}(x)$ in this game that is continuous in $x \in \mathrm{X}$.

Proof If $\alpha_i = \text{const} \geqslant 0$ ($i \in \mathbb{N}$) and $\sum_{i=1}^{N} \alpha_i > 0$, then for each $x \in \mathrm{X}$ the minimizer

$$y_{\mathrm{S}}(x) = \arg\min_{y \in \mathrm{Y}} \sum_{i=1}^{N} \alpha_i f_i(x, y) \tag{3.5.12}$$

is [152, pp. 68–69] a Slater-minimal uncertainty [79, 80] in (3.5.11). On the other hand, under the assumptions of Proposition 3.5.1, using (3.5.12) with $\alpha_j = \text{const} > 0$ and $\alpha_k = 0$ ($k \neq j,\ k \in \mathbb{N}$) leads to the desired result, see [14, p. 54]. ■

Thus, in the game Γ the *first move* belongs to the players: they choose and then report their pure strategies $x_i \in \mathrm{X}_i$ (i.e., the strategy profile $x = (x_1, \ldots, x_N) \in \mathrm{X}$) to a DM, who is "in charge of" uncertainty design. The *second move* is given to the DM—he generates the Slater minimal uncertainty $y(x) = y_S(x)$ and then reports it to each player. The *third move* is made by the players—in the induced noncooperative game without uncertainty

$$\Gamma_b = \langle \mathbb{N}, \{X_i\}_{i\in\mathbb{N}}, \{f_i(x, y_S(x))\}_{i\in\mathbb{N}} \rangle, \tag{3.5.13}$$

they find a Berge equilibrium $x^B \in X$ from the conditions

$$\max_{x\in X} f_i\left(x||x_i^B, y_S\left(x||x_i^B\right)\right) = f_i\left(x^B, y_S\left(x^B\right)\right) \quad (i \in \mathbb{N}). \tag{3.5.14}$$

However, some difficulties may arise concerning the existence of pure-strategy Berge equilibria $x^B = (x_1^B, \ldots, x_N^B)$ as game (3.5.13) evolves. (These equilibria must satisfy the system of N equalities (3.5.14)). In fact, despite the continuity of $f_i[x] = f_i(x, y_S(x))$ $(i \in \mathbb{N})$, there are numerous examples without an equilibrium x^B. Following the standard approach of mathematical game theory, we will consider the mixed extension of game (3.5.13), i.e.,

$$\tilde{\Gamma}_b = \langle \mathbb{N}, \{\mu_i\}_{i\in\mathbb{N}}, \{f_i[\mu]\}_{i\in\mathbb{N}} \rangle. \tag{3.5.15}$$

By Theorem 2.9.1, game (3.5.15) possesses Berge equilibria $\mu^B(\cdot) \in \{\mu\}$ provided the functions $f_i(x, y_S(x))$ are continuous in $x \in X$ $(i \in \mathbb{N})$. The Berge equilibria are obtained from N equalities of the form

$$\max_{\mu(\cdot)\in\{\mu\}} f_i\left[\mu||\mu_i^B\right] = f_i\left[\mu^B\right] \quad (i \in \mathbb{N}). \tag{3.5.16}$$

Next, for each compact set X_i, one considers the Borel σ-algebra of all subsets of the set X_i and chooses as a mixed strategy $\mu_i(\cdot)$ a nonnegative countably additive scalar function $\mu_i(\cdot)$ defined on this Borel σ-algebra that is normalized by unity on X_i. Denote by $\{\mu_i\}$ the set of such mixed strategies. We introduce the product measure $\mu(dx) = \mu_1(dx_1)\cdots\mu_N(dx_N)$ and the set $\{\mu\}$ in the same way as before. Finally, in (3.5.15) and (3.5.16) the expectations are the payoff functions of players, i.e.,

$$f_i[\mu] = \int_X f_i(x, y_S(x))\mu(dx) \quad (i \in \mathbb{N}).$$

Theorem 2.9.1 ensures the existence of a product measure $\mu^B(\cdot) \in \{\mu\}$ that satisfies conditions (3.5.16). Furthermore, the set of such Berge equilibrium measures $\{\mu^B\}$ is weakly compact (see Proposition 2.9.1).

We will study game (3.5.1) and associate with it the *quasi-mixed extension*

$$\langle \mathbb{N}, \{\mu_i\}_{i\in\mathbb{N}}, Y^X, \{f_i[\mu]\}_{i\in\mathbb{N}} \rangle, \tag{3.5.17}$$

where $\mathbb{N} = \{1, \ldots, N\}$;

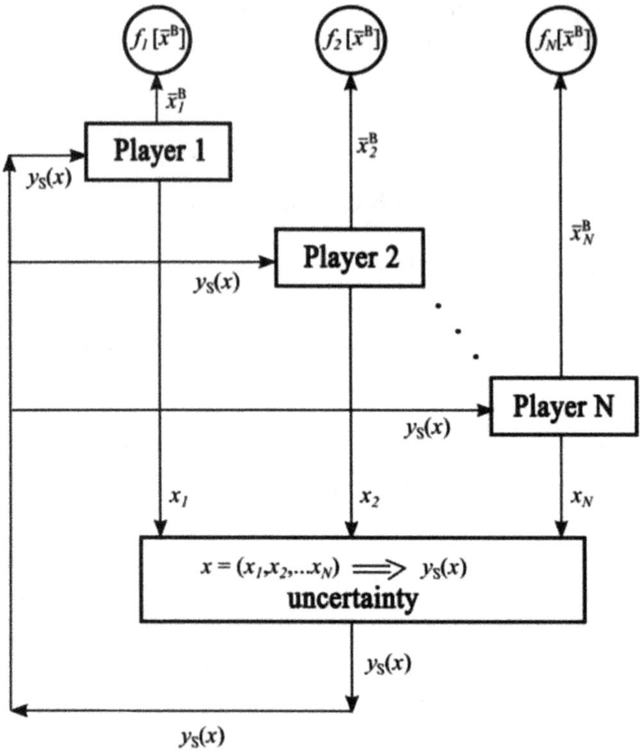

Fig. 3.10 Sequence of moves in game (3.5.1) based on SGBE

$\mathrm{X} = \prod_{i \in \mathbb{N}} \mathrm{X}_i$ is the set of pure strategy profiles $x = (x_1, \dots, x_N) \in \mathrm{X}$ in game (3.5.1);

$\{\mu_i\}$ stands for the set of mixed strategies $\mu_i(\cdot)$ of player $i \in \mathbb{N}$; a mixed strategy profile is the product measure $\mu(\cdot) = \mu_1(\cdot) \cdots \mu_N(\cdot)$;

Y^{X} is regarded as the set of uncertain factors, i.e., counter-strategies $y(x) : \mathrm{X} \to \mathrm{Y}$;

$f_i[\mu] = \int_{\mathrm{X}} f_i(x, y(x))\mu(dx)$ is the payoff function of player i in game (3.5.17), which represents the expectation of the payoff function $f_i(x, y) = f_i(x, y(x))$ in game (3.5.1) under any realizations of the strategy profile $x \in \mathrm{X}$ and continuous uncertainty $y(\cdot) \in C(\mathrm{X}, \mathrm{Y})$ (Fig. 3.10).

Definition 3.5.2 A pair $(\bar{\mu}^{\mathrm{B}}(\cdot), \widetilde{f}^{\mathrm{S}}) \in \{\mu\} \times \mathbb{R}^N$ is called a Slater-guaranteed Berge equilibrium in mixed strategies in game (3.5.1) if there exists an uncertainty, i.e., a counter-strategy $y_{\mathrm{S}}(x) : \mathrm{X} \to \mathrm{Y}$, such that

(1°) for each strategy profile $x \in \mathrm{X}$ the uncertainty $y_{\mathrm{S}}(x)$ is a Slater-minimal alternative in the N-criteria choice problem

$$\Gamma(x) = \langle\, \mathrm{Y}, \{f(x, y)\}\, \rangle,$$

that is, for each $x \in X$ the system of inequalities

$$f_i(x, y) < f_i(x, y_S(x)) \quad \forall y \in Y, \ i \in \mathbb{N},$$

is inconsistent;

(2°) the mixed strategy profile $\mu^B(\cdot) \in \{\mu\}$ is a Berge equilibrium in the mixed extension

$$\langle \mathbb{N}, \{\mu_i\}_{i \in \mathbb{N}}, \{f_i[\mu] = \int\limits_X f_i(x, y_S(x))\mu(dx)\}_{i \in \mathbb{N}} \rangle$$

of the noncooperative game without uncertainty

$$\langle \mathbb{N}, \{X_i\}_{i \in \mathbb{N}}, \{f_i(x, y_S(x)) = f_i[x]\}_{i \in \mathbb{N}} \rangle,$$

i.e., for $\mu^B(\cdot)$ all the N equalities of the form (3.5.16) hold; denote by $\{\mu^B\}$ the set of all $\mu^B(\cdot)$;

(3°) the strategy profile $\bar{\mu}^B(\cdot) \in \{\mu^B\}$ is a Slater-maximal alternative in the N-criteria choice problem

$$\langle \{\mu^B\}, \{f_i[\mu]\}_{i \in \mathbb{N}} \rangle, \tag{3.5.18}$$

i.e., for any $\mu(\cdot) \in \{\mu^B\}$ the system of N strict inequalities

$$f_i[\mu] > f_i\left[\bar{\mu}^B\right] \quad (i \in \mathbb{N})$$

is inconsistent;

(4°) the components $\widetilde{f}_i^S$ $(i \in \mathbb{N})$ of the vector $\widetilde{f}^S = (\widetilde{f}_1^S, \ldots, \widetilde{f}_N^S)$ satisfy $\widetilde{f}_i^S = f_i[\bar{\mu}^B]$ $(i \in \mathbb{N})$.

3.5.3 Existence Theorem

He that will not apply new remedies must expect new evils.
—Bacon[27]

The central result of this section—the existence of a Slater-guaranteed Berge equilibrium in mixed strategies in game (3.5.1) under standard assumptions of mathematical game theory—is established.

[27]Sir Francis Bacon, (1561–1626), was an English lawyer, statesman, and philosopher.

Theorem 3.5.1 *Consider game* (3.5.1) *under the assumptions that*

(1°) *the sets* X_i $(i \in \mathbb{N})$ *and* Y *are convex and compact;*
(2°) *the payoff functions* $f_i(x, y)$ $(i \in \mathbb{N})$ *are continuous on* $\mathrm{X} \times \mathrm{Y}$ *and there exists at least one number* $j \in \mathbb{N}$ *such that for each* $x \in \mathrm{X}$ *the function* $f_j(x, y)$ *is strictly convex in* $y \in \mathrm{Y}$.

Then there exists a Slater-guaranteed Berge equilibrium in mixed strategies in this game.

Proof Assumptions (1°) and (2°) of Theorem 3.5.1 in combination with Proposition 3.5.1 imply the existence of a continuous uncertainty $y_S(x) : \mathrm{X} \to \mathrm{Y}$ on X that is Slater minimal in the N-criteria choice problem $\Gamma(x)$ (3.5.11) for each $x \in \mathrm{X}$. Next, construct the noncooperative N-player game (3.5.13) without uncertainty. In this game, the payoff functions $f_i(x, y_S(x))$ are continuous on X as superpositions of the continuous functions $f_i(x, y)$ and $y_S(x)$. Then the mixed extension (3.5.15) of game (3.5.13) possesses a Berge equilibrium $\mu^{\mathrm{B}}(\cdot) \in \{\mu\}$. Denote by $\{\mu^{\mathrm{B}}\}$ the set of all Berge equilibria $\mu^{\mathrm{B}}(\cdot)$. This set is weakly compact, which follows from the same weak properties of $\{\mu\}$ and inequalities (3.5.16). But then the set $\mathcal{F}^{\mathrm{B}} = \left\{f[\mu^{\mathrm{B}}] | \mu^{\mathrm{B}}(\cdot) \in \{\mu^{\mathrm{B}}\}\right\}$ is also compact in $\mathbb{R}^N$, and in addition $\mathcal{F}^{\mathrm{B}} \subset \mathcal{F} = \{f[\mu] | \, \mu(\cdot) \in \{\mu\}\}$.

Consider the linear convolution $\sum_{i \in \mathbb{N}} \alpha_i f_i$, where $\alpha_i = \mathrm{const} \geq 0$ $(i \in \mathbb{N})$, defined on the set $\mathcal{F}^B$. Due to the continuity on the compact set $\mathcal{F}^{\mathrm{B}}$, there exists an N-dimensional vector $\widetilde{f}^{\mathrm{S}} = (\widetilde{f}_1^{\mathrm{S}}, \ldots, \widetilde{f}_N^{\mathrm{S}}) \in \mathcal{F}^{\mathrm{B}}$ such that

$$\max_{f \in \mathcal{F}^{\mathrm{B}}} \sum_{i \in \mathbb{N}} \alpha_i f_i = \sum_{i \in \mathbb{N}} \alpha_i \widetilde{f}_i^{\mathrm{S}}.$$

Using $\widetilde{f}^{\mathrm{S}}$, find a mixed strategy profile $\bar{\mu}^{\mathrm{B}}(\cdot) \in \in \{\mu^{\mathrm{B}}\}$ for which

$$\widetilde{f}_i^{\mathrm{S}} = f_i[\bar{\mu}^{\mathrm{B}}] \quad (i \in \mathbb{N}).$$

This profile $\bar{\mu}^{\mathrm{B}}(\cdot)$ is a Slater-maximal alternative in the N-criteria choice problem (3.5.18). Therefore, the resulting pair $(\bar{\mu}^{\mathrm{B}}(\cdot), \widetilde{f}^{\mathrm{S}}) \in \{\mu\} \times \mathbb{R}^N$ is the Slater-guaranteed Berge equilibrium in mixed strategies in game (3.5.1), as follows directly from Definition 3.5.2. ■

Remark 3.5.2 Let us discuss the game-theoretic meaning of Definition 3.5.2; recall that $f = (f_1, \ldots, f_N)$.

First, in accordance with condition (2°) of this definition, every strategy profile $x \in \mathrm{X}$ generates a vector guarantee $f(x, y_S(x))$ in pure strategies, since for all $y \in \mathrm{Y}$ all payoffs $f_i(x, y)$ cannot be simultaneously smaller than $f_i(x, y_S(x))$ $(i \in \mathbb{N})$. This expresses an analog of the inner minimum in maximin.

Second, inequalities (3.5.16) lead to

$$f_i\left[x || \mu_i^{\mathrm{B}}\right] \leqslant f_i\left[\mu^{\mathrm{B}}\right] \quad \forall x \in \mathrm{X} \; (i \in \mathbb{N}),$$

because the Dirac δ-function $\delta(x_i - \bar{x}_i)(d\bar{x}_i)$ is a probability measure from $\{\mu_i\}$ and hence [22, p. 125]

$$f\left[x||\mu_i^{\mathrm{B}}\right] \nless f\left[\mu^{\mathrm{B}}\right].$$

Hence, the mixed strategy Berge equilibrium $\mu^{\mathrm{B}}(\cdot)$ is stable against any pure strategy deviations of the coalition of size $N-1$.

Third, each vector guarantee $f(x, y_{\mathrm{S}}(x))$ in pure strategies (the Slater minimum in $\Gamma(x) = \langle \mathrm{Y}, \{f(x, y)\}\rangle$) yields a vector guarantee $f[\mu]$ in mixed strategies. Really, the system of inequalities

$$f_i^{\mathrm{e}}[x] = f_i(x, y_{\mathrm{S}}(x)) > f_i(x, y) \quad \forall\, y = \mathrm{const} \in \mathrm{Y},\ i \in \mathbb{N}, \tag{3.5.19}$$

is inconsistent for all $x \in \mathrm{X}$ if and only if, for each $x \in \mathrm{X}$ and each $y \in \mathrm{Y}$, there exists a corresponding number $j(x, y) = j \in \mathbb{N}$ such that

$$f_j(x, y_{\mathrm{S}}(x)) \leqslant f_j(x, y).$$

Integrating both sides with respect to x using an arbitrary mixed strategy profile $\mu(\cdot) \in \{\mu\}$ as the integration measure gives

$$f_j^{\mathrm{S}}[\mu] = \int\limits_{\mathrm{X}} f_j(x, y_{\mathrm{S}}(x))\mu(dx) \leqslant \int\limits_{\mathrm{X}} f_j(x, y)\mu(dx) = f_j[\mu, y] \quad \forall\, y = \mathrm{const} \in \mathrm{Y},$$

which is equivalent to the following. Each mixed strategy profile $\mu(\cdot) \in \{\mu\}$ yields the vector guarantee $f^{\mathrm{S}}[\mu] = (f_1^{\mathrm{S}}[\mu], \ldots, f_N^{\mathrm{S}}[\mu])$, because for any $y \in \mathrm{Y}$ all payoffs $f_i[\mu, y]$ cannot be simultaneously smaller than $f_i^{\mathrm{S}}[\mu]$ (in terms of component-wise comparison).

Fourth, by associating with the pure strategy game

$$\langle\, \mathbb{N}, \{\mathrm{X}_i\}_{i \in \mathbb{N}}, \{f_i(x, y_S(x)) = f_i^{\mathrm{B}}[x]\}_{i \in \mathbb{N}}\, \rangle \tag{3.5.20}$$

its mixed extension

$$\langle\, \mathbb{N}, \{\mu_i\}_{i \in \mathbb{N}}, \{f_i^{\mathrm{B}}[\mu]\}_{i \in \mathbb{N}}\, \rangle, \tag{3.5.21}$$

we have actually passed from the noncooperative game of vector guarantees (3.5.20) in pure strategies to its mixed extension, i.e., the noncooperative game of vector guarantees (3.5.21) in mixed strategies. Now, an analog of the outer maximum in maximin is a sequential application of two operations, the calculation of all Berge equilibria in game (3.5.15) and the construction of the Slater-maximal Berge equilibrium $\bar{\mu}^{\mathrm{B}}(\cdot)$ among them. Consequently, by choosing their mixed strategies and forming a mixed strategy profile $\mu(\cdot) \in \{\mu\}$, the players obtain the payoffs

$f_i[\mu, y] = \int_X f_i(x, y)\mu(dx)$ that cannot be simultaneously smaller than $f_i^S[\mu] = \int_X f_i(x, y_S(x))\mu(dx)$ $(i \in \mathbb{N})$ under any pure uncertainties $y \in Y$. Among all the Berge equilibria $\mu^B(\cdot) \in \{\mu\}$, the solution recommends that the players use the Slater-maximal measure, i.e., the strategy profile $\bar{\mu}^B(\cdot) \in \{\mu^B\}$ yielding the largest (Slater-maximal) vector payoff $\bar{f}[\bar{\mu}^B]$. As a matter of fact, this expresses an analog of the outer maximum in maximin. ■

Chapter 4
Applications to Competitive Economic Models

There is no branch of mathematics, however abstract,
which may not some day be applied to phenomena of the real world.
—Lobachevsky[1]

This chapter is devoted to a study of the equilibrium solutions (in the sense of Berge and Nash) of the Cournot and Bertrand oligopoly models. As a special case, the models with import as an uncertain disturbance are also analysed using mathematical theory of noncooperative games.

4.1 The Cournot Oligopoly Model

Never let any Government imagine that it can choose perfectly safe courses;
rather let it expect to have to take very doubtful ones, because it is found
in ordinary affairs that one never seeks to avoid one trouble without
running into another; but prudence consists in knowing how to distinguish
the character of troubles, and for choice to take the lesser evil.
—Machiavelli[2]

Berge equilibria in the Cournot oligopoly model are constructed. A detailed comparison of the Berge and Nash equilibria in this model is given. Conditions under which the players obtain higher payoffs in a Berge equilibrium than in a Nash equilibrium are established.

[1]Nikolay I. Lobachevsky, (1792–1856), was a Russian mathematician and founder of non-Euclidean geometry.

[2]Niccolò Machiavelli, (1469–1527), was an Italian Renaissance political philosopher and statesman, secretary of the Florentine republic. A quote from *The Prince*, Chapter XXI.

© Springer Nature Switzerland AG 2020

M. E. Salukvadze, V. I. Zhukovskiy, *The Berge Equilibrium: A Game-Theoretic Framework for the Golden Rule of Ethics*, Static & Dynamic Game Theory: Foundations & Applications, https://doi.org/10.1007/978-3-030-25546-6_4

4.1.1 Introduction

Start from inferiority
in order to reach superiority;
in other words, scratch your heels
rather than the back of your head.
—Kozma Prutkov[3]

In many large areas of economy (such as metallurgy, oil production and refining, electronics), the main competition takes place among several companies that dominate the market. The first models of such markets—oligopolies—were described more than a 100 years ago by Cournot [226], Bertrand [205], and Hotelling [239]. Modeling of oligopolies continues in many modern works. For instance, in 2014 the Nobel Prize in Economic Sciences was awarded to J. Tirole "for his analysis of market power and regulation in sectors with few large companies." Tirole is the author of *The Theory of Industrial Organization* [278], an excellent modern textbook on the theory of imperfect competition [157–159, 162, 163].

Publications studying the behavior of oligopolies usually proceed from the assumption that *each company is primarily concerned with its own profits*. This approach meets the concept of Nash equilibrium [258], widely adopted to model the behavior of players in a competitive market. The exact opposite of such a "selfish" equilibrium is the "altruistic" concept of Berge equilibrium: without caring about himself, each player acts (chooses strategies) so as to maximize the profits of all other market participants. This concept appeared in Russia in 1994 and was called Berge equilibrium in reference to C. Berge's monograph [202], which was originally published in French back in 1957. The first research works on the concept of Berge equilibrium belong to Vaisman and Zhukovskiy [11, 13, 302]. Once it became known outside Russia, the concept of Berge equilibrium gradually gained popularity, as witnessed by a large number of publications related to this type of equilibrium. Most of them however deal with purely theoretical issues or applications to psychology [223, 227]. To our knowledge there are only a few researchers exploring Berge equilibrium in economic problems. Perhaps, this state of affairs is to a large extent a consequence of Shubik's critical review [269] of Berge's book [202] ("…The arguments have been presented in a rather abstract manner and no attention has been paid to applications to economics. The book will be of a little direct interest to economists…"). As it turns out, things are not so black and white. In Sect. 4.1, we will consider Berge equilibrium in the Cournot oligopoly model and also its relationship to Nash equilibrium. What is important, we will exhibit the cases in which the players gain greater profits by following the concept of Berge equilibrium than by using the Nash equilibrium strategies.

[3] An English translation of a quote from [168, p. 239].

4.1.2 *Basic Notations and Definitions*

Mathematics may be defined as the subject
in which we never know what we are talking about,
nor whether what we are saying is true.
—Russel

The concepts of Berge and Nash equilibria are formalized for the Cournot oligopoly model.

Recall that we are considering a noncooperative N-player game identified with a triplet [42–45]

$$\Gamma = \langle \mathbb{N}, \{\mathrm{X}_i\}_{i\in\mathbb{N}}, \{f_i(x)\}_{i\in\mathbb{N}}\rangle. \tag{4.1.1}$$

Here $\mathbb{N} = \{1, 2, \ldots, N\}$ denotes the set of players ($N > 1$); each of N players chooses his strategy (action) $x_i \in \mathrm{X}_i \subseteq \mathbb{R}^{n_i}$ without forming coalitions with other players (as before, $\mathbb{R}^k$, $k \geqslant 1$, is the k-dimensional Euclidean space—the real arithmetical space composed of all ordered collections of k real numbers written as column vectors, with the standard scalar product and the Euclidean norm $\|\cdot\|$); their choice forms a *strategy profile*

$$x = (x_1, \ldots, x_N) \in \mathrm{X} = \prod_{i\in\mathbb{N}} \mathrm{X}_i \subseteq \mathbb{R}^n \quad (n = \sum_{i\in\mathbb{N}} n_i)$$

in this game; the payoff function $f_i(x)$, defined on the set X, numerically assesses the performance (quality) of player i ($i \in \mathbb{N}$); in the sequel, we denote $(x\|z_i) = (x_1, \ldots, x_{i-1}, z_i, x_{i+1}, \ldots, x_N)$ and $f = (f_1, \ldots, f_N)$.

Definition 4.1.1 A pair $(x^{\mathrm{e}}, f^{\mathrm{e}}) = \big((x_1^{\mathrm{e}}, \ldots, x_N^{\mathrm{e}}), (f_1(x^{\mathrm{e}}), \ldots, f_N(x^{\mathrm{e}}))\big) \in \mathrm{X} \times \mathbb{R}^N$ is called a Nash equilibrium in game (4.1.1) if

$$\max_{x_i\in\mathrm{X}_i} f_i(x^{\mathrm{e}}\|x_i) = f_i(x^{\mathrm{e}}) \quad (i \in \mathbb{N}); \tag{4.1.2}$$

in what follows, x^{e} itself will be also called a Nash equilibrium in game (4.1.1).

Definition 4.1.2 A pair $(x^{\mathrm{B}}, f^{\mathrm{B}}) = ((x_1^{\mathrm{B}}, \ldots, x_N^{\mathrm{B}}), (f_1(x^{\mathrm{B}}), \ldots, f_N(x^{\mathrm{B}}))) \in \mathrm{X} \times \mathbb{R}^N$ is called a Berge equilibrium in game (4.1.1) if

$$\max_{x\in\mathrm{X}} f_i(x\|x_i^{\mathrm{B}}) = f_i(x^{\mathrm{B}}) \quad (i \in \mathbb{N}); \tag{4.1.3}$$

by analogy, x^{B} itself will be also called a Berge equilibrium in game (4.1.1).

4.1.3 The Cournot Oligopoly and Equilibrium Strategies

Explicit forms of the Berge and Nash equilibria as well as corresponding payoffs in the Cournot oligopoly model are found.

In 1838, Cournot [226] studied a market with domination of several large players producing the same product. It was postulated in [226] that the market participants are competing with each other for the market share by the law of demand and supply. Later on, this model was called the *Cournot oligopoly*.

Thus, adhering to this pricing model, we will consider the Cournot oligopoly—a market of a homogeneous product with N players (firms). Assign numbers from 1 to N to them and denote by $\mathbb{N} = \{1, 2, \ldots, N\}$ the set of players. Let q_i be the quantity of products supplied by firm i ($i \in \mathbb{N}$) during a given time period. In the Cournot oligopoly, each quantity q_i satisfies the constraints

$$\alpha \leqslant q_i \leqslant \beta \quad (i = 1, \ldots, N), \tag{4.1.4}$$

where $\alpha > 0$ and β are constants. The right inequality $q_i \leqslant \beta$ in (4.1.4) means that the production capacity of each firm is limited, while the left inequality $\alpha \leqslant q_i$ requires from each firm to supply a guaranteed minimum quantity α in order to be admitted to the market. In other words, the market has a regulating arbitrator (e.g., government in the electricity market) that allows only large players with a guaranteed minimum supply α regardless of the market price.

Next, assume that the *production cost of player i* ($i \in \mathbb{N}$) is a linear function of the quantity q_i, i.e., can be written as $cq_i + d$, where the constants c and d specify the average variable and fixed cost, respectively. Variable cost includes, e.g., wages, raw material purchases, and depreciation of equipment, while fixed cost includes the rent of premises, land, equipment, licences, and so on.

The price p is determined by the law of supply and demand depending on the total quantity $\bar{q} = q_1 + q_2 + \cdots + q_N$ supplied by all players. Let the price p be a linear function of the total supply as follows:

$$p(\bar{q}) = a - b\bar{q}, \tag{4.1.5}$$

where $a = \text{const} > 0$ is an initial price and the positive constant b (known as the elasticity coefficient) shows the price drop in response to unit product supply.

Suppose the resulting price balances the existing demand and supply. In other words, each firm sells everything it produces. Thus, the revenue of player i ($i \in \mathbb{N}$) is

$$p(\bar{q})q_i = (a - b\bar{q})q_i = \Big[a - b\sum_{k\in\mathbb{N}} q_k\Big]q_i,$$

and its *profit* (revenue minus cost) is

$$\pi_i(q_i, \dots, q_N) = \Big[a - b \sum_{k \in \mathbb{N}} q_k\Big] q_i - (c q_i + d). \qquad (4.1.6)$$

Finally, a natural hypothesis accepted in this model is that each firm determines the quantity of its product, expecting the rational behavior of its competitors.

This economic interaction can be described by a noncooperative N-player game of the form

$$\langle \mathbb{N}, \{\mathrm{Q}_i = [\alpha, \beta]\}_{i \in \mathbb{N}}, \{\pi_i(q_1, \dots, q_N)\}_{i \in \mathbb{N}} \rangle. \qquad (4.1.7)$$

Here, like in (4.1.1), $\mathbb{N} = \{1, 2, \dots, N\}$ denotes the set of players, and $\mathrm{Q}_i = [\alpha, \beta]$ is the set of admissible strategies of player i $(i \in \mathbb{N})$. All strategy profiles have the form $q = (q_1, \dots, q_N) \in \mathrm{Q} = \mathrm{Q}_1 \times \mathrm{Q}_2 \times \dots \times \mathrm{Q}_N$, while the payoff function $\pi_i(q) = \pi_i(q_1, \dots, q_N)$ of player i is given by (4.1.6).

Proposition 4.1.1 *If $a > c$, then game* (4.1.7) *possesses a Berge equilibrium $q^{\mathrm{B}} = (q_1^{\mathrm{B}}, q_2^{\mathrm{B}}, \dots, q_N^{\mathrm{B}})$, where $q_i^{\mathrm{B}} = \alpha$ $(i \in \mathbb{N})$, and the corresponding payoffs of the players are*

$$\pi_i^{\mathrm{B}} = \pi_i(q^{\mathrm{B}}) = [a - Nb\alpha]\alpha - (c\alpha + d) = [a - c]\alpha - bN\alpha^2 - d.$$

Proof A Berge equilibrium in game (4.1.7) satisfies the system of N inequalities

$$\pi_i(q \| q_i^{\mathrm{B}}) \leqslant \pi_i(q^{\mathrm{B}}) \quad \forall\, q \in \mathrm{Q} \quad (i \in \mathbb{N}), \qquad (4.1.8)$$

where, as before, $(q \| q_i^{\mathrm{B}}) = (q_1, q_2, \dots, q_{i-1}, q_i^{\mathrm{B}}, q_{i+1}, \dots, q_N)$.

With (4.1.6), inequalities (4.1.8) can be written as

$$\begin{cases} \big[a - b\big(q_1^{\mathrm{B}} + q_2 + \dots + q_N\big)\big] q_1^{\mathrm{B}} - (c q_1^{\mathrm{B}} + d) \\ \qquad \leqslant \big[a - b\big(q_1^{\mathrm{B}} + q_2^{\mathrm{B}} + \dots + q_N^{\mathrm{B}}\big)\big] q_1^{\mathrm{B}} - (c q_1^{\mathrm{B}} + d), \\ \big[a - b\big(q_1 + q_2^{\mathrm{B}} + \dots + q_N\big)\big] q_2^{\mathrm{B}} - (c q_2^{\mathrm{B}} + d) \\ \qquad \leqslant \big[a - b\big(q_1^{\mathrm{B}} + q_2^{\mathrm{B}} + \dots + q_N^{\mathrm{B}}\big)\big] q_2^{\mathrm{B}} - (c q_2^{\mathrm{B}} + d), \\ \quad \cdot \quad \cdot \quad \cdot \quad \cdot \quad \cdot \quad \cdot \quad \cdot \quad \cdot \quad \cdot \quad \cdot \\ \big[a - b\big(q_1 + q_2 + \dots + q_N^{\mathrm{B}}\big)\big] q_N^{\mathrm{B}} - (c q_N^{\mathrm{B}} + d) \\ \qquad \leqslant \big[a - b\big(q_1^{\mathrm{B}} + q_2^{\mathrm{B}} + \dots + q_N^{\mathrm{B}}\big)\big] q_N^{\mathrm{B}} - (c q_N^{\mathrm{B}} + d). \end{cases}$$

They hold for all $q_i \in \mathrm{Q}_i$ $(i \in \mathbb{N})$.

Clearly, a Berge equilibrium in game (4.1.7) is $q^{\mathrm{B}} = (\alpha, \alpha, \dots, \alpha)$, and the corresponding payoffs π_i^{B} $(i \in \mathbb{N})$ are

$$\pi_i^{\mathrm{B}} = \pi_i(q^{\mathrm{B}}) = [a - Nb\alpha]\alpha - (c\alpha + d) = [a - c]\alpha - bN\alpha^2 - d.$$

Indeed, by reducing its supply as much as possible, each player i ($i \in \mathbb{N}$) increases the profit of all other participants of game (4.1.7). ■

Now consider game (4.1.7) with Nash equilibrium as its solution concept.

Proposition 4.1.2 *If $a > c$, then game* (4.1.7) *possesses a Nash equilibrium*

$$q^{\mathrm{e}} = (q_1^{\mathrm{e}}, q_2^{\mathrm{e}}, \ldots, q_N^{\mathrm{e}}),$$

where for each $i \in \mathbb{N}$ the equilibrium strategy is calculated by

$$q_i^{\mathrm{e}} = \begin{cases} \alpha, & \text{if } \dfrac{a-c}{b(N+1)} \leqslant \alpha, \\ \dfrac{a-c}{b(N+1)}, & \text{if } \alpha < \dfrac{a-c}{b(N+1)} < \beta, \\ \beta, & \text{if } \dfrac{a-c}{b(N+1)} \geqslant \beta. \end{cases} \tag{4.1.9}$$

The payoff of player i ($i \in \mathbb{N}$) in the Nash equilibrium is given by

$$\pi_i^{\mathrm{e}} = \pi_i(q^{\mathrm{e}}) = \begin{cases} (a-c)\alpha - bN\alpha^2 - d, & if \ \dfrac{a-c}{b(N+1)} \leqslant \alpha, \\ \dfrac{(a-c)^2}{(N+1)^2 b} - d, & if \ \alpha < \dfrac{a-c}{b(N+1)} < \beta, \\ (a-c)\beta - bN\beta^2 - d, & if \ \dfrac{a-c}{b(N+1)} \geqslant \beta. \end{cases}$$

Proof A Nash equilibrium in (4.1.7) satisfies the system of N equalities (see Definition 4.1.1)

$$\begin{aligned}
\pi_1(q^{\mathrm{e}}) &= \max_{q_1 \in [\alpha,\beta]} \pi_1(q^{\mathrm{e}} \| q_1) \\
&= \max_{q_1 \in [\alpha,\beta]} \{[a - b(q_1 + q_2^{\mathrm{e}} + \cdots + q_N^{\mathrm{e}})]q_1 - (cq_1 + d)\}, \\
\pi_2(q^{\mathrm{e}}) &= \max_{q_2 \in [\alpha,\beta]} \pi_2(q^{\mathrm{e}} \| q_2) \\
&= \max_{q_2 \in [\alpha,\beta]} \{[a - b(q_1^{\mathrm{e}} + q_2 + \cdots + q_N^{\mathrm{e}})]q_2 - (cq_2 + d)\}, \qquad (4.1.10) \\
&\cdots\cdots\cdots\cdots\cdots \\
\pi_N(q^{\mathrm{e}}) &= \max_{q_N \in [\alpha,\beta]} \pi_N(q^{\mathrm{e}} \| q_N) \\
&= \max_{q_N \in [\alpha,\beta]} \{[a - b(q_1^{\mathrm{e}} + q_2^{\mathrm{e}} + \cdots + q_N)]q_N - (cq_N + d)\}.
\end{aligned}$$

In (4.1.11), for each $i \in \mathbb{N}$ the symbol $(q^{\mathrm{e}} \| q_i)$ denotes a strategy profile q^{e} in which the strategy q_i^{e} of player i is replaced by q_i.

For each $i \in \mathbb{N}$, the function $\pi_i(q^{\mathrm{e}} \| q_i)$ achieves maximum in the variable q_i under two conditions, namely,

$$\left.\frac{\partial \pi_i(q^{\mathrm{e}} \| q_i)}{\partial q_i}\right|_{q_i = q_i^{\mathrm{e}}} = \left[a - 2bq_i - b\left(q_1^{\mathrm{e}} + \cdots + q_{i-1}^{\mathrm{e}} + q_{i+1}^{\mathrm{e}} + \cdots + q_N^{\mathrm{e}}\right) - c\right]\Big|_{q_i = q_i^{\mathrm{e}}} = 0, \quad (4.1.11)$$

$$\left.\frac{\partial^2 \pi_i(q^{\mathrm{e}} \| q_i)}{\partial q_i^2}\right|_{q_i = q_i^{\mathrm{e}}} = -2b < 0.$$

The second condition in system (4.1.12) holds since the elasticity coefficient is $b > 0$. For each $i \in \mathbb{N}$, the first equality leads to the following system of N linear equations:

$$\begin{cases} 2q_1^{\mathrm{e}} + q_2^{\mathrm{e}} + q_3^{\mathrm{e}} + \cdots + q_N^{\mathrm{e}} = \dfrac{a-c}{b}, \\ q_1^{\mathrm{e}} + 2q_2^{\mathrm{e}} + q_3^{\mathrm{e}} + \cdots + q_N^{\mathrm{e}} = \dfrac{a-c}{b}, \\ \cdots\cdots\cdots\cdots \\ q_1^{\mathrm{e}} + q_2^{\mathrm{e}} + q_3^{\mathrm{e}} + \cdots + 2q_N^{\mathrm{e}} = \dfrac{a-c}{b}. \end{cases}$$

The solution is the strategy profile

$$q^{\mathrm{e}} = (q_1^{\mathrm{e}}, q_2^{\mathrm{e}}, \ldots, q_N^{\mathrm{e}}) = \left(\frac{a-c}{(N+1)b}, \frac{a-c}{(N+1)b}, \ldots, \frac{a-c}{(N+1)b}\right).$$

Under the condition $\alpha < \dfrac{a-c}{b(N+1)} < \beta$, the resulting strategies q_i^{e} $(i \in \mathbb{N})$ maximize the functions $\pi_i(q^{\mathrm{e}} \| q_i)$ on the interval $[\alpha, \beta]$ and hence are Nash equilibrium strategies in game (4.1.7).

In the case $\dfrac{a-c}{b(N+1)} \leqslant \alpha$, the second inequality in (4.1.12) implies that each of the functions $\pi_i(q^{\mathrm{e}} \| q_i)$ $(i \in \mathbb{N})$ is monotonically decreasing on the interval $[\alpha, \beta]$. Therefore, the maxima in (4.1.11) are achieved at $q_i^{\mathrm{e}} = \alpha$ $(i \in \mathbb{N})$.

If $\dfrac{a-c}{b(N+1)} \geqslant \beta$, then for each $i \in \mathbb{N}$ the function $\pi_i(q^{\mathrm{e}} \| q_i)$ is monotonically increasing on $[\alpha, \beta]$. Accordingly, equalities (4.1.11) hold at $q_i^{\mathrm{e}} = \beta$ $(i \in \mathbb{N})$.

Combining the three cases above, we conclude that the Nash equilibrium strategies in game (4.1.7) are given by formula (4.1.9).

Let us calculate the payoffs $\pi_i(q^{\mathrm{e}})$ in the strategy profile (4.1.9) ($i \in \mathbb{N}$). Substituting expressions (4.1.9) into the payoff functions (4.1.6) yields the Nash equilibrium payoffs in game (4.1.7) for each player $i \in \mathbb{N}$. In particular, under the condition $\frac{a-c}{b(N+1)} \leqslant \alpha$, we find that

$$\pi_i^{\mathrm{e}} = \pi_i(\alpha, \alpha, \ldots, \alpha) = [a - bN\alpha]\,\alpha - (c\alpha + d) = (a-c)\alpha - bN\alpha^2 - d.$$

In the case where $\frac{a-c}{b(N+1)} \geqslant \beta$, the Nash equilibrium payoff of player i is

$$\pi_i^{\mathrm{e}} = \pi_i(\beta, \beta, \ldots, \beta) = [a - bN\beta]\,\beta - (c\beta + d) = (a-c)\beta - bN\beta^2 - d.$$

Finally, if $\alpha < \frac{a-c}{b(N+1)} < \beta$, the Nash equilibrium payoff is

$$\begin{aligned}
\pi_i^{\mathrm{e}} &= \pi_i\left(\frac{a-c}{(N+1)b}, \frac{a-c}{(N+1)b}, \ldots, \frac{a-c}{(N+1)b}\right) \\
&= \left[a - bN\frac{a-c}{(N+1)b}\right]\frac{a-c}{(N+1)b} - \left(c\frac{a-c}{(N+1)b} + d\right) \\
&= \frac{(a-c)^2}{(N+1)b} - \frac{(a-c)^2}{(N+1)b}\cdot\frac{N}{N+1} - d = \frac{(a-c)^2}{(N+1)^2 b} - d.
\end{aligned}$$

Summarizing, the payoff of player i in the Nash equilibrium profile has the form

$$\pi_i^{\mathrm{e}} = \pi_i(q^{\mathrm{e}}) = \begin{cases} (a-c)\alpha - bN\alpha^2 - d, & \text{if } \dfrac{a-c}{b(N+1)} \leqslant \alpha, \\[2ex] \dfrac{(a-c)^2}{(N+1)^2 b} - d, & \text{if } \alpha < \dfrac{a-c}{b(N+1)} < \beta, \\[2ex] (a-c)\beta - bN\beta^2 - d, & \text{if } \dfrac{a-c}{b(N+1)} \geqslant \beta. \end{cases}$$

■

4.1.4 *Comparison of Payoffs: Berge Equilibrium vs. Nash Equilibrium*

It would not be an overstatement to say that,
with the recent development of computing means and
advances in mathematical logic, contemporary mathematics
has entered a new period in which the focus of research

is shifting from an object under study towards the ways and forms to define it; from problems themselves to possible methods of solution.

—Aleksandrov[4]

In this section, the advantages provided by Berge equilibrium in real applications are demonstrated for the first time in economic research. More specifically, using the Cournot oligopoly model, the cases in which the firms obtain higher payoffs in the Berge equilibrium than in the traditional Nash equilibrium are exhibited.

Let us compare the payoffs yielded by the Berge equilibrium strategies with their counterparts in the case of Nash equilibrium. Again we consider three cases as follows.

Case I. If $\dfrac{a-c}{b(N+1)} \leqslant \alpha$, the Nash equilibrium strategies x_i^{e} defined by (4.1.9) coincide with the Berge equilibrium strategies $x_i^{\mathrm{B}} = \alpha$ $(i \in \mathbb{N})$. Hence, the players have the same payoffs in the Nash equilibrium as in the Berge equilibrium, i.e.,

$$\pi_i^{\mathrm{e}} = \pi_i^{\mathrm{B}} \quad \text{for} \quad \frac{a-c}{b(N+1)} \leqslant \alpha. \tag{4.1.12}$$

Case II. Under the inequality $\alpha < \dfrac{a-c}{b(N+1)} < \beta$, the payoff of player i $(i \in \mathbb{N})$ in the Nash equilibrium is

$$\pi_i^{\mathrm{e}} = \pi_i(q^{\mathrm{e}}) = \frac{(a-c)^2}{(N+1)^2 b} - d,$$

while the Berge equilibrium payoff is

$$\pi_i^{\mathrm{B}} = \pi_i(q^{\mathrm{B}}) = \alpha[a-c-bN\alpha] - d.$$

Their difference equals

$$\begin{aligned}\pi_i^{\mathrm{e}} - \pi_i^{\mathrm{B}} &= \left(\frac{(a-c)^2}{(N+1)^2 b} - d\right) - \Big((a-c)\alpha - bN\alpha^2 - d\Big)\\ &= bN\alpha^2 - (a-c)\alpha + \frac{(a-c)^2}{(N+1)^2 b}\\ &= bN\left(\alpha - \frac{a-c}{(N+1)b}\right)\cdot\left(\alpha - \frac{a-c}{N(N+1)b}\right).\end{aligned}$$

[4]Aleksandr D. Aleksandrov, (1912–1999), was a Soviet and Russian mathematician, physicist, philosopher, and mountaineer. An English translation of a quote from the book *Mathematics: Its Contents, Methods and Meaning*, Moscow, 1956, vol. 1, p. 59.

Because the game involves at least two players ($N > 1$) and $a - c > 0$,

$$\frac{a-c}{N(N+1)b} < \frac{a-c}{(N+1)b}.$$

Recall that the guaranteed minimum quantity and the elasticity coefficient are $\alpha > 0$ and $b > 0$. Hence, the difference $\pi_i^{\mathrm{e}} - \pi_i^{\mathrm{B}}$ is positive if

$$0 < \alpha < \frac{a-c}{N(N+1)b},$$

negative if

$$\frac{a-c}{N(N+1)b} < \alpha < \frac{a-c}{(N+1)b},$$

and zero if

$$\alpha = \frac{a-c}{N(N+1)b}.$$

Thus, for all $i \in \mathbb{N}$,

$$\begin{cases} \pi_i^{\mathrm{e}} > \pi_i^{\mathrm{B}}, & \text{if } \alpha < \dfrac{a-c}{N(N+1)b} \text{ and } \dfrac{a-c}{(N+1)b} < \beta, \\ \pi_i^{\mathrm{e}} = \pi_i^{\mathrm{B}}, & \text{if } \alpha = \dfrac{a-c}{N(N+1)b} \text{ and } \dfrac{a-c}{(N+1)b} < \beta, \\ \pi_i^{\mathrm{e}} < \pi_i^{\mathrm{B}}, & \text{if } \dfrac{a-c}{N(N+1)b} < \alpha < \dfrac{a-c}{(N+1)b} < \beta. \end{cases} \tag{4.1.13}$$

Finally, consider

Case III. In this case, $\alpha < \beta \leqslant \dfrac{a-c}{(N+1)b}$, and the Nash equilibrium strategy of firm i is to supply the maximum possible quantity of products in the market, i.e., $x_i^{\mathrm{e}} = \beta$ ($i \in \mathbb{N}$). Its profit in the Nash equilibrium is

$$\pi_i^{\mathrm{e}} = (a-c)\beta - bN\beta^2 - d.$$

Following the concept of Berge equilibrium, the player has to reduce the supplied quantity down to the guaranteed minimum, i.e., $x_i^{\mathrm{B}} = \alpha$ ($i \in \mathbb{N}$). The corresponding Berge equilibrium payoff is

$$\pi_i^{\mathrm{B}} = (a-c)\alpha - bN\alpha^2 - d.$$

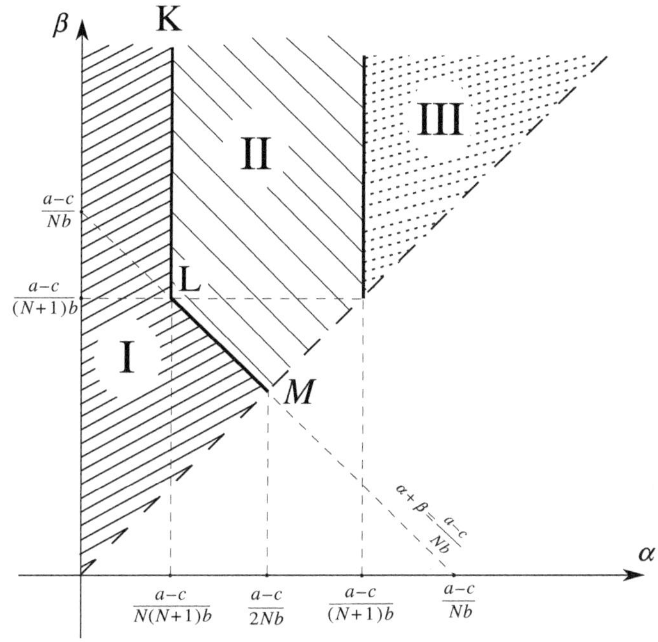

Fig. 4.1 Comparison of payoffs in Berge and Nash equilibria

Consider the difference of the players' payoffs in the Nash and Berge equilibria within the Cournot oligopoly model (4.1.7). For player $i \in \mathbb{N}$,

$$\pi_i^e - \pi_i^B = (a-c)\beta - bN\beta^2 - d - \Big[(a-c)\alpha - bN\alpha^2 - d\Big]$$
$$= (a-c)(\beta-\alpha) - bN(\beta^2-\alpha^2) = (\beta-\alpha)\cdot[a-c-bN(\beta+\alpha)]\,.$$

Since $\beta > \alpha$, the difference has the same sign as the decreasing linear function

$$a - c - bN(\alpha+\beta).$$

This function changes sign if $\alpha+\beta = \dfrac{a-c}{bN}$. Hence, the difference $\pi_i^e - \pi_i^B$ vanishes when $\alpha+\beta = \dfrac{a-c}{bN}$ (see the segment LM in Fig. 4.1). For $\alpha+\beta < \dfrac{a-c}{bN}$, the difference $\pi_i^e - \pi_i^B$ takes positive values; for $\alpha+\beta > \dfrac{a-c}{bN}$, negative values.

Consequently, in the case $\alpha < \beta \leqslant \dfrac{a-c}{(N+1)b}$, the payoffs of player i in the Berge and Nash equilibria, π_i^{B} and π_i^{e}, satisfy

$$\begin{cases} \pi_i^{\mathrm{e}} > \pi_i^{\mathrm{B}}, & \text{if } \alpha+\beta < \dfrac{a-c}{Nb} \text{ and } \alpha < \beta, \\ \pi_i^{\mathrm{e}} = \pi_i^{\mathrm{B}}, & \text{if } \alpha+\beta = \dfrac{a-c}{Nb} \text{ and } \alpha < \beta, \\ \pi_i^{\mathrm{e}} < \pi_i^{\mathrm{B}}, & \text{if } \alpha+\beta > \dfrac{a-c}{Nb} \text{ and } \alpha < \beta. \end{cases} \tag{4.1.14}$$

Bringing together all the three cases, using formulas (4.1.12)–(4.1.14) we easily obtain a complete comparison of the Berge and Nash equilibrium payoffs of player i ($i \in \mathbb{N}$) in game (4.1.7) (see Fig. 4.1). Within domain I, the payoff π_i^{e} in the Nash equilibrium is greater than its counterpart π_i^{B} in the Berge equilibrium.

Within domain II, the opposite holds, as the Berge equilibrium yields a higher payoff for player i ($i \in \mathbb{N}$) than the Nash equilibrium.

Within domain III and also on the jogged line KLM, the payoffs of all players in the two types of equilibria coincide.

Remark 4.1.1 Propositions 4.1.1 and 4.1.2 as well as Sect. 4.1.3 justify the following method for selecting solutions (as set of players' strategies) in the Cournot oligopoly model.

Step I. For the given constants a, b, c, and N, calculate the four values

$$\frac{a-c}{N(N+1)b}, \quad \frac{a-c}{2Nb}, \quad \frac{a-c}{(N+1)b}, \quad \frac{a-c}{Nb}.$$

Step II. Using these values, draw domains I–III as illustrated in Fig. 4.1.

Step III. Determine the values α^* and β^* that specify the "corridors" of admissible supply quantities q_i.

Step IV. Find the point (α^*, β^*) in Fig. 4.1 and identify the domain it belongs to.

Finally, using Propositions 4.1.1 and 4.1.2 and also the results derived in Sect. 4.1.3, write the explicit form of the equilibrium solution—the equilibrium profile and corresponding payoffs of all players.

In conclusion, let us re-emphasize a crucial aspect: the comparative analysis performed in this section breaks the dominating stereotype imposed by M. Shubik that the altruistic Berge equilibrium is fruitless for economic problems. Hopefully, these pioneering results on an efficient use of Berge equilibrium in competitive economic models will be further developed to cover other economic applications. Thus, Berge equilibrium will take a worthy place in economics, just like in psychology and sociology (e.g., see surveys in [223, 227]). Some steps towards the realization of these hopes will be made in the forthcoming sections of the book.

4.2 The Cournot Duopoly with Import

In the development of mathematics, it is easy to observe three major trends that are intrinsic to Russian mathematics as well. *The first trend* concerns an intensive elaboration of separate branches of mathematics, with further differentiation and fragmentation, the appearance of new and narrow directions of research that have specific problems and methods of analysis ... *The second trend* is, to some extent, opposite to the first, seeking for an extensive coverage of the subject of mathematics and revealing the ideas that connect its different parts for further development using some general methods... *The third trend* of modern mathematics, which is of crucial importance and actually a continuation of the best traditions of its classical development, consists in an organic link between theory and practice, in aspiration for filling the subject of mathematics with a particular content through a wide use of mathematical methods in natural and technical problems.

—Vekua*[5]

In a series of papers [73, 74], new concepts of guaranteed equilibria in noncooperative games under uncertainty were suggested. In this section, we will exhibit two types of such equilibria in the Cournot duopoly with import, which plays the role of an uncertain factor.

4.2.1 Mathematical Model

In ancient times the mathematical problems were set by the Gods, as, for example, the problem of the doubling the cube, in connection with the alteration in the dimensions of the Delphic altar. Then, in a later period, the problems were set by the Demigods: Newton, Euler, Lagrange. Today we have a third period, when problems are set by practice.

—Chebyshev*[6]

The Cournot oligopoly model with two restrictions is considered. *First*, only two players (firms) are engaged in the noncooperative game. *Second*, import is incorporated in the model as an interval uncertainty.

Recall that in the Cournot duopoly model two firms (termed players 1 and 2) are competing in a homogeneous product market. Let q_1 and q_2 be the quantities of products supplied by them during a given time period. Imagine that another firm (importer) enters the market, and both players know nothing about the intentions and quantity of products supplied by it. They can merely hypothesize that this quantity

[5]Ilya N. Vekua, (1907–1977), was a distinguished Georgian mathematician; see the Short Biographies at the end of the book. An English translation of a quote from *Nature*, 1957, no. 11.

[6]Pafnuty L. Chebyshev, (1821–1894), was the founder of the St. Petersburg mathematical school; see the Short Biographies at the end of the book. An English translation of a quote from *Encyclopedia of Mathematics*, Grave D.A., Ed., Kiev, 1912, p. 10.

has some nonnegative value $y \in [0, +\infty)$. The *production cost of player i* is a linear function of the quantity q_i $(i = 1, 2)$, i.e., can be written as $cq_i + d$, where the constants c and d specify the average variable and fixed cost, respectively. Variable cost covers, e.g., wages, raw material purchases, and depreciation of equipment, while fixed cost covers the rent of premises, land, equipment, licences, etc. The price p is determined by the law of supply and demand depending on the total quantity $\bar{q} = q_1 + q_2 + y$ supplied by all players. Let the price p be a linear function of the total supply: $p(\bar{q}) = a - b\bar{q}$, where $a = \text{const} > 0$ is an initial price and the positive constant b (the elasticity coefficient) characterizes the price drop in response to unit product supply. Suppose the resulting price balances the existing demand and supply. In other words, each firm sells everything it produces. Thus, the revenue of player 1 is [291–294]

$$p(\bar{q})q_1 = (a - b\bar{q})q_1 = [a - b\,(q_1 + q_2 + y)]\,q_1,$$

and its *profit* (revenue minus cost) is

$$\begin{aligned} \psi_1(q_1, q_2, y) &= [a - b\,(q_1 + q_2 + y)]\,q_1 - (cq_1 + d) \\ &= aq_1 - bq_1^2 - bq_1q_2 - byq_1 - cq_1 - d. \end{aligned} \tag{4.2.1}$$

The profit of player 2 has the form

$$\begin{aligned} \psi_2(q_1, q_2, y) &= [a - b\,(q_1 + q_2 + y)]\,q_2 - (cq_2 + d) \\ &= aq_2 - bq_1q_2 - bq_2^2 - byq_2 - cq_2 - d. \end{aligned} \tag{4.2.2}$$

Each firm defines the quantity of its product, expecting the rational behavior of the competitor and also any realization of the uncertain factor (the quantity supplied by the importer).

Following Germeier's principle of guaranteed result, we will assume that each player i $(i = 1, 2)$ seeks to maximize the payoff function

$$F_i(q_1, q_2, y) = \psi_i(q_1, q_2, y) + y^2 \tag{4.2.3}$$

with an appropriate choice of the quantity q_i.

The first term in (4.2.3) represents the profit of player i while the second compels him to oppose the uncertainty as much as possible.

Remark 4.2.1 The presence of y^2 in (4.2.3) can be also explained in the following way. For each player i $(i = 1, 2)$, we actually study a bi-criteria choice problem in which the first criterion is his profit $\psi_i(q_1, q_2, y)$ and the second criterion agrees with the principle of guaranteed result: player i should take decisions in response to the worst-case uncertainty that "maximally spoils his life" with largest admissible values. This recommendation leads to the second criterion y^2, also maximized by player i. Therefore, the bi-criteria choice problem arising for each player has two criteria to be maximized, the profit and y^2. The linear convolution of these criteria with positive coefficients (here both equal to 1) produces the payoff function (4.2.3). And it suffices to maximize this convolution in order to obtain the Pareto-maximal alternative in the bi-criteria choice problem.

An ordered quadruple

$$\Gamma = \langle \{1, 2\}, \{\mathrm{X}_i = [0, +\infty)\}_{i=1,2}, \mathrm{Y} = [0, +\infty), \{F_i(x, y)\}_{i=1,2} \rangle$$

forms a noncooperative two-player game under uncertainty. Here 1 and 2 are the numbers of players; their strategies are $q_i \in \mathrm{X}_i = [0, +\infty)$. By choosing specific strategies, the players construct a strategy profile $x = (q_1, q_2) \in \mathrm{X} = \mathrm{X}_1 \times \mathrm{X}_2$ in the game. Some nonnegative value of the uncertainty $y \in \mathrm{Y}$ is realized regardless of their choice. The payoff function $F_i(x, y)$ (4.2.3) of player i is defined on all pairs $(x, y) \in \mathrm{X} \times \mathrm{Y}$.

Below we will employ two solution concepts, namely, strongly-guaranteed equilibrium and Pareto-guaranteed equilibrium, which were proposed in the papers [73, 74].

4.2.2 *Strongly-Guaranteed Equilibrium*

One of the endlessly alluring aspects of mathematics
is that its thorniest paradoxes have a way of
blooming into beautiful theories.
—Davis[7]

The concept of strongly-guaranteed equilibrium (SGE) is formalized on the basis of [74]. This concept lies at the junction of maximin and Nash equilibrium. Using strategic uncertainties $y(x) : \mathrm{X} \to \mathrm{Y}$, an explicit form of the SGE in the Cournot oligopoly model with import is obtained.

Definition 4.2.1 A strongly-guaranteed equilibrium (SGE) in the game Γ is a triplet $(x^{\mathrm{N}}, F_1^{\mathrm{N}}, F_2^{\mathrm{N}}) \in \mathrm{X} \times \mathbb{R}^2$ for which there exist two functions $y^{(i)}(x) : \mathrm{X} \to \mathrm{Y}$ such that

first, for each strategy profile $x \in \mathrm{X}$ the strategic uncertainty $y^{(i)}(x) : \mathrm{X} \to \mathrm{Y}$ is unique and maximal in the bi-criteria choice problem

$$\langle \mathrm{Y}, F_i(x, y) \rangle \quad (i = 1, 2),$$

[7]Philip J. Davis, (1923–2018), was an American mathematician.

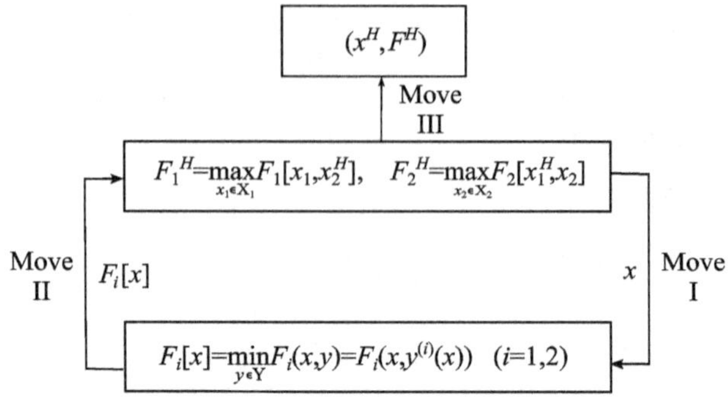

Fig. 4.2 SGE design

i.e.,

$$\min_{y \in Y} F_i(x, y) = F_i(x, y^{(i)}(x)) = F_i[x] \quad \forall\, x \in X \quad (i = 1, 2); \tag{4.2.4}$$

second, the strategy profile $x^N = (x_1^N, x_2^N)$ is a Nash equilibrium in the game of guarantees

$$\Gamma_g = \langle \{1, 2\}, \{X_i\}_{i=1,2}, \{F_i[x]\}_{i=1,2} \rangle,$$

i.e.,

$$\begin{aligned} \max_{x_1 \in X_1} F_1[x_1, x_2^N] &= F_1[x_1^N, x_2^N] = F_1^N, \\ \max_{x_2 \in X_2} F_2[x_1^N, x_2] &= F_2[x_1^N, x_2^N] = F_2^N. \end{aligned} \tag{4.2.5}$$

A hierarchical interpretation of SGE is a two-level three-stage game described as follows (see Fig. 4.2). *In the first stage (move)*, two players occupying the upper level send all their admissible strategy profiles $x = (x_1, x_2) \in X$ to the lower level.

The second stage (move) involves the lower-level player; for each $i = 1, 2$ and each strategy profile $x \in X$, this player forms the guarantees $F_i[x] \leqslant F_i(x, y)$ $\forall\, y \in Y$ $(i = 1, 2)$ and then sends them to the upper level.

In the third stage (move), the players participate in the game of guarantees

$$\Gamma_g = \langle \{1, 2\}, \{X_i\}_{i=1,2}, \{F_i[x]\}_{i=1,2} \rangle$$

and, using (4.2.5), find the Nash equilibrium and the corresponding payoffs $F_i^N = F_i[x^N]$ $(i = 1, 2)$. The resulting pair $(x^N, F^N = (F_1^N, F_2^N))$ forms the SGE of the game Γ.

The players are suggested to use the strategies (x_1^N, x_2^N), because if the actions of the two players lead to a strategy profile $x = (x_1, x_2)$, then $F_i(x, y) \geqslant F_i[x]$

$\forall\, y \in \mathrm{Y}$ $(i = 1, 2)$. In other words, the payoff of each player i cannot be smaller than its guarantee

$$F_i[x] = \min_{y \in \mathrm{Y}} F_i(x, y) \quad (i = 1, 2)$$

under any realization of the uncertainty $y \in \mathrm{Y}$. Hence, as the uncertain factor y may take arbitrary values from Y, using their strategies from the profile $x = (x_1, x_2)$ both players may firmly count on their guarantees $F_i[x]$ $(i = 1, 2)$ only. In this case, a natural approach (see Chap. 3 and the papers [73, 74]) is to choose as a solution of the game Γ the strategy profiles $x^{\mathrm{N}} \in \mathrm{X}$ that implement the Nash equilibrium in the game of guarantees Γ_{g}. Thus, we have justified the following *design method* for the strongly-guaranteed equilibrium in the game Γ.

(a) find two scalar functions $y^{(i)}(x) : \mathrm{X} \to \mathrm{Y}$ $(i = 1, 2)$ such that

$$F_i[x] = \min_{y \in \mathrm{Y}} F_i(x, y) = F_i(x, y^{(i)}(x)) \quad \forall\, x \in \mathrm{X} \quad (i = 1, 2);$$

(b) for the game of guarantees

$$\Gamma_{\mathrm{g}} = \langle \{1, 2\}, \{\mathrm{X}_i\}_{i=1,2}, \{F_i[x]\}_{i=1,2} \rangle,$$

construct the Nash equilibrium x^{N} defined by equalities (4.2.5).

The resulting triplet $(x^{\mathrm{N}}, F_1^{\mathrm{N}}, F_2^{\mathrm{N}})$ is the *strongly-guaranteed equilibrium in the game* Γ.

Proposition 4.2.1 *In the game Γ with $b > 0$, $d < \dfrac{5(a-c)^2}{49b}$, and $a > c$, the strongly-guaranteed equilibrium has the form*

$$((x_1^{\mathrm{N}}, x_2^{\mathrm{N}}), (F_1^{\mathrm{N}}, F_2^{\mathrm{N}})) = \left(\left(\frac{2(a-c)}{7b}, \frac{2(a-c)}{7b} \right), \left(\frac{5(a-c)^2}{49b}, \frac{5(a-c)^2}{49b} \right) \right).$$

Proof Apply the above-mentioned procedure to the mathematical model of the Cournot duopoly with import—the game Γ. Let us implement steps (a) and (b).

(a) For the payoff function of player 1,

$$F_1(x, y) = [a - b(x_1 + x_2 + y)]x_1 - (cx_1 + d) + y^2,$$

construct the function

$$y^{(1)}(x) = \arg\min_{y \in \mathrm{Y}} F_i(x, y) \quad \forall x \in \mathrm{X}.$$

A sufficient condition consists of

$$\left.\frac{\partial F_1(x,y)}{\partial y}\right|_{y=y^{(1)}(x)} = -bx_1 + 2y^{(1)}(x) = 0 \quad \forall x \in \mathrm{X},$$

$$\left.\frac{\partial^2 F_1(x,y)}{\partial y^2}\right|_{y=y^{(1)}(x)} = 2 > 0.$$

Hence, $y = y^{(1)}(x) = \dfrac{bx_1}{2}$, and therefore

$$F_1[x] = \min_{y\in \mathrm{Y}} F_1(x,y) = F_1(x, y^{(i)}(x)) = \left[a - b\left(\frac{3x_1}{2} + x_2\right)\right]x_1$$

$$-(cx_1 + d) + \frac{bx_1^2}{4} = ax_1 - \frac{5}{4}bx_2^2 - bx_1x_2 - (cx_1 + d),$$

$$F_2[x] = \min_{y\in \mathrm{Y}} F_2(x,y) = ax_2 - \frac{5}{4}bx_1^2 - bx_1x_2 - (cx_2 + d).$$

(b) Conditions (4.2.5) hold if

$$\left.\frac{\partial F_1[x_1, x_2^{\mathrm{N}}]}{\partial x_1}\right|_{x_1=x_1^{\mathrm{N}}} = (a-c) - \frac{5}{2}bx_1^{\mathrm{N}} - bx_2^{\mathrm{N}} = 0,$$

$$\frac{\partial^2 F_1[x_1, x_2^{\mathrm{N}}]}{\partial x_1^2} = -\frac{5}{2}b < 0,$$

$$\left.\frac{\partial F_2[x_1^{\mathrm{N}}, x_2]}{\partial x_2}\right|_{x_2=x_2^{\mathrm{N}}} = (a-c) - bx_1^{\mathrm{N}} - \frac{5}{2}bx_2^{\mathrm{N}} = 0,$$

$$\frac{\partial^2 F_2[x_1^{\mathrm{N}}, x_2]}{\partial x_2^2} = -\frac{5}{2}b < 0.$$

This leads to the following system to calculate $x^{\mathrm{N}} = (x_1^{\mathrm{N}}, x_2^{\mathrm{N}})$:

$$\begin{cases} 5x_1^{\mathrm{N}} + 2x_2^{\mathrm{N}} = 2\dfrac{a-c}{b}, \\ 2x_1^{\mathrm{N}} + 5x_2 = 2^{\mathrm{N}}\dfrac{a-c}{b}. \end{cases}$$

Then

$$x_i^{\mathrm{N}} = \frac{2(a-c)}{7b} \quad (i = 1, 2),$$

and finally

$$F_i^{\mathrm{N}} = F_i[x^{\mathrm{N}}] = \frac{5(a-c)^2}{49b} - d \quad (i = 1, 2).$$

■

4.2.3 *Pareto-Guaranteed Equilibrium*

As a rule, it happens that natural phenomena and economic processes are wider than available mathematical tools. This is a permanent motivation for further development of mathematics, its concepts and theories.
—Gnedenko[8]

As in Sect. 4.2.2, the concept of Pareto-guaranteed equilibrium in which the payoffs are not smaller than in the strongly-guaranteed equilibrium is introduced. This concept was proposed in [74] at the junction of Pareto minimum and Nash equilibrium.

The guarantees F_i^N $(i = 1, 2)$ obtained in Proposition 4.2.1 are smallest. However, the players seek for maximum payoffs and hence for as high guarantees as possible. Therefore, we will adopt the guarantees that are not smaller than their counterparts from Sect. 4.2.1.

Definition 4.2.2 A Pareto-guaranteed equilibrium (*PGE*) in the game Γ is a triplet $(x^{\mathrm{e}}, \psi_1^{\mathrm{e}}, \psi_2^{\mathrm{e}})$ for which there exists a function $y_{\mathrm{P}}(x) : \mathrm{X} \to \mathrm{Y}$ such that

first, for each strategy profile $x \in \mathrm{X}$ the function $y_{\mathrm{P}}(x)$ is a Pareto-minimal uncertainty in the bi-criteria choice problem

$$\langle\, \mathrm{Y}, \{F_i(x, y)\}_{i=1,2} \,\rangle,$$

which is derived from Γ for each fixed strategy profile $x = (x_1, x_2) \in \mathrm{X}$; i.e., under each frozen profile $x \in \mathrm{X}$ the system of inequalities

$$F_i(x, y) \leqslant F_i(x, y_{\mathrm{P}}(x)) \quad \forall\, y \in \mathrm{Y} \quad (i = 1, 2),$$

with at least one strict inequality, is inconsistent;

[8]Boris V. Gnedenko, (1912–1995), was a Soviet mathematician and student of A. N. Kolmogorov. An English translation of a quote from *Kvant*, 1977, no. 11, p. 26.

second, the strategy profile $x^{\mathrm{e}} = (x_1^{\mathrm{e}}, x_2^{\mathrm{e}})$ is a Nash equilibrium in the game (without uncertainty)

$$\langle \{1, 2\}, \{\mathrm{X}_i\}_{i=1,2}, \{F_i(x, y_{\mathrm{P}}(x))\}_{i=1,2} \rangle,$$

which is obtained from the game Γ by replacing the uncertainty y with its realization $y_{\mathrm{P}}(x)$.

In addition, x^{e} is called a *Pareto-guaranteeing equilibrium* while $F^{\mathrm{e}} = (F_1^{\mathrm{e}}, F_2^{\mathrm{e}})$, $F_i^{\mathrm{e}} = F_i(x^{\mathrm{e}}, y_{\mathrm{P}}(x^{\mathrm{e}}))$ $(i = 1, 2)$, is called the *corresponding vector guarantee of player i*.

A Pareto-guaranteed equilibrium (PGE) in the Cournot duopoly with import is a triplet $(x^{\mathrm{e}}, \psi_1^{\mathrm{e}}, \psi_2^{\mathrm{e}})$, *where the Pareto-guaranteeing strategy profile* $x^{\mathrm{e}} = (x_1^{\mathrm{e}}, x_2^{\mathrm{e}})$ *is the same as in the PGE of the game* Γ, *while* $\psi_i^e = \psi_i(x^e, y_P(x^e))$ $(i = 1, 2)$ *are the players' profits included in their guaranteed payoff* $(F_i^{\mathrm{e}} = F_i(x_1^{\mathrm{e}}, x_2^{\mathrm{e}}, y_{\mathrm{P}}(x_1^{\mathrm{e}}, x_2^{\mathrm{e}}))$ *for firm* $i, i = 1, 2)$.

4.2.3.1 Design Algorithm for Pareto-Guaranteed Equilibrium

In accordance with Definition 4.2.2, we suggest the following design algorithm for the PGE in the Cournot duopoly model with import.

Step I. Pareto inner minimum calculation: find a continuous function $y_{\mathrm{P}}(x) : \mathrm{X} \to \mathrm{Y}$ that yields the Pareto minimum in the bi-criteria choice problem

$$\langle\, \mathrm{Y} = [0, +\infty)\, , \{F_i(x, y)\}_{i=1,2}\, \rangle \quad \forall x \in \mathrm{X}, \tag{4.2.6}$$

which is obtained from the game Γ for each fixed strategy profile $x = (x_1, x_2) \in \mathrm{X}$;

Step II. Nash equilibrium calculation: find the Nash equilibrium $x^{\mathrm{e}} = (x_1^{\mathrm{e}}, x_2^{\mathrm{e}})$ in the game of guarantees (without uncertainty)

$$\langle \{1, 2\}, \{\mathrm{X}_i = [0, +\infty)\}_{i=1,2}, \{F_i(x, y_{\mathrm{P}}(x))\}_{i=1,2} \rangle, \tag{4.2.7}$$

which is obtained from the game Γ by substituting the Pareto-minimal uncertainty $y_{\mathrm{P}} = y_{\mathrm{P}}(x)$;

Step III. Calculation of the profits ψ_i^{e}: find the players' profits $\psi_i(x_1^{\mathrm{e}}, x_2^{\mathrm{e}}, y_{\mathrm{P}}(x_1^{\mathrm{e}}, x_2^{\mathrm{e}})) = \psi_i^{\mathrm{e}}$ $(i = 1, 2)$.

4.2.3.2 Pareto Inner Minimum Calculation

For the sake of compactness, we will further adopt the form

$$F_i(x, y) = \psi_i(q_1, q_2, y) + \frac{y^2}{2} \quad (i = 1, 2),$$

where $\psi_i(q_1, q_2, y)$ are given by (4.2.1) and (4.2.2).

Lemma 4.2.1 *Suppose there exist values* $\alpha, \beta > 0$ *and a scalar function* $y_P(x) : X \to Y$ *such that, for each* $x \in X$,

$$\min_{y \in Y} [\alpha F_1(x, y) + \beta F_2(x, y)] = Idem\,[y \to y_P(x)],$$

where $Idem\,[y \to y_P(x)]$ *denotes the bracketed expression with* y *replaced by* $y_P(x)$*. Then for each* $x \in X$ *the function* $y_P(x)$ *is Pareto minimal in the bi-criteria choice problem (4.2.6).*

Proof of this result can be found in almost every textbook on multicriteria optimization.

Lemma 4.2.2 *The uncertainty*

$$y_P(x_1, x_2) = \frac{b(x_1 + x_2)}{2}$$

is Pareto minimal in the bi-criteria choice problem (4.2.6) *for each strategy profile* $x = (x_1, x_2) \in [0, +\infty)^2$.

Proof Consider the function

$$\begin{aligned} F(x, y) &= F_1(x, y) + F_2(x, y) = \psi_1(x_1, x_2, y) + \psi_2(x_1, x_2, y) + y^2 \\ &= a(x_1 + x_2) - b(x_1 + x_2)^2 - by(x_1 + x_2) - c(x_1 + x_2) - 2d + y^2. \end{aligned}$$

For each fixed $x = (x_1, x_2) \in X$, the minimal value of this function is achieved at $y_P(x) = \dfrac{b(x_1 + x_2)}{2}$ because

$$\left.\frac{\partial F}{\partial y}\right|_{y=y_P(x)} = -b(x_1 + x_2) + 2y_P(x) = 0$$

and

$$\left.\frac{\partial^2 F}{\partial y^2}\right|_{y=y_P(x)} = 2 > 0.$$

Hence, by Lemma 4.2.1 with $\alpha = \beta = 1$, the uncertainty $y_P(x) = \dfrac{b(x_1 + x_2)}{2}$ is Pareto minimal in the bi-criteria choice problem (4.2.6). ■

4.2.3.3 Nash Equilibrium Calculation

Proposition 4.2.2 *For $b > 0$ and $a > c$, the Nash equilibrium in game* (4.2.7) *has the form*

$$x^{\mathrm{e}} = (x_1^{\mathrm{e}}, x_2^{\mathrm{e}}) = \left(\frac{a-c}{b(3+b)}, \frac{a-c}{b(3+b)}\right).$$

Proof Using the uncertainty $y_{\mathrm{P}}(x)$ found in Lemma 4.2.2 as well as formulas (4.2.2) and (4.2.3), we obtain

$$\begin{aligned} F_1(x, y_{\mathrm{P}}(x)) &= ax_1 - bx_1^2 - bx_1x_2 \\ &\quad - \frac{b^2(x_1+x_2)}{2}x_1 - cx_1 - d + \frac{b^2(x_1+x_2)^2}{8}, \\ F_2(x, y_{\mathrm{P}}(x)) &= ax_2 - bx_2^2 - bx_1x_2 \\ &\quad - \frac{b^2(x_1+x_2)}{2}x_2 - cx_2 - d + \frac{b^2(x_1+x_2)^2}{8}. \end{aligned} \tag{4.2.8}$$

The sufficient conditions for the existence of a Nash equilibrium $x^{\mathrm{e}} = (x_1^{\mathrm{e}}, x_2^{\mathrm{e}})$ in game (4.2.7) can be reduced to the following combined equalities and inequalities:

$$\left.\frac{\partial F_1}{\partial x_1}\right|_{x=x^{\mathrm{e}}} = a - 2bx_1^{\mathrm{e}} - bx_2^{\mathrm{e}} - \frac{b^2}{2}(2x_1^{\mathrm{e}} + x_2^{\mathrm{e}}) - c + \frac{b^2}{4}(x_1^{\mathrm{e}} + x_2^{\mathrm{e}}) = 0, \tag{4.2.9}$$

$$\left.\frac{\partial^2 F_1}{\partial x_1^2}\right|_{x=x^{\mathrm{e}}} = -2b - \frac{3b^2}{4} < 0, \tag{4.2.10}$$

$$\left.\frac{\partial F_2}{\partial x_2}\right|_{x=x^{\mathrm{e}}} = a - bx_1^{\mathrm{e}} - 2bx_2^{\mathrm{e}} - \frac{b^2}{2}(x_1^{\mathrm{e}} + 2x_2^{\mathrm{e}}) - c + \frac{b^2}{4}(x_1^{\mathrm{e}} + x_2^{\mathrm{e}}) = 0, \tag{4.2.11}$$

$$\left.\frac{\partial^2 F_2}{\partial x_2^2}\right|_{x=x^{\mathrm{e}}} = -2b - \frac{3b^2}{4} < 0. \tag{4.2.12}$$

Conditions (4.2.10) and (4.2.12) hold because $b > 0$, while (4.2.9) and (4.2.11) represent a system of two linear inhomogeneous equations with constant coefficients,

$$\begin{cases} \left(2b + \dfrac{3b^2}{4}\right)x_1^{\mathrm{e}} + \left(b + \dfrac{b^2}{4}\right)x_2^{\mathrm{e}} = a - c, \\ \left(b + \dfrac{b^2}{4}\right)x_1^{\mathrm{e}} + \left(2b + \dfrac{3b^2}{4}\right)x_2^{\mathrm{e}} = a - c. \end{cases} \tag{4.2.13}$$

The determinant of this system is

$$\Delta = \begin{vmatrix} 2b+\dfrac{3b^2}{4} & b+\dfrac{b^2}{4} \\ b+\dfrac{b^2}{4} & 2b+\dfrac{3b^2}{4} \end{vmatrix} = \left(2b+\frac{3b^2}{4}\right)^2 - \left(b+\frac{b^2}{4}\right)^2 = (3b+b^2)\left(b+\frac{b^2}{2}\right),$$

and $\Delta \neq 0$ by $b > 0$.

Calculating the determinants

$$\Delta_1 = \begin{vmatrix} a-c & b+\dfrac{b^2}{4} \\ a-c & 2b+\dfrac{3b^2}{4} \end{vmatrix} = (a-c)\begin{vmatrix} 1 & b+\dfrac{b^2}{4} \\ 1 & 2b+\dfrac{3b^2}{4} \end{vmatrix}$$

$$= (a-c)\left(2b+\frac{3b^2}{4}-b-\frac{b^2}{4}\right) = (a-c)\left(b+\frac{b^2}{2}\right),$$

$$\Delta_2 = \begin{vmatrix} 2b+\dfrac{3b^2}{4} & a-c \\ b+\dfrac{b^2}{4} & a-c \end{vmatrix} = (a-c)\begin{vmatrix} 2b+\dfrac{3b^2}{4} & 1 \\ b+\dfrac{b^2}{4} & 1 \end{vmatrix}$$

$$= (a-c)\left(2b+\frac{3b^2}{4}-b-\frac{b^2}{4}\right) = (a-c)\left(b+\frac{b^2}{2}\right)$$

we finally write the solution of system (4.2.13) as

$$x_1^e = x_2^e = \frac{\Delta_1}{\Delta} = \frac{\Delta_2}{\Delta} = \frac{(a-c)\left(b+\dfrac{b^2}{2}\right)}{(3b+b^2)\left(b+\dfrac{b^2}{2}\right)} = \frac{a-c}{b(3+b)} > 0.$$

■

Proposition 4.2.3 *The Pareto-guaranteed equilibrium in the Cournot duopoly with import is the triplet* $(x^{\mathrm{e}}, \psi_1^{\mathrm{e}}, \psi_2^{\mathrm{e}})$, *where*

$$x^{\mathrm{e}} = (x_1^{\mathrm{e}}, x_2^{\mathrm{e}}) = \left(\frac{a-c}{b(3+b)}, \frac{a-c}{b(3+b)}\right)$$

and the corresponding profit of firm i is

$$\psi_i^{\mathrm{e}} = \psi_i(x_1^{\mathrm{e}}, x_2^{\mathrm{e}}, y_P(x_1^{\mathrm{e}}, x_2^{\mathrm{e}})) = \frac{(a-c)^2}{b(3+b)^2} - d \quad (i = 1, 2).$$

Proof Taking advantage of Proposition 4.2.2, we merely have to calculate the profits ψ_i^{e} and the guaranteed payoffs F_i^{e}.

First, a direct substitution of x^{e} into $y_{\mathrm{P}}(x_1, x_2) = \dfrac{b(x_1 + x_2)}{2}$ gives $y_{\mathrm{P}}(x_1^{\mathrm{e}}, x_2^{\mathrm{e}}) = \dfrac{a-c}{3+b}$.

Second, using $x^{\mathrm{e}} = (x_1^{\mathrm{e}}, x_2^{\mathrm{e}}) = \left(\dfrac{a-c}{b(3+b)}, \dfrac{a-c}{b(3+b)}\right)$ and $y_{\mathrm{P}}(x_1^{\mathrm{e}}, x_2^{\mathrm{e}}) = \dfrac{a-c}{3+b}$ in (4.2.1)–(4.2.3), we easily find the guaranteed payoffs F_i^{e} of the players, which contain

$$\begin{aligned}
\psi_i^{\mathrm{e}} &= \psi_i(x_1^{\mathrm{e}}, x_2^{\mathrm{e}}, y_{\mathrm{P}}(x_1^{\mathrm{e}}, x_2^{\mathrm{e}})) = \left[a - b\left(x_1^{\mathrm{e}} + x_2^{\mathrm{e}} + y_{\mathrm{P}}(x_1^{\mathrm{e}}, x_2^{\mathrm{e}})\right)\right] x_i^{\mathrm{e}} - (cx_i^{\mathrm{e}} + d) \\
&= \left[a - b\left(2\frac{a-c}{b(3+b)} + \frac{a-c}{3+b}\right)\right] \cdot \frac{a-c}{b(3+b)} - \left[\frac{c(a-c)}{b(3+b)} + d\right] \\
&= \frac{(a-c)^2}{b(3+b)^2} - d \quad (i = 1, 2).
\end{aligned}$$

The guaranteed payoffs are

$$F_i^{\mathrm{e}} = F_i(x^{\mathrm{e}}, y_{\mathrm{P}}(x^{\mathrm{e}})) = \psi_i^{\mathrm{e}} + \frac{y_{\mathrm{P}}^2(x^{\mathrm{e}})}{2} = \left(\frac{a-c}{3+b}\right)^2 \frac{2+b}{2b} - d \quad (i = 1, 2). \quad \blacksquare$$

Remark 4.2.2 It follows from (4.2.8) that $\dfrac{\partial^2 F_1(x, y_{\mathrm{P}}(x))}{\partial x_2^2} = \dfrac{b^2}{4} > 0$. In accordance with Table 2.1 (row 4) and (2.10.1), the noncooperative game of guarantees

$$\langle \{1, 2\}, \{\mathrm{X}_i = \mathbb{R}^{n_i}\}_{i=1,2}, \{F_i(x, y_{\mathrm{P}}(x))\}_{i=1,2} \rangle$$

possesses no Berge equilibria. At the same time, there exists a guaranteeing Nash equilibrium in this game, and we have found its explicit form.

4.3 The Bertrand Duopoly Model

Wherever there is number, there is beauty.
—Proclus[9]

[9]Proclus, (c. 410–485 A.D.), was the last major ancient Greek philosopher.

4.3.1 Mathematical Model

A writer only begins a book. A reader finishes it.
—Johnson[10]

In 1883, Bertrand [205] suggested a price competition model for duopoly markets in which the strategy of each player is the unit price of products. Note that Bertrand continued Cournot's research [225]. The mathematical model developed by Cournot proceeds from the assumption that both players (firms) choose the quantity of supplied products while the unit price is determined by the law of supply and demand. More specifically, the unit price is established at the level of buyers demand for all products put in the market. On the other hand, Bertrand relied on a more natural behavior of economic agents, assuming that each seller chooses not the quantity of supplied products (like in the Cournot setup) but the unit price.[11]

As a rule, buyers consider the same-usage products supplied by different firms as different products. Therefore, suppose each firm enters the market with its own product but all products are substitutable.

Thus, assume there are two firms selling the same product in the market. Let *the strategy of each firm* (player) be the unit price for its product. Thus, each firm i chooses a unit price $p_i = \text{const} \geqslant 0$ $(i = 1, 2)$. The announcement of unit prices by both players leads to a *strategy profile*—a unit price vector $\vec{p} = (p_1, p_2)$. The market demand for the product of player i $(i \in \{1, 2\})$ is assumed to be a linear function of the announced unit prices, i.e.,

$$Q_1(p) = q - l_1 p_1 + l_2 p_2, \quad Q_2(p) = q - l_1 p_2 + l_2 p_1. \tag{4.3.1}$$

Here q specifies an initial demand and the elasticity coefficient $l_1 = \text{const} > 0$ shows *the demand drop* in response to a unit price increase. In turn, the elasticity coefficient $l_2 = \text{const} > 0$ characterizes the *demand rise* in response to the unit price increase of the substitute product. Denote by $c > 0$ the cost of unit product; then the profit of firm i (further referred to as *the payoff function* of player $i \in \{1, 2\}$) is given by

$$\begin{aligned} f_1(\vec{p}) &= [q - l_1 p_1 + l_2 p_2](p_1 - c), \\ f_2(\vec{p}) &= [q - l_1 p_2 + l_2 p_1](p_2 - c). \end{aligned} \tag{4.3.2}$$

[10]Samuel Johnson, byname Dr. Johnson, (1709–1784), was an English critic, biographer, essayist, poet, and lexicographer.

[11]"PRICE, n. Value, plus a reasonable sum for the wear and tear of conscience in demanding it." An ironic definition that belongs to Ambrose Gwinnett Bierce, (1842–1914), an American newspaperman and satirist.

Consequently, the mathematical model of this competitive interaction between the firms (sellers) can be described as the ordered triplet

$$\Gamma = \langle \{1, 2\}, \{P_i = (c, \beta]\}_{i=1,2}, \ \{f_i(\vec{p})\}_{i=1,2} \rangle. \tag{4.3.3}$$

This noncooperative two-player normal form game uses the following notations: $\{1, 2\}$ is the set of players; $\beta = \text{const} > c$ as a maximum unit price determined by the market (and conscience!), sometimes regardless of the player's preferences; $p_i \in (c, \beta]$ as the *strategy of player i*—the unit price of product; $\vec{p} = (p_1, p_2)$ as a *strategy profile*, i.e., the price policy established in the market; $f_i(\vec{p})$ as the *payoff function of player i*, which measures its performance in a current strategy profile $\vec{p} = (p_1, p_2) \in P = P_1 \times P_2$; an explicit form of $f_i(\vec{p})$ is given by (4.3.2).

The game Γ has several specific features, namely,

first, the maximum unit price β and cost c are assumed to be the same for both players (a quite natural hypothesis for single-product markets);

second, the rules of this game forbid the coalition $\{1, 2\}$ (which reflects the noncooperative character of the game Γ);

and third, the unit price is $p_i \geqslant c$ $(i = 1, 2)$ (otherwise, it makes no economic sense for player i to enter the market).

4.3.2 Main Notions

In 1949, 21 years old American mathematician and economist J. Nash, at that time a postgraduate of Princeton University, suggested an original concept of equilibrium [257] in noncooperative games. As repeatedly mentioned throughout the book, in 1994 J. Nash, J. Harsanyi and R. Selten were awarded the Nobel Prize in Economic Sciences. For the game Γ, this concept can be defined as follows.

Definition 4.3.1 A pair $(\vec{p}^{\rm e}, \ \vec{f}(\vec{p}^{\rm e}) = \vec{f}^{\rm e}) \in P \times \mathbb{R}^2$ is called a Nash equilibrium in the game Γ if

$$\max_{p_i \in P_i} f_i(\vec{p}^{\rm e} \parallel p_i) = f_i(\vec{p}^{\rm e}) \quad (i = 1, 2); \tag{4.3.4}$$

a strategy profile $\vec{p}^{\rm e} = (p_1^{\rm e}, p_2^{\rm e})$ that satisfies (4.3.4) is also called a Nash equilibrium.

Here and in the sequel, we use the standard notations of noncooperative games, $(\vec{p}^{\rm e} \parallel p_1) = (p_1, p_2^{\rm e})$ and $(\vec{p}^{\rm e} \parallel p_2) = (p_1^{\rm e}, p_2)$ for $\vec{p} = (p_1, p_2)$. In addition, introduce the vector $\vec{f} = (f_1, f_2) \in \mathbb{R}^2$.

This concept of equilibrium has turned out to be very attractive in economics, sociology, military sciences and other fields, causing "a whole star shower" of Nobel Prizes in Economic Sciences, which is still not exhausted. However, there are spots on the sun: condition (4.3.4) reflects selfish behavioral principles and, being guided by (4.3.4), each player seeks *to increase his own payoff only*, ignoring the interests of the others. In particular, with this concept of equilibrium each player breaks *the Golden Rule of ethics*: "Do to others as you would like them to do to you." It originates from the New Testament, see the Gospel according to St. Luke, Chapter 6:31, precepting that "And as ye would that men should do to you, do ye also to them likewise." Such an altruistic approach is implemented by the concept of Berge equilibrium.

Definition 4.3.2 A pair $(\vec{p}^{\mathrm{B}},\ \vec{f}(\vec{p}^{\mathrm{B}})) \in P \times \mathbb{R}^2$ is called a Berge equilibrium in the game Γ if

$$\max_{\vec{p}\in P} f_i(\vec{p} \parallel p_i^{\mathrm{B}}) = f_i(\vec{p}^{\mathrm{B}}) \quad (i = 1, 2); \tag{4.3.5}$$

a strategy profile $\vec{p}^{\mathrm{B}} = (p_1^{\mathrm{B}}, p_2^{\mathrm{B}})$ that satisfies (4.3.5) is also called a Berge equilibrium.

As noted above, the concept of Berge equilibrium was suggested in 1994 by Russian mathematician K. Vaisman from Orekhovo-Zuevo State Pedagogical Institute, see [11, 13, 302], in his Candidate of Sciences Dissertation. Sadly, Vaisman died on March 10, 1998, after a struggle with cancer, not reaching even the age of 36. The whole idea of this equilibrium appeared after a careful reading of C. Berge's book [202] and a brainstorming session on the advantages and drawbacks of the Nash equilibrium $\vec{p}^{\mathrm{e}}$. It consists in replacing $\vec{p}^{\mathrm{e}}$ with $\vec{p}^{\mathrm{B}}$, p_i with p_i^{B} and $\vec{p}^{\mathrm{e}} \parallel p_i$ with $\vec{p} \parallel p_i^{\mathrm{B}}$ in formula (4.3.4). Yet these simple modifications eliminate the selfish character of Nash equilibrium. Indeed, following their strategies from the Berge equilibrium $\vec{p}^{\mathrm{B}}$, the players forget about their individual interests, making every effort to increase the payoffs of the other players. Such an altruistic approach is cultivated by family relationships (of course, in tight-knit loving families!) and religious communities. The elements of this altruism are inherent to charitable associations, sponsors and kinsmen in general. Note that, on the strength of (4.3.5), application of Berge equilibria rules out military conflicts and bloodletting. This concept of equilibrium also provides a solution to the well-known Prisoner's Dilemma.

Unfortunately, the monograph [202] had an unenviable fate. Soon after publication, a well-recognized expert in game theory, M. Shubik wrote the review [269], underlining that "...no attention has been paid to applications to economics" (in our view, that is totally *unfair*) and "...the book will be of a little direct interest to economists..." (again, in our view, *unfair*). The latter judgement stimulated our paper [70] with a detailed study of the Berge and Nash equilibria in the well-known competitive Cournot oligopoly model [225], including the cases in which a Berge equilibrium yields higher payoffs than a Nash equilibrium. The same problem, now for the Bertrand duopoly model, will be treated in the current section. We will give a comparative analysis of using both types of equilibria depending on the highest unit price of the product supplied in the market.

Good surveys of Berge equilibrium can be found in [223] and [131, pp. 53–56]. As indicated by these surveys, most research works consider the properties and specific features of Berge equilibria as well as their modifications and connections to Nash equilibria. We anticipate that the incipient framework of Berge equilibria will soon reach the status of a rigorous mathematical theory. Probably, the intensive accumulation of facts will be replaced by a stage of evolutionary development. We believe that this book and the papers [70–72, 300, 303, 304] are contributing to this second stage.

4.3.3 *Explicit Design of Berge and Nash Equilibria*

Thus, the game Γ is defined, and we will find an explicit form of the Nash and Berge equilibria (see Definitions 4.3.1 and 4.3.2, respectively).

4.3.3.1 Berge Equilibrium

On the basis of Definition 4.3.2, we obtain

Proposition 4.3.1 *If* $l_2 > l_1$*, the Berge equilibrium in the game* Γ *has the form*

$$(\vec{p}^{\mathrm{B}},\ \vec{f}(\vec{p}^{\mathrm{B}})) = (p_1^{\mathrm{B}}, p_2^{\mathrm{B}}, f_1(\vec{p}^{\mathrm{B}}), f_2(\vec{p}^{\mathrm{B}})), \tag{4.3.6}$$

where $\vec{p}^{\mathrm{B}} = (p_1^{\mathrm{B}}, p_2^{\mathrm{B}})$, $p_i^{\mathrm{B}} = \beta$ $(i = 1, 2)$*, and the vector of the Berge equilibrium payoffs* $\vec{f}(\vec{p}^{\mathrm{B}}) = (f_1(\vec{p}^{\mathrm{B}}), f_2(\vec{p}^{\mathrm{B}}))$ *is*

$$f_i(\vec{p}^{\mathrm{B}}) = (l_2 - l_1)(\beta - c)^2 + (q + c(l_2 - l_1))(\beta - c) \quad (i = 1, 2). \tag{4.3.7}$$

Recall that β is the maximum unit price of the product determined by the law of supply and demand.

Proof Let p be the unit price of the product in the market. Then for each $i = 1, 2$,

$$\begin{aligned} f_i(\vec{p}) &= [q + p(l_2 - l_1)](p - c) = (l_2 - l_1)p(p - c) + q(p - c) \\ &= (l_2 - l_1)(p - c)^2 + (q + c(l_2 - l_1))(p - c), \end{aligned}$$

where $p \in (c, \beta]$. Since $l_2 > l_1$ and $\dfrac{\partial^2 f_i(\vec{p})}{\partial(p - c)^2} = 2(l_2 - l_1) > 0$, each payoff function $f_i(\vec{p})$ is strictly convex and increasing in p. Therefore, each $f_i(\vec{p})$ achieves maximum at $p = \beta$, which gives $p_1^{\mathrm{B}} = p_2^{\mathrm{B}} = \beta$. And direct substitution of this result into the payoff function yields the desired formula of the Berge equilibrium payoffs. ■

4.3.3.2 Nash Equilibrium

Using Definition 4.3.1, we arrive at

Proposition 4.3.2 *If $l_2 \neq 2l_1$, the Nash equilibrium in the game Γ has the form*

$$(\vec{p}^{\mathrm{e}},\ \vec{f}(\vec{p}^{\mathrm{e}})) = (p_1^{\mathrm{e}}, p_2^{\mathrm{e}}, f_1(\vec{p}^{\mathrm{e}}), f_2(\vec{p}^{\mathrm{e}})),$$

where

$$\vec{p}^{\mathrm{e}} = (p_1^{\mathrm{e}}, p_2^{\mathrm{e}}), \quad p_i^{\mathrm{e}} = \frac{q + l_1 c}{2l_1 - l_2} = p^N \quad (i = 1, 2), \tag{4.3.8}$$

and the vector of the Nash equilibrium payoffs is

$$f_i(\vec{p}^{\mathrm{e}}) = l_1(p^N - c)^2 = l_1\left(\frac{q + (l_2 - l_1)c}{2l_1 - l_2}\right)^2 \quad (i = 1, 2). \tag{4.3.9}$$

Proof Differentiating twice the explicit form (4.3.2) of the function $f_i(\vec{p})$ gives $\dfrac{\partial^2 f_i(\vec{p}^{\mathrm{e}} \parallel p_i)}{\partial {p_i}^2} = -2l_i < 0$. Thus, the payoff function $f_i(\vec{p})$ is strictly concave in p_i. This means that

$$\max_{p_1 \in P_1} f_1(p_1, p_2^{\mathrm{e}})$$

is achieved at p_1^{e} under the condition

$$\left.\frac{\partial f_1(p_1, p_2^{\mathrm{e}})}{\partial p_1}\right|_{p_1 = p_1^{\mathrm{e}}} = q - 2l_1 p_1^{\mathrm{e}} + l_2 p_2^{\mathrm{e}} + l_1 c = 0. \tag{4.3.10}$$

In a similar way,

$$\left.\frac{\partial f_2(p_1^{\mathrm{e}}, p_2)}{\partial p_2}\right|_{p_2 = p_2^{\mathrm{e}}} = q + l_2 p_1^{\mathrm{e}} - 2l_1 p_2^{\mathrm{e}} + l_1 c = 0.$$

Consequently, the Nash equilibrium $\vec{p}^{\mathrm{e}} = (p_1^{\mathrm{e}}, p_2^{\mathrm{e}})$ satisfies the following system of two linear inhomogeneous equations with constant coefficients:

$$\begin{cases} -2l_1 p_1^{\mathrm{e}} + l_2 p_2^{\mathrm{e}} = -(q + l_1 c), \\ l_2 p_1^{\mathrm{e}} - 2l_1 p_2^{\mathrm{e}} = -(q + l_1 c). \end{cases}$$

For $l_2 \neq 2l_1$, the solution is

$$p_1^{\mathrm{e}} = p_2^{\mathrm{e}} = \frac{q + l_1 c}{2l_1 - l_2} = p^N.$$

Moreover, it follows from (4.3.10) that

$$q - l_1 p_1^{\mathrm{e}} + l_2 p_2^{\mathrm{e}} = l_1 (p^N - c),$$

and, in terms of notations (4.3.8), $f_i(\vec{p^e}) = l_1(p^N - c)^2$. Hence, by (4.3.8) and (4.3.2), we get (4.3.9). Concluding this proof, note that the assumption $p^N > c$ is quite natural: otherwise, Nash equilibrium incurs losses. ■

4.3.4 Use of Berge Equilibrium

Remark 4.3.1 Based on formulas (4.3.7) and (4.3.9), let us construct the following auxiliary scalar function under the conditions $l_2 > l_1$ and $l_2 \neq 2l_1$:

$$\begin{aligned} F(l_1, l_2, \beta) &= f_i(\vec{p}^{\mathrm{B}}) - f_i(\vec{p}^{\mathrm{e}}) = (l_2 - l_1)(\beta - c)^2 \\ &+ [q + c(l_2 - l_1)](\beta - c) - l_1 \left(\frac{q + c(l_2 - l_1)}{2l_1 - l_2} \right)^2 \quad (i = 1, 2). \end{aligned} \tag{4.3.11}$$

Then from Propositions 4.3.1 and 4.3.2 we derive the following results.

First, if there exist positive values (l_1, l_2, β) such that $f_i(\vec{p}^{\mathrm{B}}) - f_i(\vec{p}^{\mathrm{e}}) = F(l_1, l_2, \beta) > 0$ $(<, =)$, then the Berge equilibrium with these values (l_1, l_2, β) yields a greater (smaller or the same) payoff as the Nash equilibrium.

Second, it is necessary to check that, for such positive values (l_1, l_2, β), the unit prices chosen by both players and also the maximum unit price β are not smaller than the cost (otherwise, sales become unprofitable and make no economic sense).

We will explore two cases of possible relationships between the elasticity coefficients, namely, $l_2 > l_1$ and $l_2 < l_1$.

Case I: $l_2 > l_1 > 0$.

Then the equation

$$(l_2 - l_1)(\beta - c)^2 + [q + c(l_2 - l_1)](\beta - c) - l_1 \left(\frac{q + c(l_2 - l_1)}{2l_1 - l_2} \right)^2 = 0$$

has the roots

$$\begin{aligned} (\beta - c)_{\pm} &= \frac{-[q + c(l_2 - l_1)] \pm \sqrt{[q + c(l_2 - l_1)]^2 + \dfrac{4l_1(l_2 - l_1)[q + c(l_2 - l_1)]^2}{(2l_1 - l_2)^2}}}{2(l_2 - l_1)} \\ &= \frac{-[q + c(l_2 - l_1)] \pm |q + c(l_2 - l_1)| \dfrac{l_2}{|2l_1 - l_2|}}{2(l_2 - l_1)}. \end{aligned} \tag{4.3.12}$$

Subcase Ia: $2l_1 - l_2 > 0$.

It follows from (4.3.12) that

$$(\beta^{(1)} - c)_+ = \frac{[q + c(l_2 - l_1)]\left(-1 + \dfrac{l_2}{2l_1 - l_2}\right)}{2(l_2 - l_1)} = \frac{q + c(l_2 - l_1)}{2l_1 - l_2} > 0,$$

$$(\beta - c)_- = \frac{[q + c(l_2 - l_1)]\left(-1 - \dfrac{l_2}{2l_1 - l_2}\right)}{2(l_2 - l_1)} = -\frac{[q + c(l_2 - l_1)]l_1}{(l_2 - l_1)(2l_1 - l_2)} < 0,$$

which yields

$$\beta^{(1)} = \frac{q + c(l_2 - l_1)}{2l_1 - l_2} + c = \frac{q + cl_1}{2l_1 - l_2} = p^N > c. \tag{4.3.13}$$

The graph of the function $F = F(l_1, l_2, \beta)$ for each fixed pair $(l_1, l_2) \in \{(l_1, l_2)|2l_1 > l_2 > l_1 > 0\}$ is shown in Fig. 4.3 (under the inequality $|(\beta - c)_-| > (\beta^{(1)} - c)_+$).

Combining Propositions 4.3.1 and 4.3.2 with the conditions $l_2 > l_1$ and $2l_1 > l_2$ (Fig. 4.4), Remark 4.3.1 and formula (4.3.13), we establish the following fact.

Proposition 4.3.3 *Consider the game* Γ *with* $0 < l_1 < l_2 < 2l_1$. *Then, for players* $i = 1, 2$,

$$\begin{aligned} &f_i(\vec{p}^{\mathrm{B}}) > f_i(\vec{p}^{\mathrm{e}}) \text{ if } \beta > \beta^{(1)} = \frac{q + cl_1}{2l_1 - l_2} = p^N;\\ &f_i(\vec{p}^{\mathrm{B}}) = f_i(\vec{p}^{\mathrm{e}}) \text{ if } \beta = \beta^{(1)};\\ &f_i(\vec{p}^{\mathrm{B}}) < f_i(\vec{p}^{\mathrm{e}}) \text{ if } \beta \in (c, \beta^{(1)}), \end{aligned} \tag{4.3.14}$$

where the Berge and Nash equilibria are $\vec{p}^{\mathrm{B}} = (\beta, \beta)$ *and* $\vec{p}^{\mathrm{e}} = (p_1^{\mathrm{e}}, p_2^{\mathrm{e}})$, $p_1^{\mathrm{e}} = p_2^{\mathrm{e}} = \dfrac{q + cl_1}{2l_1 - l_2}$, *respectively. The payoffs in these equilibria are given by*

$$\begin{aligned} &f_i(\vec{p}^{\mathrm{B}}) = (l_2 - l_1)(\beta - c)^2 + [q + c(l_2 - l_1)](\beta - c),\\ &f_i(\vec{p}^{\mathrm{e}}) = l_1\left(\frac{q + c(l_2 - l_1)}{2l_1 - l_2}\right)^2 \quad (i = 1, 2). \end{aligned} \tag{4.3.15}$$

Remark 4.3.2 By Proposition 4.3.1, for any (l_1, l_2) from the interior of the shaded domain in Fig. 4.4, the payoffs of both players are greater in the Berge equilibrium $\vec{p}^{\mathrm{B}}$ than in the Nash equilibrium $\vec{p}^{\mathrm{e}}$ if the maximum unit price satisfies the condition $\beta > p^N$ (smaller if $c < \beta < p^N$ and coincide if $\beta = p^N$).

Subcase Ib: $l_2 > 2l_1$.

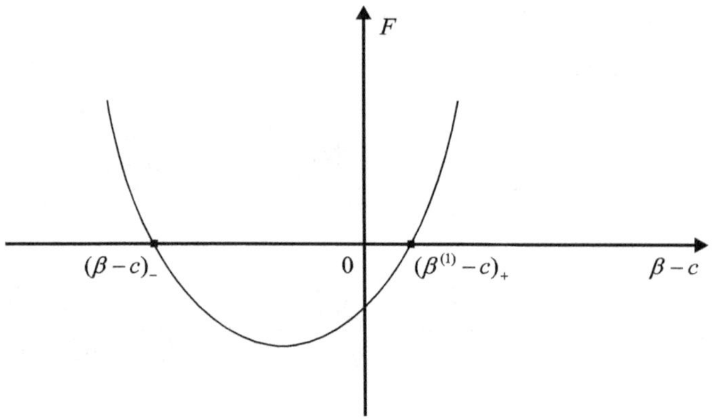

Fig. 4.3 Graph of the function $F(l_1, l_2, \beta)$ for $2l_1 > l_2 > l_1 > 0$

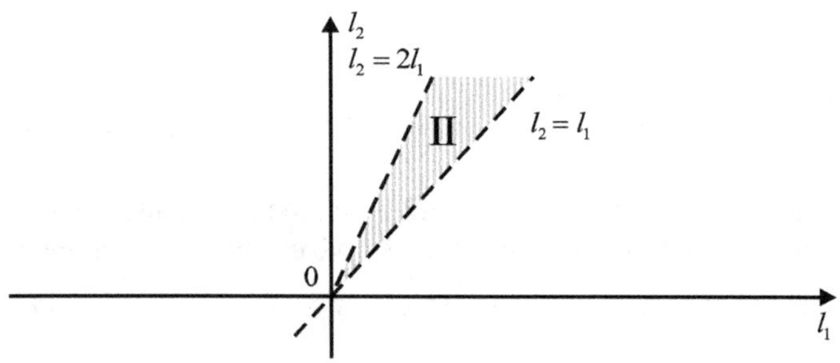

Fig. 4.4 Set of pairs $l = (l_1, l_2)$ with $0 < l_1 < l_2 < 2l_1$

Proposition 4.3.4 *Consider the game* Γ *with* $l_2 > 2l_1$. *Then, for players* $i = 1, 2$,

$$\begin{aligned} &f_i(\vec{p}^{\mathrm{B}}) > f_i(\vec{p}^{\mathrm{e}}) \text{ if } \beta > \beta^{(2)} = \frac{ql_1 + c(l_2 - l_1)^2}{(l_2 - l_1)(l_2 - 2l_1)} > c, \\ &f_i(\vec{p}^{\mathrm{B}}) = f_i(\vec{p}^{\mathrm{e}}) \text{ if } \beta = \beta^{(2)}, \\ &f_i(\vec{p}^{\mathrm{B}}) < f_i(\vec{p}^{\mathrm{e}}) \text{ if } \beta \in (c, \beta^{(2)}), \end{aligned} \tag{4.3.16}$$

and the payoffs of both players in the Berge equilibrium $\vec{p}^{\mathrm{B}} = (\beta, \beta)$ *and in the Nash equilibrium* $\vec{p}^{\mathrm{e}} = (p_1^{\mathrm{e}}, p_2^{\mathrm{e}})$, $p_i^{\mathrm{e}} = \dfrac{q + cl_1}{2l_1 - l_2}$ $(i = 1, 2)$, *again have form* (4.3.15).

As before, β denotes the maximum unit price.

Remark 4.3.3 The proof of Proposition 4.3.4 is based on
first, the implication $[l_2 > 2l_1] \Rightarrow [l_2 > l_1]$ and the two roots

$$(\beta^{(2)} - c)_+ = \frac{[q + c(l_2 - l_1)]l_1}{(l_2 - l_1)(l_2 - 2l_1)} > 0, \quad (\beta - c)_- = -\frac{q + c(l_2 - l_1)}{l_2 - 2l_1} < 0,$$

of the equation $F(l_1, l_2, p) = 0$;

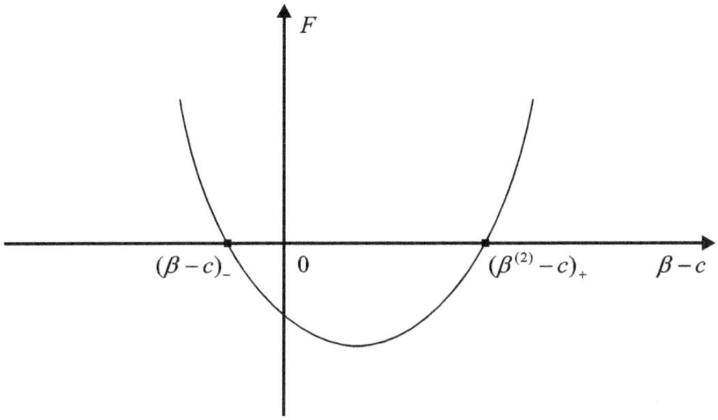

Fig. 4.5 Graph of the function $F(l_1, l_2, \beta)$ for $l_2 > 2l_1$

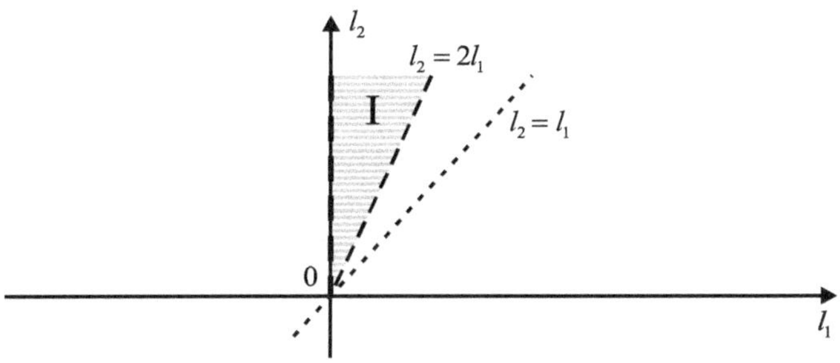

Fig. 4.6 Set of pairs $l = (l_1, l_2)$ for $l_2 > 2l_1 > 0$

second, the graph of the function $F(l_1, l_2, \beta) = f_i(\vec{p}^{\mathrm{B}}) - f_i(\vec{p}^{\mathrm{e}})$ displayed in Fig. 4.5;

third, the chain of relations

$$\beta^{(2)} = (\beta^{(2)} - c)_+ + c = \frac{ql_1 + c(l_2 - l_1)^2}{(l_2 - l_1)(l_2 - 2l_1)} > c;$$

fourth, the inequality

$$(\beta^{(2)} - c)_+ > |(\beta - c)_-| = \beta^{(1)} - c.$$

By Proposition 4.3.4, for any (l_1, l_2) from the interior of the shaded wedge in Fig. 4.6 (excluding its boundary—the lines $l_1 = 0$ and $l_2 = 2l_1$), the payoffs of both players are greater in the Berge equilibrium than in the Nash equilibrium if the maximum unit price satisfies the condition $\beta > \beta^{(2)}$ (smaller if $\beta \in (c, \beta^{(2)})$ and coincide if $\beta = \beta^{(2)}$).

Case II: $l_1 > l_2 > 0$.

There are two important subcases, $q + c(l_2 - l_1) > 0$ and $q + c(l_2 - l_1) < 0$.

Subcase IIa: $q + c(l_2 - l_1) > 0$.

Proposition 4.3.5 *If the elasticity coefficients l_1 and l_2 from* (4.3.2) *satisfy*

$$l_1 > l_2 > l_1 - \frac{q}{c}, \tag{4.3.17}$$

then the Berge equilibrium in the game Γ has the form

$$\begin{aligned}(\vec{p}^{\mathrm{B}}; \vec{f}(\vec{p}^{\mathrm{B}})) &= (\beta, \beta; f_1(\vec{p}^{\mathrm{B}}), f_2(\vec{p}^{\mathrm{B}})) \\ &= \left(\frac{q + c(l_1 - l_2)}{2(l_1 - l_2)}, \frac{q + c(l_1 - l_2)}{2(l_1 - l_2)}; \frac{[q + c(l_2 - l_1)]^2}{4(l_1 - l_2)}, \right. \\ &\qquad \left. \frac{[q + c(l_2 - l_1)]^2}{4(l_1 - l_2)}\right). \end{aligned} \tag{4.3.18}$$

Proof Conditions (4.3.17) directly imply

$$[l_1 > l_2] \Rightarrow [2l_1 - l_2 > 0], \quad \left[l_2 > l_1 - \frac{q}{c}\right] \Rightarrow [q + c(l_2 - l_1) > 0].$$

The set of the two-dimensional vectors

$$l = (l_1, l_2) \in \left\{(l_1, l_2) | \left[l_1 > l_2 > l_1 - \frac{q}{c}\right] \wedge [l_i > 0 \ (i = 1, 2)]\right\}$$

is shown in Fig. 4.7.

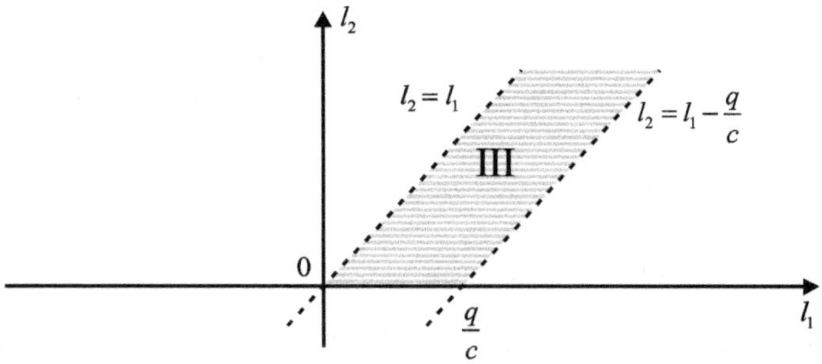

Fig. 4.7 Set of pairs $\left\{(l_1, l_2) | \left[l_1 > l_2 > l_1 - \frac{q}{c}\right] \wedge [l_i > 0 \ (i = 1, 2)]\right\}$

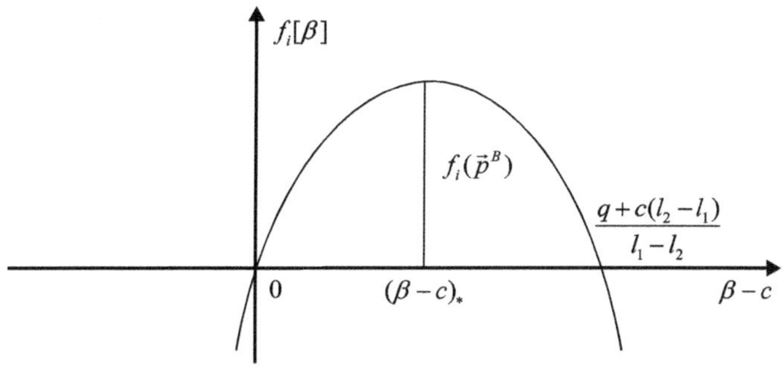

Fig. 4.8 Graph of the function $f_i[\beta]$ for $l_1 > l_2 > l_1 - \frac{q}{c}$

Recall that the two players have the same cost c and the same maximum unit price β of their products. Following (4.3.2), we introduce the two coinciding functions

$$\begin{aligned} f_i[\beta] &= [q + \beta(l_2 - l_1)](\beta - c) \\ &= (l_2 - l_1)(\beta - c)^2 + [q + c(l_2 - l_1)](\beta - c) \quad (i = 1, 2). \end{aligned} \tag{4.3.19}$$

Double differentiation of the function $f_i[\beta]$ $(i = 1, 2)$ (4.3.19) yields $\frac{d^2 f_i[\beta]}{d(\beta - c)^2} = 2(l_2 - l_1) < 0$, and hence $f_i[\beta]$ is strictly concave in $\beta - c$. The graph of $f_i[\beta]$ intersects the axis $(\beta - c)$ at the points $(\beta - c)_1 = 0$ and $(\beta - c)_2 = -\frac{q + c(l_2 - l_1)}{l_2 - l_1} > 0$. The function $f_i[\beta]$ achieves maximum at the point $(\beta - c)_* = \frac{q + c(l_2 - l_1)}{2(l_1 - l_2)}$, and the maximum value is $f_i[\beta_*] = \frac{[q + c(l_2 - l_1)]^2}{4(l_1 - l_2)}$, where $\beta_* = (\beta - c)_* + c = \frac{q + c(l_1 - l_2)}{2(l_1 - l_2)}$ (see Fig. 4.8).

Thus, under conditions (4.3.17), the criterion $f_i[\beta]$ determining the maximum unit price β_* takes the maximum value $f_i[\beta_*] = \frac{[q + c(l_2 - l_1)]^2}{4(l_1 - l_2)}$ at the point $\beta_* = \frac{q + c(l_1 - l_2)}{2(l_1 - l_2)}$. This concludes the proof of Proposition 4.3.5. ■

Combining Propositions 4.3.5 and 4.3.2, we get the following result.

Proposition 4.3.6 *Consider the game* Γ *in which the elasticity coefficients* $l_i = \text{const} > 0$ $(i = 1, 2)$ *satisfy*

$$l_1 > l_2 > l_1 - \frac{q}{c}.$$

Then both players obtain higher payoffs in the Berge equilibrium $\vec{p}^{\mathrm{B}}$ *than in the Nash equilibrium, i.e.,*

$$f_i(\vec{p}^{\mathrm{B}}) > f_i(\vec{p}^{\mathrm{e}}) \quad (i = 1, 2), \tag{4.3.20}$$

where the Berge equilibrium payoff is given by (4.3.18) *and the Nash equilibrium payoff by* (4.3.8) *and* (4.3.9).

Proof Due to the implications $[l_1 > l_2] \Rightarrow [2l_1 > l_2] \Rightarrow [2l_1 - l_2 > 0]$ and Proposition 4.3.2, there exists the Nash equilibrium $(\vec{p}^{\mathrm{e}}; \vec{f}(\vec{p}^{\mathrm{e}}))$ defined by formulas (4.3.8) and (4.3.9). Next, recall the two implications $[l_1 > l_2] \Rightarrow [l_1 - l_2 > 0]$ and $\left[l_2 > l_1 - \frac{q}{c}\right] \Rightarrow [q + c(l_2 - l_1) > 0]$ from the proof of Proposition 4.3.5, which ensure the existence of the Berge equilibrium (4.3.18) in the game Γ. A direct comparison of (4.3.18) and (4.3.9) yields

$$\begin{aligned} f_i(\vec{p}^{\mathrm{B}}) - f_i(\vec{p}^{\mathrm{e}}) &= \frac{[q + c(l_2 - l_1)]^2}{4(l_1 - l_2)} - l_1 \frac{[q + c(l_2 - l_1)]^2}{(2l_1 - l_2)^2} \\ &= \frac{[q + c(l_2 - l_1)]^2 {l_2}^2}{4(l_1 - l_2)(2l_1 - l_2)^2} > 0 \ \ (i = 1, 2), \end{aligned}$$

i.e., (4.3.20) holds. ■

Remark 4.3.4 Proposition 4.3.6 has a dual character: it can be applied to analyze the real competitive markets described by the Bertrand duopoly model and also to design such markets.

4.3.4.1 First Application

Step 1. For a real competitive market, identify *the numerical values* of the following parameters:
the elasticity coefficients l_1^* and l_2^*;
the cost c;
the initial demand q;
and the maximum unit price β.

Step 2. Using the values c and q, draw Fig. 4.7 in the first quadrant of the plane $\{l_1, l_2\}$.

Step 3. Answer the two questions:

(a) Does the point $l^* = (l_1^*, l_2^*)$ belong to the interior of the shaded domain in Fig. 4.7?

(b) Does the value β coincide with $\dfrac{q + c(l_1^* - l_2^*)}{2(l_1^* - l_2^*)}$?

Step 4. If both answers are affirmative, the players should choose their strategies from the Berge equilibrium $\vec{p}^{\mathrm{B}} = (p_1^{\mathrm{B}}, p_2^{\mathrm{B}}) = (\beta, \beta) = \left(\frac{q+c(l_1^*-l_2^*)}{2(l_1^*-l_2^*)}, \frac{q+c(l_1^*-l_2^*)}{2(l_1^*-l_2^*)}\right)$, thereby obtaining the payoffs $f_i(\vec{p}^{\mathrm{B}}) = \frac{[q+c(l_2^*-l_1^*)]^2}{4(l_1^*-l_2^*)}$ $(i = 1, 2)$, which are greater than the payoffs $f_i(\vec{p}^{\mathrm{e}}) = l_1\left(\frac{q+c(l_2^*-l_1^*)}{2l_1^*-l_2^*}\right)^2$ $(i = 1, 2)$ in the Nash equilibrium $\vec{p}^{\mathrm{e}} = \left(\frac{q+l_1^*c}{2l_1^*-l_2^*}, \frac{q+l_1^*c}{2l_1^*-l_2^*}\right)$.

Second application is the *design* of competitive markets.

Step 1. Using the *desired numerical values* of the cost c and initial demand q, draw the bisecting line $l_2 = l_1$ in the first quadrant of the plane $\{l_1, l_2\}$ as well as its translate to the right by q/c (see Fig. 4.9).
Thus, in the first quadrant of the plane $\{l_1, l_2\}$, we obtain a track with the deadlock $[0, q/c]$ on the axis l_1, as illustrated in Fig. 4.9.

Step 2. Adjust the maximum unit price β to the value $\frac{q+c(l_1-l_2)}{2(l_1-l_2)}$ with economic, governmental and other regulatory measures.

Step 3. Then for all points $l = (l_1, l_2)$ inside the shaded track in Fig. 4.9 (which is formed by the two half-lines $l_2 = l_1$, $l_2 = l_1 - \frac{q}{c}$, $l_i > 0$ $(i = 1, 2)$ and the deadlock $\left[0, \frac{q}{c}\right]$ on the axis l_1), the players should choose their strategies $\beta = \frac{q+c(l_1-l_2)}{2(l_1-l_2)}$ from the Berge equilibrium $\vec{p}^{\mathrm{B}} = (\beta, \beta)$, as the corresponding payoffs $\frac{[q+c(l_2-l_1)]^2}{4(l_1-l_2)}$ are higher that their counterparts $l_1\left(\frac{q+c(l_1-l_2)}{2l_1-l_2}\right)^2$ in the Nash equilibrium $\vec{p}^{\mathrm{e}} = \left(\frac{q+cl_1}{2l_1-l_2}, \frac{q+cl_1}{2l_1-l_2}\right)$).

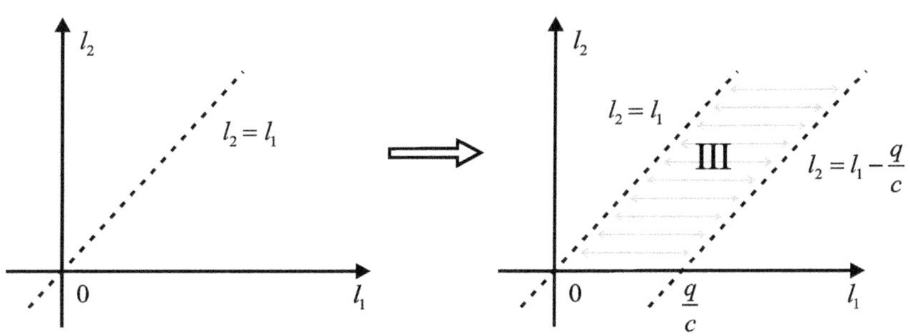

Fig. 4.9 Construction of the domain $\left\{l = (l_1, l_2) | l_1 > l_2 > l_1 - \frac{q}{c}\right\}$

Subcase IIb: $\left[l_2 < l_1 - \frac{q}{c}\right] \Leftrightarrow [q + c(l_2 - l_1) < 0]$.

Proposition 4.3.7 *Consider the game Γ with $l_2 < l_1 - \frac{q}{c}$. Then, for any $\beta > c$,*

$$f_i(\vec{p}^{\mathrm{e}}) > 0 > f_i(\vec{p}^{\mathrm{B}}) \quad (i = 1, 2), \tag{4.3.21}$$

where the payoffs $f_i(\vec{p}^{\mathrm{e}})$ and strategies p_i^{e} in the Nash equilibrium are given by (4.3.9) *and* (4.3.8), *respectively.*

Proof First of all, note that

$$\left[l_2 < l_1 - \frac{q}{c}\right] \Rightarrow [l_2 < l_1] \Rightarrow [l_2 - 2l_1 < 0].$$

In addition, $q + c(l_2 - l_1) < 0$. Then the set $\left\{l = (l_1, l_2) | 0 < l_2 < l_1 - \frac{q}{c}\right\}$ is the interior of an acute angle that adjoins the axis l_1 (see Fig. 4.10).

Wedge IV is formed by the rays $l_2 = 0$ and $l_2 = l_1 - \frac{q}{c}$ emanating from the vertex $(\frac{q}{c}, 0)$.

For any $l = (l_1, l_2) \in \operatorname{int} \mathrm{IV}$, we have $l_2 < l_1$ and $q + c(l_2 - l_1) < 0$. These inequalities in combination with (4.3.2) yield

$$f_i(\beta, \beta) = (l_2 - l_1)(\beta - c)^2 + [q + c(l_2 - l_1)](\beta - c) < 0 \quad \forall \beta > c \quad (i = 1, 2),$$

which proves the right-hand inequality in (4.3.21).

Further, for any $l = (l_1, l_2) \in \operatorname{int} \mathrm{IV}$, by the inequalities $2l_1 - l_2 \neq 0$, $q + c(l_2 - l_1) < 0$ and Proposition 4.3.2,

$$f_i(\vec{p}^{\mathrm{e}}) = l_1 \left(\frac{q + c(l_2 - l_1)}{2l_1 - l_2}\right)^2 > 0 \quad (i = 1, 2).$$

This proves the left-hand inequality in (4.3.21). ■

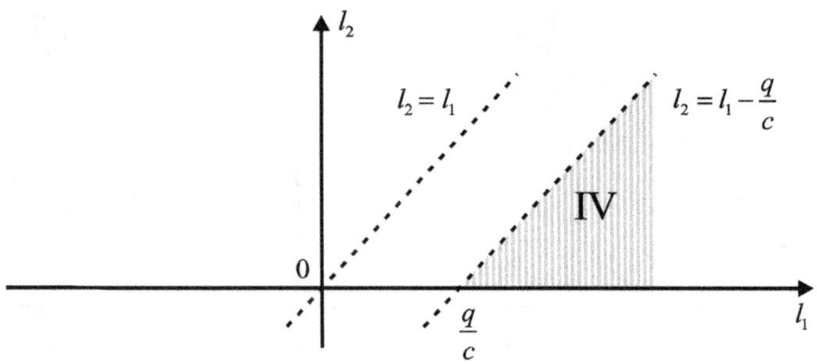

Fig. 4.10 Set of pairs $l = (l_1, l_2) \in \left\{(l_1, l_2) | 0 < l_2 < l_1 < l_1 - \frac{q}{c}\right\}$

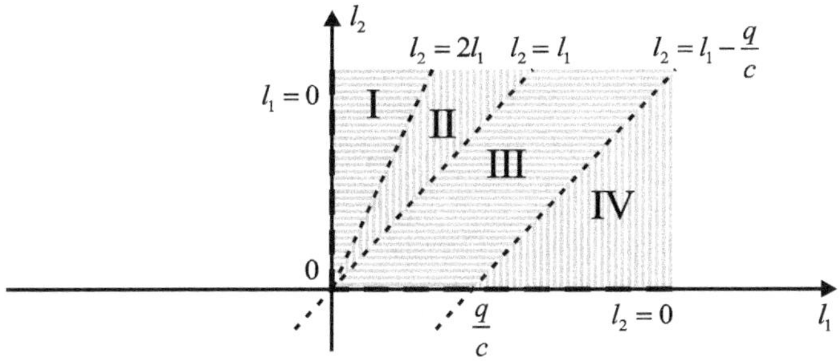

Fig. 4.11 Choice of better equilibrium

Remark 4.3.5 In accordance with Proposition 4.3.7, under the conditions $l_2 < l_1 - \frac{q}{c}$ and $\beta > c$, both players should prefer the Nash equilibrium to the Berge equilibrium (of course, if the point $l = (l_1, l_2)$ falls *inside* the shaded wedge in Fig. 4.10).

Remark 4.3.6 Propositions 4.3.3–4.3.7 allow one to choose a better equilibrium in terms of payoffs (Berge or Nash) depending on the location of the point $l = (l_1, l_2)$ in the first quadrant of the plane $\{l_1, l_2\}$ and the maximum unit price β only if this point belongs to the *interior* of the shaded domains I–IV in Figs. 4.4, 4.6, 4.7, and 4.10 (see Fig. 4.11).

However, which equilibrium (NE or BE) should be chosen if the point $l = (l_1, l_2)$ lies on the boundaries of domains I–IV? This question will be treated in the next section.

4.3.5 Choice of Appropriate Equilibrium on the Boundaries of the Constructed Domains

Let the point l be on the *boundaries* of the shaded domains—see the dash-and-dot lines in Fig. 4.11, which are formed by the rays $l_1 = 0, l_2 = 2l_1, l_2 = l_1, l_2 = l_1 - \frac{q}{c}$, and $l_2 = 0$ in the first quadrant ($l_i \geqslant 0$, $i = 1, 2$). Which equilibrium is a better choice then?

4.3.5.1 Boundary $l_1 = 0$

Proposition 4.3.8 *Consider the game* Γ *with* $l_1 = 0$. *Then for any* $\beta > c$,

$$f_i(\vec{p}^{\mathrm{B}}) > f_i(\vec{p^{\mathrm{e}}}) \quad (i = 1, 2),$$

i.e., the Berge equilibrium $\vec{p}^{\mathrm{B}} = (\beta, \beta)$ *is preferable to the Nash equilibrium for*

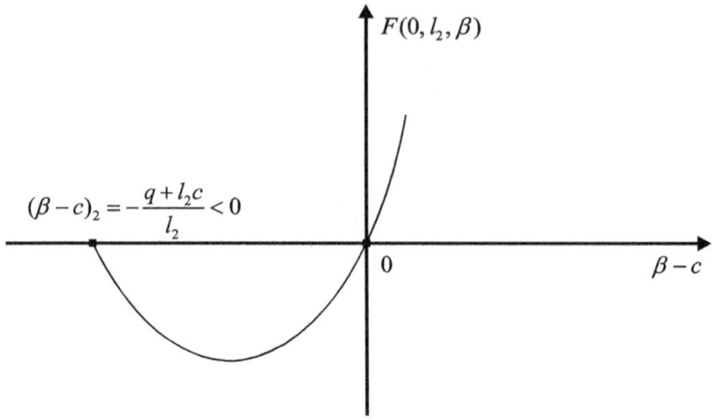

Fig. 4.12 Graph of the function $F(0, l_2, \beta)$ for $l_1 = 0$

both players, having the payoffs

$$f_i(\vec{p}^{\mathrm{B}}) = l_2(\beta - c)^2 + [q + l_2 c](\beta - c).$$

Proof For $l_1 = 0$, the auxiliary function (4.3.11) takes the form

$$F(0, l_2, \beta) = l_2(\beta - c)^2 + [q + cl_2](\beta - c).$$

Then $\dfrac{\partial F(0, l_2, \beta)}{\partial(\beta - c)} = 2l_2(\beta - c) + [q + cl_2]$ and $\dfrac{\partial^2 F(0, l_2, \beta)}{\partial(\beta - c)^2} = 2l_2 > 0$. Hence, the function $F(0, l_2, \beta)$ is strictly convex in the variable $\beta - c$ and increasing for $\beta - c > 0$ (see Fig. 4.12). In addition, the graph of this function intersects the axis $(\beta - c)$ at the two points $(\beta - c)_1 = 0$ and $(\beta - c)_2 = -\dfrac{q + l_2 c}{2l_2} < 0$.

The conclusion follows from Fig. 4.12 and the inequality $F(0, l_2, \beta) = f_i(\vec{p}^{\mathrm{B}}) - f_i(\vec{p}^{\mathrm{e}}) > 0 \; \forall \beta > c$. ■

4.3.5.2 Boundary $l_1 = l_2 > 0$

Proposition 4.3.9 *Consider the game* Γ *with* $l_2 = l_1$. *Then for each* $i = 1, 2$,

$$f_i(\vec{p}^{\mathrm{B}}) > f_i(\vec{p}^{\mathrm{e}}) \text{ if } \beta > c + \frac{q}{l_1},$$
$$f_i(\vec{p}^{\mathrm{B}}) < f_i(\vec{p}^{\mathrm{e}}) \text{ if } \beta < c + \frac{q}{l_1},$$
$$f_i(\vec{p}^{\mathrm{B}}) = f_i(\vec{p}^{\mathrm{e}}) \text{ if } \beta = c + \frac{q}{l_1},$$

where $\vec{p}^{\mathrm{B}} = (\beta, \beta)$, $\vec{p}^{\mathrm{e}} = (p_1^{\mathrm{e}}, p_2^{\mathrm{e}})$, *and* $p_1^{\mathrm{e}} = p_2^{\mathrm{e}} = \dfrac{q + l_1 c}{l_1}$.

Proof For $l_1 = l_2$, the auxiliary function (4.3.11) takes the form

$$F(l_1, l_2, \beta) = q(\beta - c) - l_1 \frac{q^2}{(2l_1 - l_2)^2} = q[\beta - c - \frac{q}{l_1}].$$

Hence,

$$F(l_1, l_2, \beta) = f_i(\vec{p}^{\mathrm{B}}) - f_i(\vec{p}^{\mathrm{e}}) = q\left[\beta - c - \frac{q}{l_1}\right],$$

which leads to the desired result by Remark 4.3.1. ■

4.3.5.3 Boundary $l_2 = l_1 - \frac{q}{c}$

Proposition 4.3.10 *Consider the game* Γ *with* $l_2 = l_1 - \frac{q}{c}$. *Then both players have the highest payoffs in the Nash equilibrium* $\vec{p}^{\mathrm{e}} = (p_1^{\mathrm{e}}, p_2^{\mathrm{e}}) = (c, c)$, *which here coincides with the Berge equilibrium.*

Proof Using (4.3.11) and the implication

$$\left[l_2 = l_1 - \frac{q}{c}\right] \Rightarrow [q + c(l_2 - l_1) = 0],$$

we have

$$F(l_1, l_2, \beta) = f_i(\vec{p}^{\mathrm{B}}) - f_i(\vec{p}^{\mathrm{e}}) = (l_2 - l_1)(\beta - c)^2 = -\frac{q}{c}(\beta - c)^2 < 0,$$

where $l_2 - l_1 = -\frac{q}{c}$.

For $q + c(l_2 - l_1) = 0$, vector (4.3.9) has components $f_i(\vec{p^e}) = 0$ $(i = 1, 2)$. Thus, it is necessary to extend the set of admissible strategies with the additional point $\beta = c$. Then $\vec{p}^{\mathrm{B}} = (c, c) = \vec{p^e}$. ■

4.3.5.4 Boundary $l_2 = 0$

Once again, formula (4.3.11) with $l_2 = 0$ yields

$$F(l_1, 0, \beta) = -l_1(\beta - c)^2 + (q - cl_1)(\beta - c) - \frac{(q - cl_1)^2}{4l_1}.$$

Since $\dfrac{\partial F(l_1, 0, \beta)}{\partial(\beta - c)} = -2l_1(\beta - c) + [q - cl_1]$ and $\dfrac{\partial^2 F(l_1, 0, \beta)}{\partial(\beta - c)^2} = -2l_1 < 0$, the function $F(l_1, 0, \beta)$ is concave in the variable $\beta - c$. We will analyze two subcases, $q > cl_1$ and $q < cl_1$.

4.3.5.5 Subcase $q > cl_1$

Proposition 4.3.11 *Consider the game* Γ *with* $l_2 = 0$ *and* $q > cl_1$. *Then for each* $i = 1, 2$,

$$\begin{aligned} f_i(\vec{p}^{\,\mathrm{B}}) &= f_i(\vec{p}^{\,\mathrm{e}}) \text{ for } \beta = \frac{q + cl_1}{2l_1} = p^{\mathrm{N}}, \\ f_i(\vec{p}^{\,\mathrm{B}}) &< f_i(\vec{p}^{\,\mathrm{e}}) \text{ for } [\beta > c] \wedge [\beta \neq p^{\mathrm{N}}], \end{aligned}$$

where $\vec{p^e} = \left(\dfrac{q + l_1 c}{2l_1}, \dfrac{q + l_1 c}{2l_1}\right)$, $f_i(\vec{p^e}) = \dfrac{(q - l_1 c)^2}{4l_1}$, $\vec{p}^{\mathrm{B}} = (\beta, \beta)$, *and* $f_i(\vec{p}^{\mathrm{B}}) = -l_1(\beta - c)^2 + (q - cl_1)(\beta - c)$.

Proof The graph of $F(l_1, 0, \beta)$ touches the axis $(\beta - c)$ at the unique point $\beta - c = \dfrac{q - cl_1}{2l_1} > 0$, since

$$F(l_1, 0, \beta) = -l_1\left[(\beta - c) - \frac{q - cl_1}{2l_1}\right]^2 \leqslant 0.$$

Due to this fact as well as the implication $[q > cl_1] \Rightarrow [q - cl_1 > 0]$ and the strict concavity of $F(l_1, 0, \beta)$ in $\beta - c$, the function $F(l_1, 0, \beta)$ has the graph presented in Fig. 4.13.

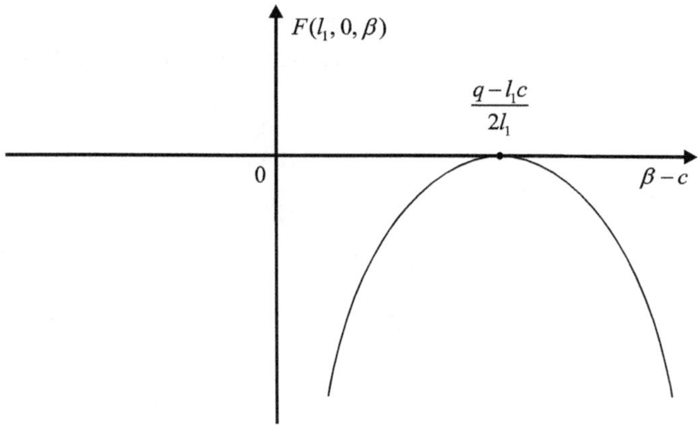

Fig. 4.13 Graph of the function $F(l_1, 0, \beta)$ for $l_2 = 0$ and $q > cl_1$

Thus, we have shown that $F(l_1, 0, \beta) < 0$ for $\beta - c \neq \dfrac{q - cl_1}{2l_1}$ and $F(l_1, 0, \beta) = 0$ only for $\beta = \dfrac{q - cl_1}{2l_1} + c = \dfrac{q + cl_1}{2l_1} = p_i^e$ $(i = 1, 2)$. In combination with Remark 4.3.1, this concludes the proof of Proposition 4.3.11. ■

4.3.5.6 Subcase $q < cl_1$

Proposition 4.3.12 *Consider the game* Γ *with* $l_2 = 0$ *and* $q < cl_1$. *Then for each* $i = 1, 2$,

$$f_i(\vec{p}^{\mathrm{e}}) > f_i(\vec{p}^{\mathrm{B}})\ \forall \beta > c,$$

where, like in Proposition 4.3.11, $\vec{p}^{\mathrm{e}} = \left(\dfrac{q + l_1 c}{2l_1}, \dfrac{q + l_1 c}{2l_1}\right)$, $f_i(\vec{p}^{\mathrm{e}}) = \dfrac{(q - l_1 c)^2}{4l_1}$, $\vec{p}^{\mathrm{B}} = (\beta, \beta)$, *and* $f_i(\vec{p}^{\mathrm{B}}) = -l_1(\beta - c)^2 + (q - cl_1)(\beta - c)$.

Proof The conclusion is immediate from (4.3.11), the fact that

$$F(l_1, 0, \beta) = -l_1\left[(\beta - c) - \frac{q - cl_1}{2l_1}\right]^2 < 0$$

for all $\beta > c$ and Remark 4.3.1. The corresponding graph of $F(l_1, 0, \beta)$ is shown in Fig. 4.14. ■

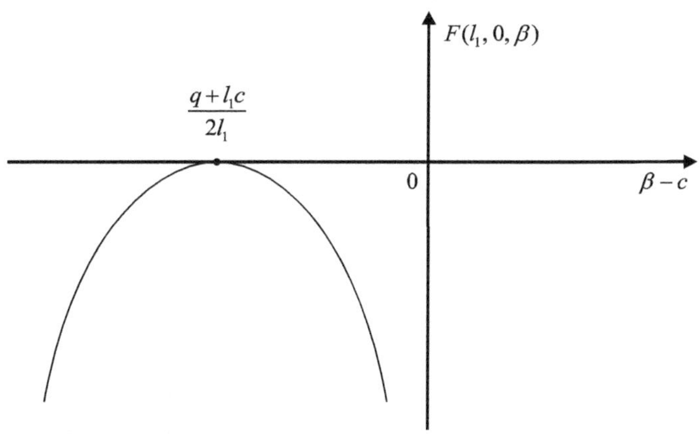

Fig. 4.14 Graph of the function $F(l_1, 0, \beta)$ for $l_2 = 0$ and $q < cl_1$

Remark 4.3.7 For the boundary $l_2 = 2l_1$, complete results have not been obtained so far.

4.3.6 Compromising Behavioral Principles for Higher Benefits

As is well-known, in any noncooperative game (of course, including the Bertrand duopoly model) the players choose their strategies in order to maximize their payoffs. For the time being, our analysis is confined to a couple of alternatives, namely,

first, the selfish Nash equilibrium (NE);

second, the altruistic Berge equilibrium (BE).

We are eliminating the concept of active equilibrium and the equilibrium in threats and counter-threats as its special case, comparing NE and BE only.

1. Here a major role can be played by the so-called psychological factor, a common assumption in theory of noncooperative games. Nash equilibrium (Proposition 4.3.2) is a recipe for an extremely self-centered player who cares about individual interests only. On the other hand, a convinced altruist should use the strategy dictated by the Golden Rule of ethics, i.e., Berge equilibrium (Proposition 4.3.1).
2. Our idea consists in neglecting the psychological factor by a simple comparison of the payoffs in the NE and BE. This can be implemented using the following procedure, dictated by the the content of the current Sect. 4.3.

Step 1. Using the numerical values of the parameters q, c, l_1, and l_2, find the three values

$$\beta^{(1)} = \frac{q + cl_1}{2l_1 - l_2}, \quad \beta^{(2)} = \frac{ql_1 + c(l_2 - l_1)^2}{(l_2 - l_1)(l_2 - 2l_1)},$$

$$f_i(\vec{p^e}) = l_1\left[\frac{q + (l_2 - l_1)c}{2l_1 - l_2}\right]^2 \quad (i = 1, 2).$$

Step 2. In the first quadrant of the plane $\{l_1, l_2\}$, draw the half-lines $l_1 = 0$, $l_2 = l_1$, $l_2 = 2l_1$, $l_2 = l_1 - \frac{q}{c}$, and $l_2 = 0$ to exhibit the domains I–IV (see Fig. 4.11).

Step 3. Find out which of the domains I–IV contains the point $l^* = (l_1, l_2)$. Further steps are devoted to the choice of an appropriate solution concept (Nash or Berge equilibrium) as well as to the explicit-form calculation of the corresponding strategies and payoffs of the players. These issues are completely settled by two elements of the model: *first,* the maximum unit price β determined by the law of supply and demand in the market; *second,* the domain that contains the point l^*.

(a) Let $l^* \in \mathrm{I}$; then by Proposition 4.3.4,

(a1) for $\beta > \beta^{(2)}$, both players should choose the BE $\vec{p}^{\mathrm{B}} = (\beta, \beta)$, obtaining the payoffs

$$f_i(\vec{p}^{\mathrm{B}}) = (l_2 - l_1)(\beta - c)^2 + [q + c(l_2 - l_1)](\beta - c) \quad (i = 1, 2); \tag{4.3.22}$$

(a2) for $c < \beta < \beta^{(2)}$, both players should choose the NE $\vec{p}^{\mathrm{e}} = (p_1^N, p_2^N)$, $p_i^N = \dfrac{q + cl_1}{2l_1 - l_2}$ $(i = 1, 2)$, obtaining payoffs

$$f_i(\vec{p}^{\mathrm{e}}) = l_1 \left[\frac{q + (l_2 - l_1)c}{2l_1 - l_2} \right]^2 \quad (i = 1, 2); \tag{4.3.23}$$

(a3) for $\beta = \beta^{(2)}$, the payoffs of both players in the BE $\vec{p}^{\mathrm{B}}$ and in the NE $\vec{p}^{\mathrm{e}}$ coincide.

(b) Let $l^* \in \mathrm{II}$; then by Proposition 4.3.3,

(b1) for $\beta > \beta^{(1)}$, both players should choose the BE $\vec{p}^{\mathrm{B}} = (\beta, \beta)$, obtaining payoffs (4.3.22);

(b2) for $c < \beta < \beta^{(1)}$, both players should choose the NE $\vec{p}^{\mathrm{e}}$, obtaining payoffs (4.3.23);

(b3) for $\beta = \beta^{(1)}$, the payoffs of both players in the BE and NE are $l_1 \left[\dfrac{q + (l_2 - l_1)c}{2l_1 - l_2} \right]^2$.

(c) Let $l^* \in \mathrm{III}$; then by Proposition 4.3.6 both players should choose the BE $\vec{p}^{\mathrm{B}} = (\beta, \beta)$ for all $\beta > c$, obtaining payoffs (4.3.22).

(d) Let $l^* \in \mathrm{IV}$; then by Proposition 4.3.7 both players should choose the NE $\vec{p}^{\mathrm{e}}$ for all $\beta > c$, obtaining payoffs (4.3.23).

If the point $l^* = (l_1, l_2)$ falls on the boundaries of domains I–IV, then an appropriate concept of equilibrium should be chosen using the following table

$l_1 = 0,\ l_2 > 0$	Proposition 4.3.8
$l_2 = l_1 > 0$	Proposition 4.3.9
$l_2 = l_1 - q/c > 0$	Proposition 4.3.10
$l_1 > 0,\ l_2 = 0, q > cl_1$	Proposition 4.3.11
$l_1 > 0,\ l_2 = 0,\ q < cl_1$	Proposition 4.3.12

Or simply by using Fig. 4.15 as a guide for Sect. 4.3.

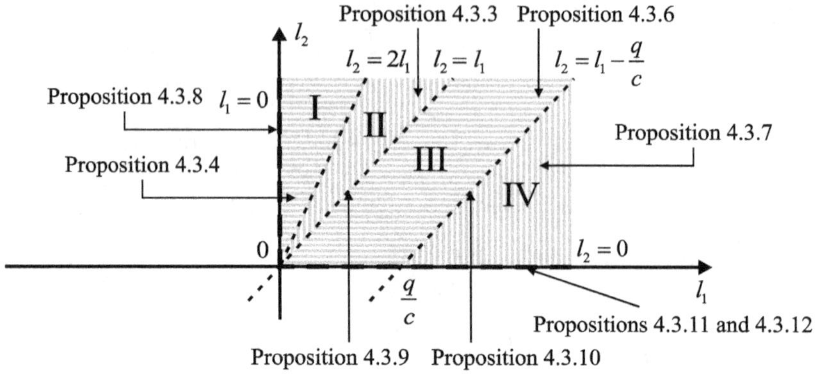

Fig. 4.15 Graph of the function $F(l_1, 0, \beta)$ for $l_2 = 0$ and $q < cl_1$

4.4 The Bertrand Model with Import

Gain cannot be made without some person's loss.
—Syrus[12]

The Bertrand price competition model of the previous section is augmented with import as a disturbing factor (uncertainty). Only a Pareto-guaranteed Nash equilibrium is constructed because Pareto-guaranteed Berge equilibria do not exist (see Remark 4.4.1).

4.4.1 Mathematical Model

Not he who has much is rich, but he who gives much.
—Fromm[13]

Generally, buyers consider the same-usage products supplied by different firms as different products. Such products are often substitutable. For example, two brands of identical-composition apple juice (sometimes, even produced from the same concentrate in the same facilities, yet for different sellers) can be positioned as different products intended for different groups of buyers. Nevertheless, a buyer may painlessly switch from one brand to the other.

Consider two firms in the market, 1 and 2, that supply products A and B, respectively. Assume A and B are interchangeable: whenever the products of one firm are unavailable, a buyer purchases the product of the other firm without

[12]Publilius Syrus, (flourished first century B.C.), was a Latin mime writer and artist.

[13]Erich Fromm, (1900–1980), was a German-born American psychoanalyst and social philosopher.

discomfort. The strategy of each firm (player in the market) is the unit price for its product. Let p_1 and p_2 be the unit prices announced by players 1 and 2, respectively. After price announcement, the demand for each product is determined by the law of supply and demand. Assume that the demand has the linear form, i.e.,

$$Q_1(p_1, p_2) = q - l_1 p_1 + l_2 p_2, \quad Q_2(p_1, p_2) = q - l_1 p_2 + l_2 p_1.$$

Here q specifies an initial demand and the elasticity coefficient $l_1 > 0$ shows the demand drop in response to a unit increase in price. In turn, the elasticity coefficient $l_2 > 0$ characterizes the demand rise in response to a unit increase in price of the substitute product.

Denote by c the cost of product unit, the same for both firms. Then the profits of the firms are described by the functions [31]

$$\begin{aligned} f_1(p_1, p_2) &= (q - l_1 p_1 + l_2 p_2)(p_1 - c), \\ f_2(p_1, p_2) &= (q - l_1 p_2 + l_2 p_1)(p_2 - c). \end{aligned} \tag{4.4.1}$$

The mathematical model of this competitive interaction between the firms is the noncooperative game without uncertainty

$$\Gamma = \langle \{1, 2\}, \{\mathrm{P}_i = [0, +\infty)\}_{i=1,2}, \{f_i(p_1, p_2)\}_{i=1,2} \rangle.$$

The strategy of player i $(i = 1, 2)$ is a unit price $p_i \geqslant 0$ announced by it without negotiations with the competitor. The payoff functions of the players are defined by (4.4.1).

Due to the strict concavity of $f_i(p_1, p_2)$ in p_i (which follows from $\dfrac{\partial^2 f_i}{\partial p_i^2} = -2l_1 < 0$), the sufficient conditions for the existence of a strategy profile p_i^{e} that maximizes $f_i(p_1, p_2)$ in p_i can be written as the system of two linear equations

$$\begin{cases} \left.\dfrac{\partial f_1(p_1, p_2)}{\partial p_1}\right|_{p^{\mathrm{e}}} = q + l_1 c - 2l_1 p_1^{\mathrm{e}} + l_2 p_2^{\mathrm{e}} = 0, \\ \left.\dfrac{\partial f_2(p_1, p_2)}{\partial p_2}\right|_{p^{\mathrm{e}}} = q + l_1 c - 2l_1 p_2^{\mathrm{e}} + l_2 p_1^{\mathrm{e}} = 0. \end{cases} \tag{4.4.2}$$

For $l_2 \neq 2l_1$, the unique solution of system (4.4.2) has the form

$$p_1^{\mathrm{e}} = p_2^{\mathrm{e}} = \frac{q + l_1 c}{2l_1 - l_2},$$

thus taking positive values for $l_2 < 2l_1$.

This solution is the Nash equilibrium in the game Γ, and the corresponding payoffs of the players are

$$f_1^{\mathrm{e}} = f_2^{\mathrm{e}} = l_1 \left[\frac{q - c(l_1 - l_2)}{2l_1 - l_2} \right]^2.$$

4.4.2 Consideration of Import

> The development of modern mathematical science
> is characterized by the growing role of algebraic methods,
> their wide penetration into many of its branches.
> —Bogolyubov[14]

An approach to account for import in the Bertrand duopoly model is suggested.

Now, in addition to the two firms (players 1 and 2) that supply the interchangeable products A and B, suppose that another firm (importer of a similar product C, e.g., a third brand of apple juice) enters the market. Suppose both players know nothing about its intentions and pricing policy. Perhaps, the importer will try to capture a market share, or mess with the market participants by cutting the unit price, or do something else. The importer will announce some unit price y, which acts as prior uncertainty in this model (the only reasonable hypothesis is that this price takes a value $y \geqslant 0$).

Choosing its strategy (the unit price p_i), each firm i must then take into account the competitor's actions and also any possible realization of the uncertainty y (the price announced by the importer).

Firms 1 and 2 announce their unit prices p_1 and p_2, respectively, thereby forming a strategy profile $(p_1, p_2) \in \mathrm{X} = \mathrm{X}_1 \times \mathrm{X}_2$, i.e., a current situation in the market. Simultaneously with and regardless of the players' choice, a specific value of the uncertainty y is realized. Let the resulting demand depend linearly on the prices with the elasticity coefficients l_1 (the demand drop in response to a unit price increase) and l_2 (the demand rise in response to a unit price increase of the substitute product). Then the demand function for the products of firm 1 is given by

$$Q_1(p_1, p_2, y) = q - l_1 p_1 + l_2 y + l_2 p_2 = q - l_1 p_1 + l_2(p_2 + y).$$

In the same way, the demand function for the products of firm 2 has the form

$$Q_2(p_1, p_2, y) = q - l_1 p_2 + l_2(p_1 + y).$$

As before, we will assume both brands have the same unit cost c. Then the profit[15] of firm 1 is given by

$$f_1(p_1, p_2, y) = Q_1(p_1, p_2, y)(p_1 - c) = [q - l_1 p_1 + l_2(p_2 + y)]\,(p_1 - c), \tag{4.4.3}$$

---

[14]Nikolay N. Bogolyubov, (1909–1992), was a Soviet mathematician and theoretical physicist who contributed to quantum field theory, classical and quantum statistical mechanics, and theory of dynamical systems. An English translation of a quote from *Vestnik Akad. Nauk SSSR*, 1966, no. 7, p. 38.

[15]"And gain is gain, however small." From *Paracelsus* by Robert Browning, (1812–1889), a major English poet of the Victorian age.

and that of firm 2 by

$$f_2(p_1, p_2, y) = Q_2(p_1, p_2, y)(p_2 - c) = [q - l_1 p_2 + l_2(p_1 + y)]\,(p_2 - c). \tag{4.4.4}$$

Without any agreement, both firms announce their unit prices, seeking to maximize their individual profits. Choosing an appropriate strategy, each firm must consider any possible realization of the uncertain factor y (the nonnegative import price).

As a mathematical model of this duopoly one can take the noncooperative game under uncertainty given by

$$\langle \{1, 2\}, \{\mathrm{P}_i\}_{i=1,2}, \mathrm{Y}, \{\Phi_i(p_1, p_2, y)\}_{i=1,2} \rangle. \tag{4.4.5}$$

Here 1 and 2 denote the numbers of players (firms); the strategy of each firm i is a unit price $p_i \in \mathrm{P}_i = [0, +\infty)$ $(i = 1, 2)$; the uncertain factor y takes arbitrary values from the set $\mathrm{Y} = [0, +\infty)$; in accordance with Germeier's principle of guaranteed result, the payoff function $\Phi_i(p_1, p_2, y)$ of player i $(i = 1, 2)$ is composed of its profit $f_i(p_1, p_2, y)$ and the term $\frac{y^2}{2}$ (see Remark 4.2.1), i.e.,

$$\begin{aligned} \Phi_1(p_1, p_2, y) &= [q - l_1 p_1 + l_2(p_2 + y)]\,(p_1 - c) + \frac{y^2}{2}, \\ \Phi_2(p_1, p_2, y) &= [q - l_1 p_2 + l_2(p_1 + y)]\,(p_2 - c) + \frac{y^2}{2}. \end{aligned} \tag{4.4.6}$$

The component $\frac{y^2}{2}$ in the payoff functions (4.4.6) compels each player to oppose the uncertainty as much as possible.

Thus, the game runs as follows. Without forming coalition, the players independently choose their individual strategies p_i $(i = 1, 2)$, which result in a strategy profile $p = (p_1, p_2) \in \mathrm{P} = \mathrm{P}_1 \times \mathrm{P}_2$ in this game. Some nonnegative value of the uncertainty $y \in \mathrm{Y}$ is realized simultaneously with and independently of their choice. The payoff function $\Phi_i(p_1, p_2, y)$ (4.4.6) of player i is defined on all pairs $(p, y) \in \mathrm{P} \times \mathrm{Y}$. For a realized pair (p, y), the value of this function gives the payoff of player i and the value of $f_i(p_1, p_2, y)$ the corresponding profit.

A Pareto-guaranteed equilibrium (PGE) of game (4.4.5) is a triplet $(p^{\mathrm{e}}, \Phi_1^{\mathrm{e}}, \Phi_2^{\mathrm{e}})$ *for which there exists a function* $y_{\mathrm{P}}(p) : \mathrm{P} \to \mathrm{Y}$ *such that*

first, for each fixed $p \in \mathrm{P}$, *the function* $y_{\mathrm{P}}(p) : \mathrm{P} \to \mathrm{Y}$ *is the Pareto-minimal uncertainty in the bi-criteria choice problem*

$$\langle \mathrm{Y}, \{\Phi_i(p, y)\}_{i=1,2} \rangle,$$

which is obtained from (4.4.5) *for each fixed strategy profile* $p = (p_1, p_2) \in \mathrm{P}$;

second, the strategy profile $p^{\mathrm{e}} = (p_1^{\mathrm{e}}, p_2^{\mathrm{e}})$ *is a Nash equilibrium in the game without uncertainty*

$$\langle \{1, 2\}, \{\mathrm{P}_i\}_{i=1,2}, \{\Phi_i(p, y_{\mathrm{P}}(p))\}_{i=1,2}\rangle,$$

which is obtained from game (4.4.5) *by replacing the uncertainty* y *with its realization* $y_{\mathrm{P}}(p)$.

In this case, p^{e} is a *Pareto-guaranteeing strategy profile*, while the corresponding *vector guarantee* is $\Phi_i^{\mathrm{e}} = \Phi_i(p^{\mathrm{e}}, y_P(p^{\mathrm{e}}))$.

A Pareto-guaranteed equilibrium (PGE) in the Bertrand duopoly with import is a triplet $(p^{\mathrm{e}}, f_1^{\mathrm{e}}, f_2^{\mathrm{e}})$, *where the Pareto-guaranteeing strategy profile* $p^{\mathrm{e}} = (p_1^{\mathrm{e}}, p_2^{\mathrm{e}})$ *is the same as in game* (4.4.5) *and* $f_i^{\mathrm{e}} = f_i(p^{\mathrm{e}}, y_{\mathrm{P}}(\mathrm{p}^{\mathrm{e}}))$ *is the corresponding profit of firm* i.

In the sequel, we will design the Pareto-guaranteed equilibrium in the Bertrand duopoly with import using an algorithm similar to the one suggested in Sect. 4.2.3.

More specifically, to get the PGE in the Bertrand duopoly model with import, it is necessary to perform the following steps:

(1) find a continuous scalar function $y_{\mathrm{P}} = y_{\mathrm{P}}(p)$ yielding the Pareto minimum in the bi-criteria choice problem

$$\langle \mathrm{Y} = [0, +\infty), \{\Phi_i(p, y)\}_{i=1,2}\rangle, \tag{4.4.7}$$

which is obtained from (4.4.5) for each fixed strategy profile $p = (p_1, p_2) \in \mathrm{P}$;

(2) construct a strategy profile $p^{\mathrm{e}} = (p_1^{\mathrm{e}}, p_2^{\mathrm{e}})$ yielding a Nash equilibrium in the game without uncertainty

$$\langle \{1, 2\}, \{\mathrm{P}_i = [0, +\infty)\}_{i=1,2}, \{\Phi_i(p, y_{\mathrm{P}}(p))\}_{i=1,2}\rangle, \tag{4.4.8}$$

which is obtained from game (4.4.5) by substituting the Pareto-minimal uncertainty $y_{\mathrm{P}} = y_{\mathrm{P}}(p)$;

(3) calculate the corresponding prices $p_1 = p_1^{\mathrm{e}}$ and $p_2 = p_2^{\mathrm{e}}$ as well as the profit $f_i(p_1^{\mathrm{e}}, p_2^{\mathrm{e}}, y_{\mathrm{P}}(p_1^{\mathrm{e}}, p_2^{\mathrm{e}})) = f_i^{\mathrm{e}}$ of each player i $(i = 1, 2)$.

4.4.3 Calculation of Inner Pareto Minimum

Calculus is the outcome of a dramatic intellectual struggle which has lasted for twenty-five hundred years.
—Courant

The Pareto-minimal guaranteeing uncertainty in the Bertrand duopoly model is constructed.

The following result holds.

Proposition 4.4.1 *The uncertainty*

$$y_{\mathrm{P}}(p_1, p_2) = l_2 \frac{2c - p_1 - p_2}{2}$$

is Pareto-minimal in the bi-criteria choice problem (4.4.7) *for each strategy profile* $p = (p_1, p_2) \in [0, +\infty)^2$.

Proof Consider the function

$$\begin{aligned} F(p, y) &= \Phi_1(p, y) + \Phi_2(x, y) = f_1(p_1, p_2, y) + f_2(p_1, p_2, y) + y^2 \\ &= [q - l_1 p_1 + l_2(p_2 + y)]\,(p_1 - c) + [q - l_1 p_2 + l_2(p_1 + y)]\,(p_2 - c) + y^2. \end{aligned}$$

For each fixed $p = (p_1, p_2) \in \mathrm{P}$, the minimum value of this function is achieved for $y_{\mathrm{P}}(p) = l_2 \frac{2c - p_1 - p_2}{2}$ if

$$\left.\frac{\partial F}{\partial y}\right|_{y = y_{\mathrm{P}}(p)} = l_2(p_1 - c) + l_2(p_2 - c) + 2y_{\mathrm{P}}(p) = 0$$

and

$$\left.\frac{\partial^2 F}{\partial y^2}\right|_{y = y_{\mathrm{P}}(p)} = 2 > 0.$$

Hence, by Lemma 4.2.1 with $\alpha = \beta = 1$, the uncertainty $y_{\mathrm{P}}(p) = l_2 \frac{2c - p_1 - p_2}{2}$ is Pareto-minimal in the bi-criteria choice problem (4.4.7) for each fixed strategy profile $p = (p_1, p_2) \in \mathrm{P}$. ∎

4.4.4 *Design of Nash Equilibrium*

A mathematician, like a painter or a poet, is a maker of patterns.
—Hardy[16]

A Nash equilibrium in the game of guarantees is constructed.

Proposition 4.4.2 *For* $l_1 > \frac{1}{8}$ *and* $l_2 > 0$, *the Nash equilibrium in game* (4.4.8) *has the form*

$$p^{\mathrm{e}} = (p_1^{\mathrm{e}}, p_2^{\mathrm{e}}) = \left(\frac{q + c(l_1 + l_2^2)}{2l_1 - l_2 + l_2^2}, \frac{q + c(l_1 + l_2^2)}{2l_1 - l_2 + l_2^2} \right).$$

[16]Godfrey Harold Hardy, (1877–1947), was a leading English pure mathematician whose work was mainly in analysis and number theory.

Proof Substituting the uncertainty $y_P(p)$ from Proposition 4.4.1 into (4.4.6) yields

$$\Phi_1(p, y_P(p)) = \left[q - l_1 p_1 + l_2 p_2 + l_2^2 \frac{2c - p_1 - p_2}{2}\right](p_1 - c)$$
$$+ l_2^2 \frac{(2c - p_1 - p_2)^2}{8},$$
$$\Phi_2(p, y_P(p)) = \left[q - l_1 p_2 + l_2 p_1 + l_2^2 \frac{2c - p_1 - p_2}{2}\right](p_2 - c)$$
$$+ l_2^2 \frac{(2c - p_1 - p_2)^2}{8}.$$

The sufficient conditions for the existence of a Nash equilibrium $p^{\mathrm{e}} = (p_1^{\mathrm{e}}, p_2^{\mathrm{e}})$ in game (4.4.8) can be written as

$$\left.\frac{\partial \Phi_1(p_1, p_2^{\mathrm{e}})}{\partial p_1}\right|_{p_1 = p_1^{\mathrm{e}}} = (-l_1 - \frac{l_2^2}{2})(p_1^{\mathrm{e}} - c) + q - l_1 p_1^{\mathrm{e}} + l_2 p_2^{\mathrm{e}}$$
$$+ l_2^2 \frac{2c - p_1^{\mathrm{e}} - p_2^{\mathrm{e}}}{2} - \frac{l_2^2}{4}(2c - p_1^{\mathrm{e}} - p_2^{\mathrm{e}}) = 0, \quad (4.4.9)$$

$$\left.\frac{\partial^2 \Phi_1(p_1, p_2^{\mathrm{e}})}{\partial p_1^2}\right|_{p_1 = p_1^{\mathrm{e}}} = -2l_1 - \frac{3}{4} l_2^2 < 0, \quad (4.4.10)$$

$$\left.\frac{\partial \Phi_2(p_1^e, p_2)}{\partial p_2}\right|_{p_2 = p_2^e} = \left(-l_1 - \frac{l_2^2}{2}\right)(p_2^{\mathrm{e}} - c) + q - l_1 p_2^{\mathrm{e}} + l_2 p_1^{\mathrm{e}}$$
$$+ l_2^2 \frac{2c - p_1^{\mathrm{e}} - p_2^{\mathrm{e}}}{2} - \frac{l_2^2}{4}(2c - p_1^{\mathrm{e}} - p_2^{\mathrm{e}}) = 0, \quad (4.4.11)$$

$$\left.\frac{\partial^2 \Phi_2(p_1^{\mathrm{e}}, p_2)}{\partial p_2^2}\right|_{p_2 = p_2^{\mathrm{e}}} = -2l_1 - \frac{3}{4} l_2^2 < 0. \quad (4.4.12)$$

Conditions (4.4.10) and (4.4.12) hold because $l_2 > 0$ and $l_1 > \frac{1}{8}$. Equalities (4.4.9) and (4.4.11) represent a system of two inhomogeneous linear equations with constant coefficients

$$\begin{cases} \left(2l_1 + \frac{3l_2^2}{4}\right) p_1^{\mathrm{e}} + \left(\frac{l_2^2}{4} - l_2\right) p_2^{\mathrm{e}} = q + c(l_1 + l_2^2), \\ \left(\frac{l_2^2}{4} - l_2\right) p_1^{\mathrm{e}} + \left(2l_1 + \frac{3l_2^2}{4}\right) p_2^{\mathrm{e}} = q + c(l_1 + l_2^2). \end{cases} \quad (4.4.13)$$

The determinant of this system is

$$\Delta = \begin{vmatrix} 2l_1 + \frac{3l_2^2}{4} & \frac{l_2^2}{4} - l_2 \\ \frac{l_2^2}{4} - l_2 & 2l_1 + \frac{3l_2^2}{4} \end{vmatrix} = \left(2l_1 + \frac{3l_2^2}{4}\right)^2 - \left(\frac{l_2^2}{4} - l_2\right)^2$$

$$= (2l_1 - l_2 + l_2^2)\left(2l_1 + l_2 + \frac{l_2^2}{2}\right),$$

and hence $\Delta \neq 0$ since $l_1 > \frac{1}{8}$ and $l_2 > 0$.

Finally, calculating the determinants

$$\Delta_1 = \begin{vmatrix} q + c(l_1 + l_2^2) & \frac{l_2^2}{4} - l_2 \\ q + c(l_1 + l_2^2) & 2l_1 + \frac{3l_2^2}{4} \end{vmatrix} = \left[q + c(l_1 + l_2^2)\right] \begin{vmatrix} 1 & \frac{l_2^2}{4} - l_2 \\ 1 & 2l_1 + \frac{3l_2^2}{4} \end{vmatrix}$$

$$= \left[q + c(l_1 + l_2^2)\right]\left(2l_1 + \frac{3l_2^2}{4} - \frac{l_2^2}{4} + l_2\right) = \left[q + c(l_1 + l_2^2)\right]\left(2l_1 + l_2 + \frac{l_2^2}{2}\right)$$

and

$$\Delta_2 = \begin{vmatrix} 2l_1 + \frac{3l_2^2}{4} & q + c(l_1 + l_2^2) \\ \frac{l_2^2}{4} - l_2 & q + c(l_1 + l_2^2) \end{vmatrix} = \left[q + c(l_1 + l_2^2)\right] \begin{vmatrix} 2l_1 + \frac{3l_2^2}{4} & 1 \\ \frac{l_2^2}{4} - l_2 & 1 \end{vmatrix}$$

$$= \left[q + c(l_1 + l_2^2)\right]\left(2l_1 + \frac{3l_2^2}{4} - \frac{l_2^2}{4} + l_2\right) = \left[q + c(l_1 + l_2^2)\right]\left(2l_1 + l_2 + \frac{l_2^2}{2}\right),$$

we obtain the solution of system (4.4.13) in the form

$$p_i^e = \frac{\Delta_i}{\Delta} = \frac{\left[q + c(l_1 + l_2^2)\right]\left(2l_1 + l_2 + \frac{l_2^2}{2}\right)}{(2l_1 - l_2 + l_2^2)\left(2l_1 + l_2 + \frac{l_2^2}{2}\right)} = \frac{q + c(l_1 + l_2^2)}{2l_1 - l_2 + l_2^2} \quad (i = 1, 2).$$

■

4.4.5 Calculation of the Corresponding Profits

For the Bertrand duopoly model, the explicit-form profits of firms in the Pareto-guaranteed equilibrium are derived.

First, a direct substitution of $p^{\mathrm{e}} = (p_1^{\mathrm{e}}, p_2^{\mathrm{e}})$ into the expression $y_{\mathrm{P}}(p) = l_2 \dfrac{2c - p_1 - p_2}{2}$ shows that $y_{\mathrm{P}}(p_1^{\mathrm{e}}, p_2^{\mathrm{e}}) = l_2 \dfrac{cl_1 - cl_2 - q}{2l_1 - l_2 + l_2^2}$.

Second, using

$$p^{\mathrm{e}} = (p_1^{\mathrm{e}}, p_2^{\mathrm{e}}) = \left(\frac{q + c(l_1 + l_2^2)}{2l_1 - l_2 + l_2^2}, \frac{q + c(l_1 + l_2^2)}{2l_1 - l_2 + l_2^2} \right)$$

and $y_{\mathrm{P}}(p_1^{\mathrm{e}}, p_2^{\mathrm{e}}) = l_2 \dfrac{cl_1 - cl_2 - q}{2l_1 - l_2 + l_2^2}$ in (4.4.3) and (4.4.4), one finds the corresponding profits of the players:

$$\begin{aligned} f_1^{\mathrm{e}} &= f_1(p_1^{\mathrm{e}}, p_2^{\mathrm{e}}, y_{\mathrm{P}}(p_1^{\mathrm{e}}, p_2^{\mathrm{e}})) \\ &= \left[q - l_1 p_1^{\mathrm{e}} + l_2(p_2^{\mathrm{e}} + y_P(p_1^{\mathrm{e}}, p_2^{\mathrm{e}}))\right](p_1^{\mathrm{e}} - c) = l_1 \frac{[q + c(l_2 - l_1)]^2}{(2l_1 - l_2 + l_2^2)^2}, \end{aligned}$$

$$\begin{aligned} f_2^{\mathrm{e}} &= f_2(p_1^{\mathrm{e}}, p_2^{\mathrm{e}}, y_{\mathrm{P}}(p_1^{\mathrm{e}}, p_2^{\mathrm{e}})) \\ &= \left[q - l_1 p_2^{\mathrm{e}} + l_2(p_1^{\mathrm{e}} + y_P(p_1^{\mathrm{e}}, p_2^{\mathrm{e}}))\right](p_2^{\mathrm{e}} - c) = l_1 \frac{[q + c(l_2 - l_1)]^2}{(2l_1 - l_2 + l_2^2)^2}. \end{aligned}$$

Consequently, we have established

Proposition 4.4.3 *Consider the Bertrand duopoly model with import in which* $l_1 > \dfrac{1}{8}$ *and* $l_2 > 0$. *The Pareto-guaranteed equilibrium in this model is the triplet* $(p^{\mathrm{e}}, f_1^{\mathrm{e}}, f_2^{\mathrm{e}})$, *where*

$$p^{\mathrm{e}} = (p_1^{\mathrm{e}}, p_2^{\mathrm{e}}) = \left(\frac{q + c(l_1 + l_2^2)}{2l_1 - l_2 + l_2^2}, \frac{q + c(l_1 + l_2^2)}{2l_1 - l_2 + l_2^2} \right),$$

and the corresponding profit of firm i is given by the formula

$$f_i^e = l_1 \frac{[q + c(l_2 - l_1)]^2}{(2l_1 - l_2 + l_2^2)^2} \quad (i = 1, 2).$$

■

Remark 4.4.1 In the game of guarantees (4.4.8), $\dfrac{\partial^2 \Phi_1}{\partial p_2^2} = \dfrac{\partial^2 \Phi_2}{\partial p_1^2} = \dfrac{l_2^2}{4} > 0$. Since game (4.4.8) is linear-quadratic, in the notations of Table 2.1 we obtain the case $C_1 > 0$ and $A_1 > 0$. In accordance with rows 3 and 4 of this table, the game of guarantees (4.4.8) has no guaranteed Berge equilibria.

Chapter 5
New Approaches to the Solution of Noncooperative Games and Multicriteria Choice Problems

The way out of trouble is never as simple as the way in.
—Howe[1]

This chapter considers three new approaches to important problems of mathematical game theory and multicriteria choice, which are described in four sections (5.1–5.4). The first approach ensures payoff increase with simultaneous risk reduction in the Savage–Niehans sense in multicriteria choice problems (Sect. 5.1) and noncooperative games (Sect. 5.2). The second approach allow us to stabilize coalitional structures in cooperative games without side payments under uncertainty (Sect. 5.3). The third approach serves to integrate the selfish Nash equilibrium with the altruistic Berge equilibrium. Note that the investigations in Sects. 5.2–5.4 involve a special Germeier convolution of criteria and calculation of its saddle point in mixed strategies.

5.1 A New Approach to Optimal Solutions of Multicriteria Choice Problems: Consideration of Savage–Niehans Risk

Wer wagt, gewinnt.[2]

Prologue

This section introduces an original approach to multicriteria choice problems under uncertainty: a decision maker (DM) seeks not only to *increase* the guaranteed values of each criterion, but also to *reduce* the guaranteed risk

[1]Edgar Watson Howe, (1853–1937), was an American editor, novelist, and essayist.

[2]German "Those who risk win." This is an analog of the English proverb "Nothing ventured, nothing gained."

© Springer Nature Switzerland AG 2020

M. E. Salukvadze, V. I. Zhukovskiy, *The Berge Equilibrium: A Game-Theoretic Framework for the Golden Rule of Ethics*, Static & Dynamic Game Theory: Foundations & Applications, https://doi.org/10.1007/978-3-030-25546-6_5

of such increase. The approach lies at the junction of multicriteria choice problems [53, 55, 152] and the Savage–Niehans principle of minimax regret (risk) [85, 268]. More specifically, we will employ the notion of a weakly efficient estimate and the Germeier theorem [213, p. 66] from the theory of multicriteria choice problems and an estimated value of the regret function as the Savage–Niehans risk from the principle of minimax regret [66]. Considerations are restricted to interval-type uncertainty, i.e., the DM merely knows the limits of a range of values, without any probabilistic characteristics. We suggest a new concept—the *Slater-maximal strongly-guaranteed solution in outcomes and risks* (SGOR)—and establish its existence under standard assumptions of mathematical programming (continuous criteria, compact strategy sets and compact uncertainty [58–65]). As a possible application, the SGOR in the diversification problem of a deposit into sub-deposits in different currencies is calculated in explicit form.

Introduction

Consider a multicriteria choice problem under uncertainty (MCPU)

$$\Gamma_c = \langle \mathbb{N}, X, Y, f(x, y) \rangle,$$

where $\mathbb{N} = \{1, \dots, N\}$ $(N \geq 2)$ denotes the set of numbers assigned to the elements $f_i(x, y)$ of a vector criterion $f(x, y) = (f_1(x, y), \dots, f_N(x, y))$; $X \subset \mathbb{R}^n$ is the set of alternatives x; $Y \subset \mathbb{R}^m$ forms the set of *interval* uncertainty y. For Savage–Niehans risk function design, we will also use the *strategic* uncertainties $y(x) : X \to Y$, denoting their set by Y^X.

At a conceptual level, it is often assumed that the DM in the problem Γ_c seeks for an alternative $x \in X$ that maximizes the values of all criteria (outcomes) under any realization of the uncertainty $y \in Y$. In Sect. 5.1, we will also take into account N new criteria—the risks posed by increasing these outcomes. Thus, the problem setup will include N additional criteria, i.e., the Savage–Niehans risk functions associated with outcome increase [172].

Thus, the next section will justify in mathematical terms the new design method of alternatives in the MCPUs that simultaneously "hits two targets," namely, achieving higher guarantees of all outcomes under smaller risks posed by them.

5.1.1 The Savage–Niehans Principle of Minimax Regret

Payer de sa personne.[3]

In 1939 A. Wald, a Romanian mathematician who emigrated to the USA in 1938, introduced the maximin principle, also known as the principle of guaranteed outcome [282]. This principle allows one to find a guaranteed outcome in a single-criterion choice problem under uncertainty (SCPU). Almost a decade later, German mathematician J. Niehans (1948) and American mathematician, economist, and statistician L. Savage (1951) suggested the principle of minimax regret (PMR) for building guaranteed risks in the SCPUs [268]. In the modern literature, this principle is also referred to as the Savage risk or the Niehans–Savage criterion. Interestingly, during World War II Savage worked as an assistant of J. von Neumann, which surely contributed to the appearance of the PMR. Note that the authors of two most remarkable dissertations in economics and statistics are annually awarded the Savage Prize, which was established in 1971 in the USA.

For the single-criterion problem $\Gamma_1 = \langle X, Y, \phi(x, y)\rangle$, the PMR is to construct a pair $(x^{\mathrm{r}}, R_{\phi}^{\mathrm{r}}) \in X \times \mathbb{R}$ that satisfies the chain of equalities [3, 4, 156]

$$R_{\phi}^{\mathrm{r}} = \max_{y\in Y} R_{\phi}(x^{\mathrm{r}}, y) = \min_{x\in X}\max_{y\in Y} R_{\phi}(x, y), \tag{5.1.1}$$

where *the Savage–Niehans risk function* [268] has the form

$$R_{\phi}(x, y) = \max_{z\in X} \phi(z, y) - \phi(x, y). \tag{5.1.2}$$

The value R_{ϕ}^{r} given by (5.1.1) is called *the Savage–Niehans risk* in the problem Γ_1. The risk function $R_{\phi}(x, y)$ assesses the difference between the realized value of the criterion $\phi(x, y)$ and its best-case value $\max_{z\in X}\phi(z, y)$ from the DM's view. Obviously, the DM strives to reduce $R_{\phi}(x, y)$ as much as possible with an appropriately chosen alternative $x \in X$, naturally expecting the strongest opposition from the uncertainty in accordance with the principle of guaranteed result (see formula (5.1.1)). Therefore, adhering to (5.1.1)–(5.1.2), the DM is an *optimist* seeking for the best-case value $\max_{x\in X}\phi(x, y)$. In contrast, the pessimistic DM is oriented towards the worst-case result—the Wald maximin solutions $\big(x^0, \phi^0 = \max_{x\in X}\min_{y\in Y}\phi(x, y) = \min_{y\in Y}\phi(x^0, y)\big)$.

In the sequel, we will consider that the DM in the problem Γ_{c} is an optimist: for each element $f_i(x, y)$ $(i \in \mathbb{N})$ of the vector criterion $f(x, y)$, he forms a corresponding Savage–Niehans risk function

$$R_i(x, y) = \max_{z\in X} f_i(z, y) - f_i(x, y) \quad (i \in \mathbb{N}). \tag{5.1.3}$$

[3] French "Put oneself at risk."

Note two important aspects as follows. *First*, each criterion $f_i(x, y)$ from Γ_c has its own risk $R_i(x, y)$ (see (5.1.3)). *Second*, the DM tries to choose alternatives $x \in X$ so as to reduce all risks $R_i(x, y)$, expecting any realization of the strategic uncertainty $y(\cdot) \in Y^X$, $y(x) : X \to Y$.

Remark 5.1.1 The models Γ_c arise naturally, e.g., in economics: a seller in a market is interested in maximizing his profits under import uncertainty.

For a survey of other possible uncertainties, we refer to the books [138, 53, pp. 19–32] and other publications [69, 75–78].

The uncertainties present in the problem Γ_1 lead to the sets

$$\phi(x, Y) = \{\phi(x, y) |\, \forall y \in Y\},$$

which are induced by an alternative $x \in X$. The set $\phi(x, Y)$ can be reduced using risks. What is a proper comprehension of risk? A well-known Russian expert in optimization, T. Sirazetdinov, claims that today there is no rigorous mathematical definition of risk. The monograph [185, p. 15] even suggested sixteen possible concepts of risk. Most of them require statistical data on uncertainty. However, in many cases the DM does not possess such information for objective reasons. Precisely these situations will be studied in the current section and Sect. 5.1.

Thus, here *risks* will be understood as *possible deviations of realized values from the desired ones*. Note that this definition (in particular, the Savage–Niehans risk) is in good agreement with the conventional notion of microeconomic risk; e.g., see [182, pp. 40–50].

Risk management is a topical problem of economics: in 1990, H. Markowitz was awarded the Nobel Prize in Economic Sciences "for having developed the theory of portfolio choice." In this chapter, the idea of his approach will be extended to the multicriteria choice problems and conflicts under uncertainty. In publications on microeconomics (e.g., see [182, p. 103; 183, p. 5]), all decision makers are divided into three categories: *risk-averse*, *risk-neutral*, and *risk-seeking*. In Sect. 5.1, the DM is assumed to be a risk-neutral person and, of course, an optimist [86–90].

5.1.2 Strong Guarantees and Transition from Γ_c to 2N-Criteria Choice Problem

The first blow is half the battle.
—English proverb

For each of the N criteria $f_i(x, y)$ $(i \in \mathbb{N})$, construct the corresponding risk function $R_i(x, y)$ using formulas (5.1.3), thereby extending the MCPU Γ_c to the $2N$-criteria choice problem

$$\langle \mathbb{N}, X, Y, \{f_i(x, y), -R_i(x, y)\}_{i \in \mathbb{N}} \rangle. \tag{5.1.4}$$

In (5.1.4), the sets $\mathbb{N}$, X, and Y are the same as in Γ_c, while the vector criterion $f(x, y)$ has an additional term in the form of the N-dimensional vector $-R(x, y) = (-R_1(x, y), \dots, -R_N(x, y))$. Here the minus sign reflects a uniform effect of any alternative $x \in X$ on each criterion $f_i(x, y) (i \in \mathbb{N})$. More specifically, in problem (5.1.4) the DM chooses an alternative $x \in X$ in order to *increase* as much as possible the value of each element $f_i(x, y)$ and $-R_i(x, y)$ $(i \in \mathbb{N})$ of the two N-dimensional vectors $f(x, y)$ and $-R(x, y)$. Moreover, the DM must expect any realization of the uncertainty $y \in Y$ (note that an increase of $-R_i(x, y)$ is equivalent to a decrease of $R_i(x, y)$ due to the minus sign and $R_i(x, y) \geq 0$).

Now, consider the *strong guarantees* of criteria. In a series of papers [73, 74], the authors suggested three methods to take the uncertain factors into account—an analog of saddle point [74] and two analogs of maximin [73], namely, strong and vector guarantees. Note that strong guarantee is used below, while vector guarantee was applied in [92, 94, 95, 97, 295].

Definition 5.1.1 A scalar function $f_i[x]$ is called a strong guarantee of a criterion $f_i(x, y) : X \to Y$ if, for each $x \in X$,

$$f_i[x] \leq f_i(x, y) \quad \forall y \in Y \quad (i \in \mathbb{N}).$$

Remark 5.1.2 Obviously, the function $f_i[x] = \min_{y \in Y} f_i(x, y)$ $\forall x \in X$ is a strong guarantee of $f_i(x, y)$. Hence, we have an explicit design method for the strong guarantees of all $2N$ criteria from (5.1.4).

Let us find the strong guarantees $R_i[x]$ of the risk functions $R_i(x, y)$ given by (5.1.3). This will be done in three steps as follows.

First, define

$$\psi_i(y) = \max_{z \in X} f_i(z, y) \quad \forall y \in Y \quad (i \in \mathbb{N}).$$

Second, construct the Savage–Niehans risk function

$$R_i(x, y) = \psi_i(y) - f_i(x, y) \quad (i \in \mathbb{N}).$$

Third, calculate the strong guarantee $\min_{y \in Y}[-R_i(x, y)]$, i.e.,

$$R_i[x] = \max_{y \in Y} R_i(x, y) \quad (i \in \mathbb{N}).$$

Note that the DM seeks to minimize the risk $R_i(x, y)$ with an appropriate alternative $x \in X$ under any realization of the uncertainty $y \in Y$.

Whenever the functions $f_i[x]$ and $-R_i[x]$ described in the remark exist, they are strong guarantees of $f_i(x, y)$ and $-R_i(x, y)$, respectively. Indeed, for each $x \in X$, we have the implications

$$\big[f_i[x] = \min_{y\in Y} f_i(x, y)\big] \Rightarrow \big[f_i[x] \le f_i(x, y) \quad \forall y \in Y\big],$$

$$\big[-R_i[x] = \min_{y\in Y}(-R_i(x, y))\big] \Rightarrow \big[-R_i[x] \le -R_i(x, y) \quad \forall y \in Y\big].$$

The existence of $f_i[x]$ and $R_i[x]$ follows from a well-known result in operations research, which was mentioned earlier.

Lemma 5.1.1 (see [136, p. 54]) *If the sets X and Y are compact and the criteria $f_i(x, y)$ are continuous on $X \times Y$, then the functions $f_i[x] = \min_{y\in Y} f_i(x, y)$ and $\psi_i[y] = \max_{z\in X} f_i(z, y)$ are continuous on X and Y, respectively.*

From this point onwards, $\text{comp}\,\mathbb{R}^n$ stands for the set of all compact sets from space $\mathbb{R}^n$. In addition, if $f_i(x, y)$ is continuous on $X \times Y$, we will write $f_i(x, y) \in C(X \times Y)$.

Remark 5.1.3 If in the MCPU Γ_c the criterion $f_i(x, y) \in C(X \times Y)$, $X \in comp\,\mathbb{R}^n$, and $Y \in \text{comp}\,\mathbb{R}^m$, then the Savage–Niehans risk function $R_i(x, y)$ $(i \in \mathbb{N})$ defined by (5.1.3) is continuous on $X \times Y$. Indeed, the continuity of $\psi_i[y] = \max_{z\in X} f_i(z, y)$ follows from Lemma 5.1.1, and hence by (5.1.3) the function $R_i(x, y) = \psi[y] - f_i(x, y)$ $(i \in \mathbb{N})$ is also continuous.

Remark 5.1.4 The Savage–Niehans risk function (5.1.3) characterizes the deviation of the criterion $f_i(x, y)$ from the desired value $\max_{z\in X} f_i(z, y)$. This stimulates the DM's choice of an alternative $x \in X$ that would *reduce* as much as possible the difference $R_i(x, y)$ from (5.1.3) or, equivalently, *maximize* $-R_i(x, y)$.

Let us associate with the initial MCPU Γ_c the $2N$-criteria choice problem (5.1.4). Once again, at a conceptual level the DM in problem (5.1.4) seeks for an alternative $x \in X$ under which all the $2N$ criteria $f_i(x, y)$ and $-R_i(x, y)$ $(i \in \mathbb{N})$ would take the *greatest* values possible under any realization of the uncertainty $y \in Y$.

5.1.3 Formalization of a Guaranteed Solution in Outcomes and Risks for Problem Γ_c

Universalia sunt realia.[4]

The MCPUs are well-described in the literature (in particular, we refer to the monograph [295]). The specifics of the interval-type uncertainty y figuring in the problem Γ_c compel the DM to use in (5.1.4) the available information (the limits

[4]This Latin phrase expresses a main postulate of realism: universals exist in reality and independently from consciousness.

of the range of values). In this section, our analysis will be confined to the strong guarantees $f_i[x]$ and $-R_i[x]$ of the criteria $f_i(x, y)$ and $-R_i(x, y)$, respectively. Therefore, it seems natural to pass from the MCPU (5.1.4) to the multicriteria choice problem of guarantees without uncertainty

$$\Gamma^{\mathrm{g}} = \langle X, \{f_i[x], -R_i[x]\}_{i\in\mathbb{N}}\rangle.$$

The criteria $f_i[x]$ and $-R_i[x]$ in Γ^{g} are closely related in terms of optimization: the criterion $R_i[x]$ is used for assessing the DM's risk posed by the outcome $f_i[x]$ so that an increase in the difference $f_i[x] - R_i[x]$ leads to a higher guaranteed outcome $f_i[x]$ and (or) a lower guaranteed risk $R_i[x]$. Conversely, a decrease in this difference leads to a lower guaranteed outcome $f_i[x]$ and (or) a higher risk $R_i[x]$. The DM is interested in the maximization of $f_i[x]$ with simultaneous minimization of $R_i[x]$ for each $i \in \mathbb{N}$. Therefore, we will associate with the original $2N$-criteria choice problem Γ^{g} the auxiliary N-criteria choice problem

$$\Gamma^{\mathrm{a}} = \langle X, \{F_i[x] = f_i[x] - R_i[x]\}_{i\in\mathbb{N}}\rangle. \tag{5.1.5}$$

For a formalization of the optimal solution in guaranteed outcomes and risks for the problem Γ_{c}, we will use a concept of vector maximum from the theory of multicriteria choice problems [152]. A first optimal solution of this type was introduced in 1909 by Italian economist and sociologist V. Pareto, (1848–1923), and subsequently it because known as a Pareto maximum.

The analysis below will employ the concept of Slater maximum, which includes the Pareto maximum as a particular case. Perhaps this concept appeared in the Russian literature after the translation [191] of a paper by Hurwitz.

Definition 5.1.2 An alternative $x^{\mathrm{S}} \in X$ is called *Slater-maximal (*weakly efficient*)* in the N-criteria choice problem *(5.1.5)* if the system of strict inequalities

$$F_i[x] > F_i[x^{\mathrm{S}}] \quad (i \in \mathbb{N})$$

is inconsistent for any $x \in X$.

Remark 5.1.5 By Definition 5.1.2, an alternative $x^* \in X$ is not Slater-maximal in problem (5.1.5) if there exists an alternative $\overline{x} \in X$ satisfying the N inequalities

$$F_i[\overline{x}] > F_i[x^*] \quad (i \in \mathbb{N}).$$

Proposition 5.1.1 (Sufficient Conditions) *If*

$$\min_{i\in\mathbb{N}} F_i[x^{\mathrm{S}}] = \max_{x\in X} \min_{i\in\mathbb{N}} F_i[x], \tag{5.1.6}$$

then the alternative $x^{\mathrm{S}} \in X$ is Slater-maximal in problem (5.1.5).

Proof By equality (5.1.6) and Remark 5.1.5, for any alternative $x \in X$ there exists a number $j \in \mathbb{N}$ such that $\left[F_j[x] \le F_j[x^S]\right] \Rightarrow \left[\text{the system of inequalities } F_j[x] > F_j[x^S]\ (i \in \mathbb{N}) \text{ is inconsistent}\right] \Rightarrow \left[x^S \text{ is Slater-maximal in problem (5.1.5)}\right]$. ■

Theorem 5.1.1 (Existence) *If $f_i(\cdot) \in C(X \times Y)$ and the sets X and Y are compact, then there exists a Slater-maximal alternative $x^S \in X$ in problem* (5.1.5).

Proof Using Lemma 5.1.1,

$$\left[f_i(\cdot) \in C(X \times Y),\ i \in \mathbb{N}\right] \Rightarrow \left[f_i[x] \in C(X),\ i \in \mathbb{N}\right],$$

and, in accordance with Remark 5.1.3, $R_i(\cdot) \in C(X \times Y)$ $(i \in \mathbb{N})$. Then, again by Lemma 5.1.1, $\min_{i \in \mathbb{N}} F_i[x] = \min_{i \in \mathbb{N}}(f_i[x] - R_i[x]) \in C(X)$ $(i \in \mathbb{N})$. Since the continuous function $\min_{i \in \mathbb{N}} F_i[x]$ defined on the compact set X achieves maximum its at some point $x^S \in X$, we arrive at (5.1.6), and now the conclusion follows from Proposition 5.1.1. ■

Definition 5.1.3 A triplet $(x^S, f[x^S], R[x^S])$ is called a strongly-guaranteed solution in outcomes and risks *(SGOR)* of the *MCPU* Γ_c if

(1) $f_i[x] = \min_{y \in Y} f_i(x, y)$, $R_i[x] = \max_{y \in Y} R_i(x, y)$ $(i \in \mathbb{N})$;
(2) the alternative x^S is Slater-maximal in problem (5.1.5).

Recall that

$$f[x] = (f_1[x], \dots, f_N[x]), \quad R[x] = (R_1[x], \dots, R_N[x]),$$
$$R_i[x] = \max_{y \in Y} R_i(x, y), \ R_i(x, y) = \max_{z \in X} f_i(z, y) - f_i(x, y) \ (i \in \mathbb{N}). \tag{5.1.7}$$

Why is the strongly-guaranteed solution in outcomes and risks (SGOR) a good solution of the MCPU Γ_c?

First, it provides an answer to the indigenous Russian question: "What is to be done?" (See, e.g., N. Chernyshevsky's book with the same title and [155].) The decision maker is suggested to choose the alternative x^S from the triplet $(x^S, f[x^S], R[x^S])$.

Second, for all $i \in \mathbb{N}$, this alternative x^S yields outcomes $f_i(x^S, y)$ that are not smaller than $f_i[x^S]$ with a risk $R_i(x^S, y)$ not exceeding $R_i[x^S]$ under any realization of the uncertainty $y \in Y$. In other words, x^S establishes lower bounds on the outcomes realized under $x = x^S$ and also upper bounds on the risks posed by them.

Third, the situation x^S implements the largest (Slater-maximal) outcomes and corresponding "minus" risks, i.e., there is no other situation $x \neq x^S$ in which all outcome guarantees $f_i[x^S]$ would increase and, at the same time, all risk guarantees $R_i[x^S]$ would decrease.

In fact, the second and third properties considered together give some analog of the maximin alternative in the single-criterion problem Γ_1 under uncertainty if *the inner minimum and outer maximum* in maximin are replaced by $\min_{y \in Y} F_i(x, y)$ $(i \in \mathbb{N})$ and Slater maximum, respectively. There are *two lines* of further investigations in this field. In accordance with the first direction, one should substitute Slater maximality with Pareto, Borwein, Geoffrion optimality or conical optimality, and then establish connections between such different solutions. The second direction proceeds from the DM's desire for higher profits under the lowest guarantees in the sense of Definition 5.1.2. Consequently, it is possible to replace scalar minimum (from the inner minimum in maximin) by one of the listed vector minima, thereby increasing the guarantees for some $i \in \mathbb{N}$.

Also, it seems interesting to build a bridge between such solutions; some research efforts were made in the monograph [295].

Remark 5.1.6 Definition 5.1.2 suggests a constructive method of SGOR design. It consists of four steps as follows.

Step 1. Using $f_i(x, y)$, find $\max_{z \in X} f_i(z, y) = \psi_i[y]$ and construct the Savage–Niehans risk function $R_i(x, y) = \psi_i[y] - f_i(x, y)$ for the criterion $f_i(x, y)$ $(i \in \mathbb{N})$.

Step 2. Evaluate the outcome guarantees $f_i[x] = \min_{y \in Y} f_i(x, y)$ and also the risk guarantees $R_i[x] = \max_{y \in Y} R_i(x, y)$ $(i \in \mathbb{N})$.

Step 3. For the auxiliary N-criteria choice problem of guarantees Γ^{a}, calculate the Slater-maximal alternative x^{S}. At this step, we may take advantage of Proposition 5.1.2 or perform transition to the concept of Pareto optimality. For the sake of completeness, we recall this concept.

Definition 5.1.4 An alternative $x^{\mathrm{P}} \in X$ is called Pareto-maximal (efficient) in problem (5.1.5) if for any alternatives $x \in X$ the system of inequalities

$$F_i[x] \geqslant F_i[x^{\mathrm{P}}] \quad (i \in \mathbb{N}),$$

with at least one strict inequality, is inconsistent.

Note that, *first*, by Definitions 5.1.2 and 5.1.3, every Pareto-maximal alternative is also Slater-maximal (the converse generally fails); *second*, by Karlin's lemma [152, p. 71], an alternative $x^{\mathrm{P}} \in X$ that satisfies the condition

$$\max_{x \in X} \sum_{i \in \mathbb{N}} \alpha_i F_i[x] = \sum_{i \in \mathbb{N}} \alpha_i F_i[x^{\mathrm{P}}] \tag{5.1.8}$$

for some $\alpha_i = \mathrm{const} > 0$ is Pareto-maximal for problem (5.1.5).

For the bi-criteria choice problem, letting $\alpha_1 = \alpha_2 = 1$ in (5.1.8) gives the equality

$$\max_{x \in X}(F_1[x] + F_2[x]) = F_1[x^{\mathrm{S}}] + F_2[x^{\mathrm{S}}] \tag{5.1.9}$$

for obtaining a Pareto-maximal (hence, Slater-maximal) alternative x^{S}.

Step 4. Using x^{S}, evaluate the guarantees $f_i[x^{\mathrm{S}}]$ and $R_i[x^{\mathrm{S}}]$ $(i \in \mathbb{N})$ and compile the two N-dimensional vectors $f[x^{\mathrm{S}}] = (f_1[x^{\mathrm{S}}], \ldots, f_N[x^{\mathrm{S}}])$ and $R[x^{\mathrm{S}}] = (R_1[x^{\mathrm{S}}], \ldots, R_N[x^{\mathrm{S}}])$.

The resulting triplet $(x^{\mathrm{S}}, f[x^{\mathrm{S}}], R[x^{\mathrm{S}}])$ is the desired SGOR, which complies with Definition 5.1.3, i.e., for each criterion $f_i(x, y)$ $(i \in \mathbb{N})$ the alternative x^{S} leads to a guaranteed outcome $f_i[x^{\mathrm{S}}]$ with a guaranteed Savage–Niehans risk $R_i[x^{\mathrm{S}}]$.

5.1.4 Risks and Outcomes for Diversification of a Deposit into Sub-deposits in Different Currencies

Verba docent, exempla trahunt.[5]

As mentioned earlier, in economics all decision makers are divided [101–104] into three categories: risk-averse, risk-neutral, and risk-seeking. In this section, we will solve the problem of diversification of a one-year deposit into sub-deposits in rubles and foreign currency for a risk-neutral person. Note that a similar problem was addressed in the paper [100, p. 9], and the results established therein differ from those below. The case is that the Slater solutions generally form a set of distinct elements. Like in [100], the analysis in this section involves different elements of the same set.

Let us proceed to the diversification problem. The amount of money in a deposit diversified into two sub-deposits (in rubles and foreign currency) accumulated by the end of the year can be represented as $\phi(x, y) = x(1 + r) + \dfrac{(1 - x)}{k}(1 + d)y$; see [96, pp. 58–60] and also the explanations below. This leads to the single-criterion choice problem $\Gamma_1 = \langle X, Y, \phi(x, y)\rangle$, which was studied in [100]. In particular, the guaranteed solutions for risk-averse, risk-neutral, and risk-seeking persons were found. In contrast to the paper [100] dealing with the *single-criterion* choice problem with the criterion $\phi(x, y)$, in this section we will consider a *bi-criteria* analog of the problem Γ_1 with the criteria

$$f_1(x) = x(1 + r), \quad f_2(x, y) = \frac{(1 - x)}{k}(1 + d)y. \tag{5.1.10}$$

Our intention is to apply the mathematical methods described in Sect. 5.1.3.

The first criterion concerns the annual income for the sub-deposit in rubles from an investment x, while the second concerns the annual income for the sub-deposit in foreign currency from the residual investment $1 - x$. In formula (5.1.10), r and d denote the interest rates for the sub-deposits in rubles and foreign currency, respectively; k and y are the exchange rates (to the ruble) at the beginning and at the

[5] Latin "Words instruct, illustrations lead."

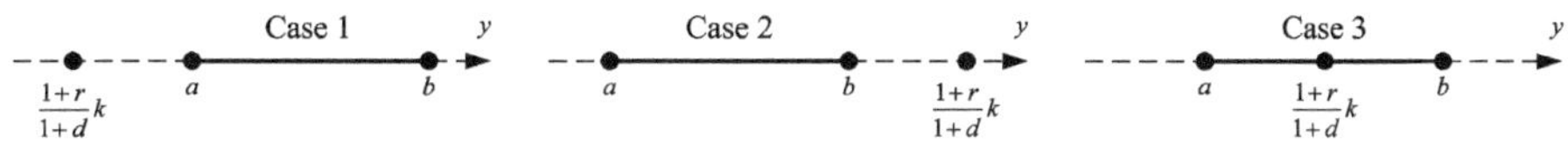

Fig. 5.1 Three possible arrangements of the point $k\frac{1+r}{1+d}$ and the interval $[a, b]$ on axis y

end of the year, respectively; finally, $x \in [0, 1]$ specifies a proportion in which the main deposit is divided into the sub-deposits. Thus, x is the part corresponding to the ruble sub-deposit, while the other part $1 - x$ is converted into foreign currency, $\frac{1-x}{k}$, and then allocated to the corresponding sub-deposit. At the end of the year, it is converted back into rubles, $\frac{(1-x)}{k}(1 + d)y$; the resulting amount of money makes up $f_1(x) + f_2(x, y)$. The decision maker (depositor) has to determine the part x under which the resulting amount of money is as large as possible. It must be taken in account that the future exchange rate y is usually unknown. However, we will assume a range of its possible fluctuations, i.e., $y \in [a, b]$, where the constants $b > a > 0$ are given or a priori known.

The mathematical model of the bi-criteria deposit diversification problem can be written as an ordered triplet

$$\Gamma_2 = \langle X = [0, 1], Y = [a, b], \{f_i(x, y)\}_{i=1,2}\rangle, \tag{5.1.11}$$

where the functions $f_i(x, y)$ are defined by (5.1.10); the set $X = [0, 1]$ consists of the DM's alternatives x; $Y = [a, b]$ is the set of uncertainties y; finally, $f_i(x, y)$ denote the DM's utility functions (criteria), and their values are called *outcomes*. In the terminology of operations research, Γ_2 is a single-criterion choice problem under uncertainty. The DM's desire to take into account the existing uncertain factors has a close connection with risk—"possible deviations of some variables from the desired values" [185, p. 18]. We will use the Savage–Niehans risk function. For problem (5.1.11), consider three cases as illustrated in Fig. 5.1, namely,

1. $k\frac{1+r}{1+d} \leq a$;
2. $k\frac{1+r}{1+d} \geq b$;
3. $a < k\frac{1+r}{1+d} < b$.

Cases 1 and 2 Recall that Γ_2 is a bi-criteria problem under uncertainty. We will solve it using Definition 5.1.4, which is based on the concept of *Pareto optimality*.[6]

[6]Cases 1 and 2 will be considered with Pareto optimality, and case 3 with Proposition 5.1.1.

Proposition 5.1.2 *In cases* 1 *and* 2, *the* SGOR *in the problem* Γ_2 *has the explicit form*

$$(x^S, f[x^S], R[x^S]) = (x^S; f_1[x^S], f_2[x^S]; R_1[x^S], R_2[x^S])$$
$$= \begin{cases} \left(0; 0, \dfrac{1+d}{k}a; 1+r, 0\right), & \text{if } \dfrac{1+r}{1+d}k \le a, \\ \\ \left(1; 1+r, 0; 0, \dfrac{1+d}{k}b\right), & \text{if } \dfrac{1+r}{1+d}k \ge b. \end{cases} \tag{5.1.12}$$

That is, in case 1 (Fig. 5.1), *the* DM *invests everything in the foreign currency sub-deposit, obtaining with zero risk (surely!) a guaranteed minimum amount of* $\frac{1+d}{k}a$ *at the end of the year; in case* 2, *he invests everything in the ruble sub-deposit, obtaining with zero risk a guaranteed minimum amount of* $1 + r$ *at the end of the year. In both cases, the guaranteed minimum amounts are obtained with zero risk under any exchange rate fluctuations* $y \in [a, b]$ *during the year.*

Proof We carry out the proof in two steps. In the first step, following Remark 5.1.6, we construct the resulting $2N$-criteria choice problem of guarantees Γ^g and then the N-criteria choice problem (5.1.5). In the second step, for this problem (5.1.5), we find the Slater-maximal alternative x^S using Proposition 5.1.1 and then calculate the explicit form of the SGOR for the bi-criteria choice problem (5.1.11).

First Step In (5.1.11), the criteria are given by

$$f_1(x, y) = f_1(x) = x(1+r), \quad f_2(x, y) = \frac{(1-x)}{k}(1+d)y.$$

Sub-step 1 Using (5.1.3), construct the Savage–Niehans risk function

$$\begin{aligned} R_1(x, y) &= \left[\max_{z\in[0,1]} f_1(z)\right] - (1+r)x \\ &= (1+r) - x(1+r) = (1-x)(1+r), \\ R_2(x, y) &= \left[\max_{z\in[0,1]} f_2(z, y)\right] - (1-x)\frac{1+d}{k}y \\ &= \frac{1+d}{k}y - (1-x)\frac{1+d}{k}y = xy\frac{1+d}{k}. \end{aligned}$$

Sub-step 2 Now, calculate the strong guarantees in outcomes and risks

$$f_1[x] = \min_{y\in[a,b]} x(1+r) = x(1+r),$$

$$f_2[x] = \min_{y\in[a,b]} (1-x)\frac{1+d}{k}y = (1-x)\frac{1+d}{k}a,$$

$$R_1[x] = \max_{y \in [a,b]} R_1(x, y) = (1 - x)(1 + r),$$

$$R_2[x] = \max_{y \in [a,b]} R_2(x, y) = x \frac{1 + d}{k} b.$$

Sub-step 3 The quad-criteria choice problem of guarantees takes the form

$$\Gamma^{\mathrm{g}} = \langle X = [0, 1], \{f_i[x], -R_i[x]\}_{i=1,2} \rangle.$$

Step 2 also allows us to define the criteria

$$\begin{aligned} F_1[x] &= f_1[x] - R_1[x] \\ &= x(1 + r) - (1 - x)(1 + r) = (2x - 1)(1 + r), \\ F_2[x] &= f_2[x] - R_2[x] \\ &= (1 - x)\frac{1 + d}{k} a - x \frac{1 + d}{k} b = \frac{1 + d}{k} a - \frac{1 + d}{k}(a + b)x \end{aligned}$$

in the auxiliary bi-criteria choice problem (5.1.5)

$$\Gamma^{\mathrm{a}} = \langle X = [0, 1], \{F_i[x]\}_{i=1,2} \rangle.$$

Second Step

Sub-step 4 Maximize the sum of criteria

$$\max_{[0,1]}(F_1[x] + F_2[x]) = F_1[x^{\mathrm{S}}] + F_2[x^{\mathrm{S}}].$$

The resulting Pareto-maximal (*ergo*, Slater-maximal) alternative x^{S} is

$$F[x^{\mathrm{S}}] = \max_{[0,1]} F[x], \tag{5.1.13}$$

where

$$\begin{aligned} F[x] = F_1[x] + F_2[x] &= (2x - 1)(1 + r) + \frac{1 + d}{k} a - \frac{1 + d}{k}(a + b)x \\ &= x\big[2(1 + r) - \frac{1 + d}{k}(a + b)\big] - (1 + r) + \frac{1 + d}{k} a \\ &= \frac{1 + d}{k}\big\{[2\gamma - (a + b)]x - \gamma + a\big\}, \end{aligned}$$

and $\gamma = \dfrac{1 + r}{1 + d} k$. The function $F[x]$ under maximization is linear in x and defined on the interval $[0, 1]$. Therefore, it achieves maximum at one of the endpoints of this interval, i.e., either at $x = 0$, or at $x = 1$. For $x = 0$, we have $F[0] = \frac{1+d}{k}(a-\gamma)$; for $x = 1$, $F[1] = \frac{1+d}{k}(\gamma - b)$.

Lemma 5.1.2 *The implication*

$$\left[a \geq \gamma\right] \Rightarrow \left[F[0] > F[1]\right]$$

is valid.

Proof Indeed,

$$[a \geq \gamma] \Leftrightarrow \left[[\frac{a+a}{2} \geq \gamma] \geq [\frac{a+b}{2} > \gamma]\right] \Rightarrow [a - \gamma > \gamma - b]$$

$$\Rightarrow \left[F[0] = \frac{1+d}{k}(a-\gamma) > F[1] = \frac{1+d}{k}(\gamma - b)\right].$$

■

In a similar fashion, we can easily establish

Lemma 5.1.3

$$\left[\gamma \geq b\right] \Rightarrow \left[F[0] < F[1]\right].$$

Proof Indeed,

$$[\gamma \geq b] \Leftrightarrow \left[\gamma \geq \frac{b+b}{2}\right] \Rightarrow \left[\gamma > \frac{b+a}{2}\right] \Rightarrow [\gamma - b > a - \gamma]$$

$$\Rightarrow \left[F[1] = \frac{1+d}{k}(\gamma - b) > F[0] = \frac{1+d}{k}(a-\gamma)\right].$$

■

Sub-step 5 By Lemmas 5.1.2 and 5.1.3, the maximum in (5.1.13) is achieved

(a) at $x^{\mathrm{S}} = 0$ if $a \geq \gamma$;
(b) at $x^{\mathrm{S}} = 1$ if $\gamma \geq b$.

The corresponding guarantees are calculated using this result and Sub-step 2: $f_1[0] = 0$, $f_2[0] = \frac{1+d}{k}a$, $R_1[0] = 1 + r$, and $R_2[0] = 0$; in addition, $f_1[1] = 1 + r$, $f_2[1] = 0$, $R_1[1] = 1$, and $R_2[1] = \frac{1+d}{k}b$. Recall that $\gamma = \frac{1+r}{1+d}k$, and the proof of Proposition 5.1.2 is complete. ■

Let us make a few of remarks. *First*, $R_1[0] = 1 + r$ means the risk with which $f_1[0] = 0$ "does not reach" the largest outcome $f_1[1] = 1+r$ (the Savage–Niehans risk). The value $R_2[1] = \frac{1+d}{k}b$ has a similar meaning. *Second*, Proposition 5.1.2 was proved in the paper [115] using a different technique.

Finally, consider case 3. Here we will utilize, *first*, the results of Sub-step 3 of Proposition 5.1.2, in particular, the bi-criteria choice problem

$$\Gamma^{\mathrm{a}} = \langle X = [0, 1], \{F_i[x]\}_{i=1,2}\rangle,$$

where

$$\begin{aligned} F_1[x] &= (2x-1)(1+r), \\ F_2[x] &= \frac{1+d}{k}a - \frac{1+d}{k}(a+b)x; \end{aligned} \tag{5.1.14}$$

second, the sufficient conditions (5.1.6) for the existence of the alternative x^{S} (see Proposition 5.1.1), writing them for the deposit diversification problem (5.1.11) as

$$\min_{i=1,2} F_i[x^{\mathrm{S}}] = \max_{x\in[0,1]} \min_{i=1,2} F_i[x].$$

Proposition 5.1.3 *If $a < \frac{1+r}{1+d}k < b$, the* SGOR *in the problem Γ_2 has the form*

$$\begin{aligned} &\left(x^{\mathrm{S}},\ f[x^{\mathrm{S}}],\ R[x^{\mathrm{S}}]\right) = \left(x^{\mathrm{S}};\ f_1[x^{\mathrm{S}}],\ f_2[x^{\mathrm{S}}];\ R_1[x^{\mathrm{S}}],\ R_2[x^{\mathrm{S}}]\right) \\ &= \left(\frac{\gamma+a}{2\gamma+a+b};\ \frac{(\gamma+a)(1+r)}{2\gamma+a+b},\ \frac{\gamma+b}{2\gamma+a+b}\frac{1+d}{k}a;\right. \\ &\left.(1+r)\frac{\gamma+b}{2\gamma+a+b},\ b\frac{1+d}{k}\frac{\gamma+a}{2\gamma+a+b}\right). \end{aligned} \tag{5.1.15}$$

Proof Draw the graphs of the two functions $F_1[x]$ and $F_2[x]$ from (5.1.14). These functions are linear in x and defined on the interval $[0, 1]$ (a compact set!); see Fig. 5.2.

In Fig. 5.2 the function $\min_{i=1,2}\{F_1[x], F_2[x]\}$ is indicated by the bold line, see the angle ABC. For $\max_{x\in[0,1]}\min_{i=1,2}\{F_1[x], F_2[x]\}$, the point B satisfies the equality

$$F_1[x^{\mathrm{S}}] = F_2[x^{\mathrm{S}}]$$

or, using (5.1.14),

$$x^S\left[2(1+r) + \frac{1+d}{k}(a+b)\right] = 1 + r + \frac{1+d}{k}a.$$

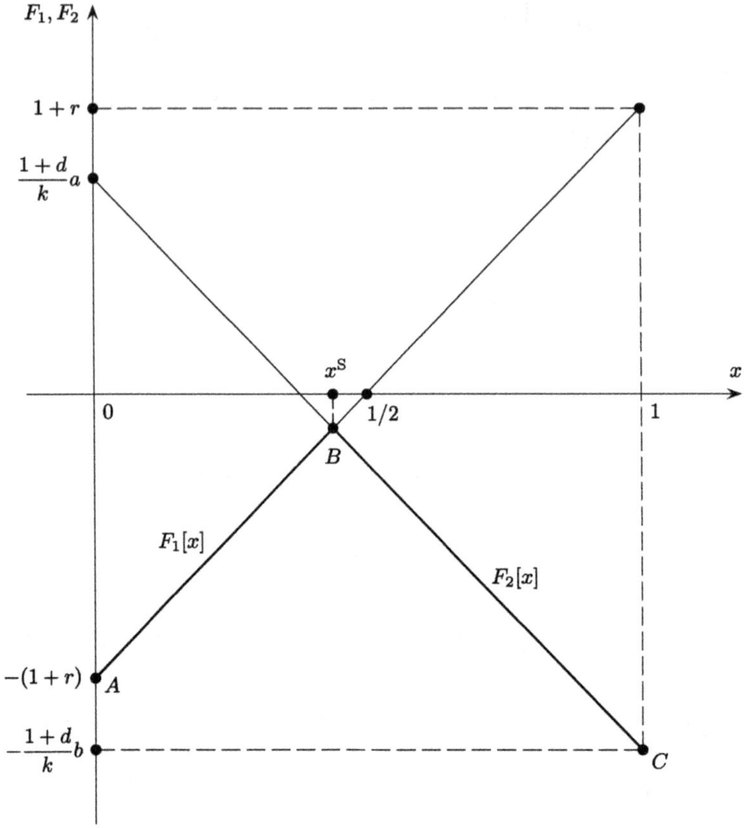

Fig. 5.2 Graphs of functions defined by (5.1.14)

With the notation $\gamma = \frac{1+r}{1+d}k$, it can be written as

$$x^S [2\gamma + a + b] = \gamma + a,$$

which gives

$$x^S = \frac{\gamma + a}{2\gamma + a + b}, \quad 1 - x^S = \frac{\gamma + b}{2\gamma + a + b}.$$

Using the formulas of Sub-step 2 (the proof of Proposition 5.1.2), we calculate the strong guarantees in outcomes and risks:

$$f_1[x^S] = (1+r)\frac{\gamma + a}{2\gamma + a + b}, \quad f_2[x^S] = \frac{1+d}{k}a\frac{\gamma + b}{2\gamma + a + b},$$
$$R_1[x^S] = (1+r)\frac{\gamma + b}{2\gamma + a + b}, \quad R_2[x^S] = \frac{1+d}{k}b\frac{\gamma + a}{2\gamma + a + b}.$$

■

Thus, we have established the following result (see Proposition 5.1.3). If $a < \frac{1+r}{1+d}k < b$, the strongly-guaranteed solution in outcomes and risks of the deposit diversification problem (Definition 5.1.2) has form (5.1.15). It suggests the DM to invest the part $\frac{\gamma+a}{2\gamma+a+b}$ in the ruble sub-deposit and the residual part $\frac{\gamma+b}{2\gamma+a+b}$ in the foreign currency sub-deposit. At the end of the year, the DM will obtain the amount $(1+r)\frac{\gamma+a}{2\gamma+a+b}$ for the ruble sub-deposit with the Savage–Niehans risk $(1+r)\frac{\gamma+b}{2\gamma+a+b}$ and the amount $\frac{1+d}{k}a\frac{\gamma+b}{2\gamma+a+b}$ (after conversion in rubles) for the foreign currency sub-deposit with the Savage–Niehans risk $\frac{1+d}{k}b\frac{\gamma+a}{2\gamma+a+b}$ under any exchange rate fluctuations $y \in [a, b]$ during the year.

Remark 5.1.7 If $\frac{1+r}{1+d}k \leqslant a$ (case 1), the DM is recommended to invest everything in the foreign currency sub-deposit, because at the end of the year he will obtain the guaranteed minimum income $\frac{1+d}{k}a$ with zero risk (Proposition 5.1.2).

If $\frac{1+r}{1+d}k \geqslant b$ (case 2), the DM is recommended to invest everything in the ruble sub-deposit, which will yield him the guaranteed minimum income $1 + r$ with zero risk at the end of the year (Proposition 5.1.2).

5.2 A New Approach to Optimal Solutions of Noncooperative Games: Accounting for Savage–Niehans Risk

He who takes chances does not risk.[7]

The novelty of the approach presented below is that each person in a conflict (player) seeks not only to increase his payoff but also to reduce his risk, taking into account a possible realization of any uncertainty from a given admissible set. A new concept, the so-called strongly-guaranteed Nash equilibrium in payoffs and risks, is introduced and its existence in mixed strategies is proved under standard assumptions of the theory of noncooperative games, i.e., compactness and convexity of the sets of players' strategies and continuity of the payoff functions.

5.2.1 Principia Universalia

Consider the noncooperative N-player normal-form game under uncertainty

$$\langle \mathbb{N}, \{\mathrm{X}_i\}_{i\in\mathbb{N}}, \mathrm{Y}, \{f_i(x, y)\}_{i\in\mathbb{N}}\rangle, \tag{5.2.1}$$

[7] An English translation of a statement from [26].

where $\mathbb{N} = \{1, 2, \ldots, N \geq 2\}$ denotes the set of players; each player $i \in \mathbb{N}$ chooses and uses a *pure strategy* $x_i \in \mathrm{X}_i \subset \mathbb{R}^{n_i}$ $(i \in \mathbb{N})$, which yields a *strategy profile* $x = (x_1, \ldots, x_N) \in \mathrm{X} = \prod_{i\in\mathbb{N}} \mathrm{X}_i \subseteq \mathbb{R}^n$ $(n = \sum_{i\in\mathbb{N}} n_i)$; regardless of the players' actions, an uncertainty $y \in \mathrm{Y} \subset \mathbb{R}^m$ is realized in game (5.2.1); *the payoff function* $f_i(x, y)$ of player i is defined on the pairs $(x, y) \in \mathrm{X} \times \mathrm{Y}$, and its value is called the *payoff* of player i.

At a conceptual level, the goal of each player in the standard setup considered before was to choose his strategy so as to achieve as great payoff as possible.

The middle of the twentieth century was a remarkable period for the theory of noncooperative games. In 1949, 21 year old Princeton University postgraduate J. Nash suggested and proved the existence of a solution [257] that subsequently became known as the *Nash equilibrium*: a strategy profile $x^{\mathrm{e}} \in \mathrm{X}$ is called a *Nash equilibrium* in a game $\langle \mathbb{N}, \{\mathrm{X}_i\}_{i\in\mathbb{N}}, \{f_i[x]\}_{i\in\mathbb{N}}\rangle$ if

$$\max_{x_i\in\mathrm{X}_i} f_i[x^{\mathrm{e}}||x_i] = f_i[x^{\mathrm{e}}] \quad (i \in \mathbb{N}),$$

where $[x^{\mathrm{e}}||x_i] = [x_1^{\mathrm{e}}, \ldots, x_{i-1}^{\mathrm{e}}, x_i, x_{i+1}^{\mathrm{e}}, \ldots, x_N^{\mathrm{e}}]$.

This concept (and the approach driven by it) has become invaluable for resolving global (and other) problems in economics, social and military sciences. After 45 years, in 1994, J. Nash together with R. Selten and J. Harsanyi were awarded the Nobel Prize in Economic Sciences "for their pioneering analysis of equilibria in the theory of non-cooperative games." In 1951, American mathematician, economist and statistician L. Savage, who worked as a statistics assistant for J. von Neumann during World War II, proposed [268] the principle of minimax regret (the Savage–Niehans risk). In particular, for a single-criterion choice problem under uncertainty $\Gamma = \langle \mathrm{X}, \mathrm{Y}, \varphi(x, y)\rangle$, the principle of minimax regret can be written as

$$\min_{x\in\mathrm{X}} \max_{y\in\mathrm{Y}} R(x, y) = \max_{y\in\mathrm{Y}} R(x^{\mathrm{e}}, y) = R, \tag{5.2.2}$$

where the Savage–Niehans risk function [268] has the form

$$R(x, y) = \max_{z\in\mathrm{X}} \varphi(z, y) - \varphi(x, y). \tag{5.2.3}$$

The value $R(x, y)$ is called *the Savage–Niehans risk* in a single-criterion choice problem Γ. It describes the risk of decision makers while choosing an alternative x (the difference between the desired value of the criterion $\max_{x\in\mathrm{X}} \varphi(x, y)$ and the realized value $\varphi(x, y)$). Note that a decision maker seeks to reduce precisely this risk as much as possible by choosing $x \in \mathrm{X}$. In fact, the combination of the concept of Nash equilibrium with the principle of minimax regret is the fundamental idea of Sect. 5.2. Such an approach matches the desire of each player not only to increase his payoff, but also to reduce his risk while realizing this desire.

In this context, two questions arise naturally:

1. How can we combine the two objectives of each player (payoff increase with simultaneous risk reduction) using only one criterion?
2. How can we combine these actions (alternatives) in a single strategy profile, in such a way that uncertainty is also accounted for?

5.2.2 How Can We Combine the Objectives of Each Player to Increase the Payoff and Simultaneously Reduce the Risk?

There's no great loss without some small gain.
—Laura Ingalls Wilder[8]

Construction of Savage–Niehans Risk Function

Recall that, in accordance with the principle of minimax regret, the risk of player i is defined by the value of the Savage–Niehans risk function [183, 184, 186]

$$R_i(x, y) = \max_{z \in \mathrm{X}} f_i(z, y) - f_i(x, y), \tag{5.2.4}$$

where $f_i(x, y)$ denotes the payoff function of player i in game (5.2.1). Thus, to construct the risk function $R_i(x, y)$ for player i, first we have to find the dependent maximum

$$\max_{x \in \mathrm{X}} f_i(x, y) = f_i[y]$$

for all $y \in \mathrm{Y}$. To calculate $f_i[y]$, in accordance with the theory of two-level hierarchical games, it is necessary to assume the *discrimination* of the lower-level player, who forms the uncertainty $y \in \mathrm{Y}$ and sends this information to the upper level for constructing counterstrategies $x^{(i)}(y) : \mathrm{Y} \to \mathrm{X}$ so that

$$\max_{x \in \mathrm{X}} f_i(x, y) = f_i(x^{(i)}(y), y) = f_i[y] \ \ \forall\, y \in \mathrm{Y}.$$

The set of such counterstrategies is denoted by X^{Y}. (Actually, this set consists of n-dimensional vector functions $x(y) : \mathrm{Y} \to \mathrm{X}$ with the domain of definition Y and the codomain X.) Thus, to construct the first term in (5.2.4) at the upper level of the

[8] Laura Elizabeth Ingalls Wilder (1867–1957) was an American writer known for the Little House on the Prairie series of children's books.

hierarchy, we have to solve N single-criterion problems of the form

$$\langle \mathrm{X}^{\mathrm{Y}}, \mathrm{Y}, f_i(x, y)\rangle \quad (i \in \mathbb{N}),$$

for each uncertainty $y \in \mathrm{Y}$; here X^{Y} is the set of counterstrategies $x(y) : \mathrm{Y} \to \mathrm{X}$, i.e., the set of pure uncertainties $y \in \mathrm{Y}$. The problem itself consists in determining the scalar functions $f_i[y]$ defined by the formula

$$f_i[y] = \max_{x(\cdot)\in \mathrm{X}^{\mathrm{Y}}} f_i(x, y) \quad \forall\, y \in \mathrm{Y}.$$

After that, the Savage–Niehans risk functions are constructed by formula (5.2.4).

Continuity of Risk Function, Guaranteed Payoffs and Risks

Ad informandum.[9]

Hereinafter, the collection of all compact sets of Euclidean space $\mathbb{R}^k$ is denoted by comp $\mathbb{R}^k$, and if a scalar function $\psi(x)$ on the set X is continuous, we write $\psi(\cdot) \in C(\mathrm{X})$.

The main role in this section will be played by the following result.

Proposition 5.2.1 *If* $\mathrm{X} \in \mathrm{comp}\, \mathbf{R}^n$, $\mathrm{Y} \in \mathrm{comp}\, \mathbf{R}^m$, *and* $f_i(\cdot) \in C(\mathrm{X}\times\mathrm{Y})$, *then*

(a) the maximum function $\max_{x\in\mathrm{X}} f_i(x, y)$ *is continuous on* Y;
(b) the minimum function $\min_{y\in\mathrm{Y}} f_i(x, y)$ *is continuous on* X.

These assertions can be found in most monographs on game theory, operations research, systems theory, and even in books on convex analysis [46].

Corollary 5.2.1 *If in game* (5.2.1) *the sets* $\mathrm{X} \in \mathrm{comp}\, \mathbb{R}^n$ *and* $\mathrm{Y} \in \mathrm{comp}\, \mathbb{R}^m$ *and the functions* $f_i(\cdot) \in C(\mathrm{X}\times\mathrm{Y})$, *then the Savage–Niehans risk function* $R_i(x, y)$ $(i \in \mathbb{N})$ *is continuous on* $\mathrm{X} \times \mathrm{Y}$.

Indeed, by Proposition 5.2.1 the first term in (5.2.4) is continuous on Y and a difference of continuous functions is itself continuous for all $(x, y) \in \mathrm{X} \times \mathrm{Y}$.

Let us proceed with *guaranteed payoffs and risks* in game (5.2.1). In a series of publications [73, 74], three different ways to account for uncertain factors of decision-making in conflicts under uncertainty were proposed. Our analysis below will be confined to one of them presented in [74]. The method that will be applied in this section consists in the following. Each payoff function $f_i(x, y)$ in game (5.2.1) is associated with its *strong guarantee* $f_i[x] = \min_{y\in\mathrm{Y}} f_i(x, y)$ $(i \in \mathbb{N})$. As a consequence, choosing their strategies from a strategy profile $x \in \mathrm{X}$, the players ensure a payoff $f_i[x] \le f_i(x, y)$ $\forall y \in \mathrm{Y}$ to each player i, i.e., under any realized uncertainty $y \in \mathrm{Y}$. Such a strongly-guaranteed payoff $f_i[x]$ seems natural for the *interval uncertainties* $y \in \mathrm{Y}$ addressed in the book, because no additional probabilistic characteristics of y (except for information on the admissible set $\mathrm{Y} \subset \mathbb{R}^m$) are available. An example of such uncertainties can be the length of women's

[9]Latin "To inform."

skirts [18]. For a clothing factory, production planning for a next year heavily affects its future profits; however, in view of the vagaries of fashion and female logic dictating fashion trends, availability of any probabilistic characteristics would be hardly expected. In such problems, it is possible to establish only some obvious limits of length variations. Proposition 5.2.1, in combination with Corollary 5.2.1 as well as the continuity of $f_i(x, y)$ and $R_i(x, y)$ on $\mathrm{X} \times \mathrm{Y}$, leads to the following result.

Proposition 5.2.2 *If in game* (5.2.1) *the sets* X_i $(i \in \mathbb{N})$ *and* Y *are compact and the payoff functions* $f_i(x, y)$ *are continuous on* $\mathrm{X} \times \mathrm{Y}$, *then the guaranteed payoffs*

$$f_i[x] = \min_{y \in \mathrm{Y}} f_i(x, y) \quad (i \in \mathbb{N}) \tag{5.2.5}$$

and the guaranteed risks

$$R_i[x] = \max_{y \in \mathrm{Y}} R_i(x, y) \quad (i \in \mathbb{N}) \tag{5.2.6}$$

are scalar functions that are continuous on X.

Remark 5.2.1 *First,* the meaning of the guaranteed payoff $f_i[x]$ from (5.2.5) is that, for any $y \in \mathrm{Y}$, the realized payoffs $f_i(x, y)$ are not smaller than $f_i[x]$. In other words, using his own strategies from a strategy profile $x \in \mathrm{X}$ in game (5.2.1), each player ensures a payoff $f_i(x, y)$ of at least $f_i[x]$ under any uncertainty $y \in \mathrm{Y}$ $(i \in \mathbb{N})$. Therefore, the guaranteed payoff $f_i[x]$ gives *a lower bound* for all possible payoffs $f_i(x, y)$ occurring when the uncertainty y runs through all admissible values from Y. *Second,* the guaranteed risk $R_i[x]$ also gives *an upper bound* for all Savage–Niehans risks $R_i(x, y)$ that can be realized under any uncertainties $y \in \mathrm{Y}$. Indeed, from (5.2.6) it immediately follows that

$$R_i(x, y) \leqslant R_i[x] \quad \forall\, y \in \mathrm{Y} \quad (i \in \mathbb{N}).$$

Thus, adhering to his strategy x_i from a strategy profile $x \in \mathrm{X}$, player $i \in \mathbb{N}$ obtains a guarantee in the payoff $f_i[x]$, because $f_i[x] \leq f_i(x, y)$ $\forall y \in \mathrm{Y}$, and simultaneously a guarantee in the risk $R_i[x] \geq R_i(x, y)$ $\forall y \in \mathrm{Y}$.

Transition from Game (5.2.1) to a Noncooperative Game with Two-Component Payoff Function

The new mathematical model of a noncooperative N-player game under uncertainty with a two-component payoff function of each player in the form

$$G = \langle \mathbb{N}, \{\mathrm{X}_i\}_{i \in \mathbb{N}}, \mathrm{Y}, \{f_i(x, y), -R_i(x, y)\}_{i \in \mathbb{N}} \rangle.$$

matches the desire of each player to increase his payoff and simultaneously reduce his risk. Here $\mathbb{N}$, X_i and Y are the same as in game (5.2.1); the novelty consists in the transition from the one-component payoff function $f_i(x, y)$ of each player

i to the two-component counterpart $\{f_i(x, y), -R_i(x, y)\}$, where $R_i(x, y)$ denotes the Savage–Niehans risk function for player i. Recall that $R_i(x, y)$ figures in the game G with the minus sign, as in this case player i seeks to *increase* both criteria simultaneously by an appropriate choice of his strategy $x_i \in \mathrm{X}_i$. In this model, we expect any uncertainty $y \in \mathrm{Y}$ to occur. Since $R_i(x, y) \geq 0$ for all $(x, y) \in \mathrm{X} \times \mathrm{Y}$, an increase of $-R_i(x, y)$ is equivalent to a reduction of $R_i(x, y)$.

Since the game G involves interval uncertainties $y \in \mathrm{Y}$ only (the only available information is the range of their variation), each player $i \in \mathbb{N}$ should focus on the guaranteed payoffs $f_i[x]$ from (5.2.5) and the guaranteed risks $R_i[x]$ from (5.2.6). This approach allows one to pass from the game G to the game of guarantees

$$G^{\mathrm{g}} = \langle \mathbb{N}, \{\mathrm{X}_i\}_{i\in\mathbb{N}}, \{f_i[x], -R_i[x]\}_{i\in\mathbb{N}} \rangle,$$

in which each player $i \in \mathbb{N}$ chooses his strategy $x_i \in \mathrm{X}_i$ so as to simultaneously maximize both criteria $f_i[x]$ and $-R_i[x]$. By "freezing" the strategies of all players in G^{g} except for x_i, we arrive at the bi-criteria choice problem

$$G_i^{\mathrm{g}} = \langle \mathrm{X}_i, \{f_i[x], -R_i[x]\} \rangle$$

for each player i. Recall that, in the bi-criteria choice problem G_i^{g}, the strategies of all players except for player i are considered to be fixed ("frozen"), and this player i chooses his strategy $x_i \in \mathrm{X}_i$ so that for $x_i = x_i^{\mathrm{S}}$ the maximum possible values of $f_i[x]$ and $-R_i[x]$ are simultaneously realized. Right here it is necessary to answer the first of the two major questions formulated at the end of Sect. 5.2.1.

How Can We Combine the Objectives of Each Player (Increase Payoff and Simultaneously Reduce Realized Risk) Using Only One Criterion?

Duo quum faciunt idem, non est idem.[10]

To answer this question, we will apply the concept of vector optimum—the Pareto efficient solution—proposed in 1909 by Italian economist and sociologist Pareto [263].

In what follows, for the choice problem G_i^{g}, introduce the notations $f_i[x_i] = f_i[x]$ and $R_i[x_i] = R_i[x]$ for the frozen strategies of all players except for the strategy x_i of player i. Then the problem $G_i^{\mathrm{g}} = \langle \mathrm{X}_i, \{f_i[x], -R_i[x]\} \rangle$ can be transformed into

$$\langle \mathrm{X}_i, \{f_i[x_i], -R_i[x_i]\} \rangle. \tag{5.2.7}$$

[10] Latin "When two do the same thing, it is not the same thing." This phrase belongs to Terence, Latin in full Publius Terentius Afer, (195–159? B.C.), after Plautus the greatest Roman comic dramatist. See *The Brothers* V. 3.

Proposition 5.2.3 *If in problem (5.2.7) there exist a strategy* $x_i^e \in X_i$ *and a value* $\sigma_i \in (0, 1)$ *such that* x_i^e *maximizes the scalar function*

$$\Phi_i[x_i] = f_i[x_i] - \sigma_i R_i[x_i], \tag{5.2.8}$$

i.e.,

$$\Phi_i[x_i^e] = \max_{x_i \in X_i} (f_i[x_i] - \sigma_i R_i[x_i]), \tag{5.2.9}$$

then x_i^e *is the Pareto-maximal alternative in (5.2.7); in other words, for any* $x_i \in X_i$ *the system of two inequalities*

$$f_i[x_i] \geqslant f_i[x_i^e], \quad -R_i[x_i] \geqslant -R_i[x_i^e],$$

with at least one strict inequality, is inconsistent.

Reductio ad Absurdum Assume on the contrary that the strategy x_i^e yielded by (5.2.9) is not the Pareto-maximal alternative in problem (5.2.7). Then there exists a strategy $\bar{x}_i \in X_i$ of player i such that the system of two inequalities

$$f_i[\bar{x}_i] \geqslant f_i[x_i^e], \quad -R_i[\bar{x}_i] \geqslant -R_i[x_i^e],$$

with at least one strict inequality, is consistent.

Multiply both sides of the first inequality by $1-\sigma_i > 0$ and of the other inequality by $\sigma_i > 0$ and then add separately the left- and right-hand sides of the resulting inequalities to obtain

$$(1 - \sigma_i) f_i[\bar{x}_i] - \sigma_i R_i[\bar{x}_i] > (1 - \sigma_i) f_i[x_i^e] - \sigma_i R_i[x_i^e]$$

or, taking into account (5.2.8),

$$\Phi_i[\bar{x}_i] > \Phi_i[x_i^e].$$

This strict inequality contradicts (5.2.9), and the conclusion follows. ■

Remark 5.2.2 The combination of criteria (5.2.5) and (5.2.6) in form (5.2.8) is of interest for two reasons. First, even if for $\bar{x}_i \neq x_i^e$ we have an increase of the guaranteed result $f_i[\bar{x}_i] > f_i[x_i^e]$, then due to the Pareto maximality of x_i^e and the fact that $R_i[x_i] \geq 0$ such an improvement of the guaranteed payoff $f_i[\bar{x}_i] > f_i[x_i^e]$ inevitably leads to an increase of the guaranteed risk $R_i[\bar{x}_i] > R_i[x_i^e]$; conversely, for the same reasons, a reduction of the guaranteed risk $R_i[\bar{x}_i] < R_i[x_i^e]$ leads to a reduction of the guaranteed payoff $f_i[\bar{x}_i] < f_i[x_i^e]$ (both cases are undesirable for player i). Therefore, the replacement of the bi-criteria choice problem (5.2.7) with the single-criterion choice problem $\langle X_i, f_i[x_i] - \sigma_i R_i[x_i]\rangle$ matches the desire of player i to increase $f_i[x_i]$ and simultaneously reduce $R_i[x_i]$.

Second, since $R_i[x_i] \geq 0$, an increase of the difference $f_i[x_i] - \sigma_i R_i[x_i]$ also matches the desire of player i to increase the guaranteed payoff $f_i[x]$ and simultaneously reduce the guaranteed risk $R_i[x]$.

5.2.3 *Formalization of Guaranteed Equilibrium in Payoffs and Risks for Game (5.2.1)*

Omnis determinatio est negatio.[11]

Punctum Saliens[12]

Now, let us answer the second question from Sect. 5.2.1: how can we combine the efforts of all N players in a single strategy profile taking into account the existing interval uncertainty? To do this, from game (5.2.1) we will pass sucessively to noncooperative games Γ_1, Γ_2 and Γ_3, where

$$\begin{aligned}\Gamma_1 &= \langle \mathbb{N}, \{\mathrm{X}_i\}_{i\in\mathbb{N}}, \mathrm{Y}, \{f_i(x,y), -R_i(x,y)\}_{i\in\mathbb{N}}\rangle,\\ \Gamma_2 &= \langle \mathbb{N}, \{\mathrm{X}_i\}_{i\in\mathbb{N}}, \{f_i[x], -R_i[x]\}_{i\in\mathbb{N}}\rangle,\\ \Gamma_3 &= \langle \mathbb{N}, \{\mathrm{X}_i\}_{i\in\mathbb{N}}, \{\Phi_i[x] = f_i[x] - \sigma_i R_i[x]\}_{i\in\mathbb{N}}\rangle.\end{aligned}$$

In all these three games, $\mathbb{N} = \{1, 2, \ldots, N \geq 2\}$ is the set of players; $x_i \in \mathrm{X}_i \subset \mathbb{R}^{n_i}$ $(i \in \mathbb{N})$ denote the strategies of player i; $x = (x_1, \ldots, x_N) \in \mathrm{X} = \prod_{i\in\mathbb{N}} \mathrm{X}_i \subset \mathbf{R}^n$ $(n = \sum_{i\in\mathbb{N}} n_i)$ forms a strategy profile; $y \in \mathrm{Y} \subset \mathbf{R}^m$ are uncertainties; the payoff function $f_i(x, y)$ of each player $i \in \mathbb{N}$ is defined on the pairs $(x, y) \in \mathrm{X}\times\mathrm{Y}$; in (5.2.4), $R_i(x, y)$ denotes the Savage–Niehans risk function of player i; finally, $\sigma_i \in (0, 1)$ $(i \in \mathbb{N})$ are some constants. In the game Γ_1, the payoff function of player i becomes two-component as the difference between the payoff function $f_i(x, y)$ of player i from (5.2.1) and the risk function $R_i(x, y)$ from (5.2.4).

In the game Γ_2, the payoff function $f_i(x, y)$ and the risk function $R_i(x, y)$ are replaced with their guarantees $f_i[x] = \min_{y\in\mathrm{Y}} f_i(x, y)$ and $R_i[x] = \max_{y\in\mathrm{Y}} R_i(x, y)$, respectively. Finally, in the game Γ_3, the linear combination of the guarantees $f_i[x]$ and $-R_i[x]$ (see Proposition 5.2.3) is used instead of the payoff function of player i.

Internal Instability of the Set of Nash Equilibria

A bird may be known by its song.
—English proverb

[11] Latin "All determination is negation." This phrase belongs to Benedict de Spinoza, (1632–1677), a Dutch Jewish philosopher.

[12] This Latin word combination identifies a starting point, an origin, a source, the heart of the matter.

Consider a noncooperative N-player game in pure strategies (a non-zero-sum game of guarantees) of the form

$$\Gamma = \langle \mathbb{N}, \{\mathrm{X}_i\}_{i\in\mathbb{N}}, \{\Phi_i[x]\}_{i\in\mathbb{N}}\rangle. \tag{5.2.10}$$

Each player i chooses and uses his *pure strategy* $x_i \in \mathrm{X}_i \subset \mathbb{R}^{n_i}$ without making coalitions with other players, thereby forming a strategy profile $x = (x_1, \dots, x_N) \in \mathrm{X} = \prod_{i\in\mathbb{N}} \mathrm{X}_i \subseteq \mathbb{R}^n$ $(n = \sum\limits_{i\in\mathbb{N}} n_i)$; a payoff function $\Phi_i[x]$ is defined for each $i \in \mathbb{N}$ on the set of strategy profiles X, and its value is called the *payoff* of player i. Below, we will again use the notations $[x^{\mathrm{e}}||x_i] = [x_1^{\mathrm{e}}, \dots, x_{i-1}^{\mathrm{e}}, x_i, x_{i+1}^{\mathrm{e}}, \dots, x_N^{\mathrm{e}}]$ and $\Phi = (\Phi_1, \dots, \Phi_N)$.

Definition 5.2.1 A strategy profile $x^{\mathrm{e}} = (x_1^{\mathrm{e}}, \dots, x_N^{\mathrm{e}}) \in \mathrm{X}$ is called a Nash equilibrium in game (5.2.10) if

$$\max_{x_i\in\mathrm{X}_i} \Phi_i[x^{\mathrm{e}}\|x_i] = \Phi_i[x^{\mathrm{e}}] \quad (i \in \mathbb{N}); \tag{5.2.11}$$

denote by X^{e} the set $\{x^{\mathrm{e}}\}$ of Nash equilibria in game (5.2.12).

Let us analyze the internal instability of X^{e}. A subset $\mathrm{X}^* \subset \mathbf{R}^n$ is *internally unstable* if there exist at least two strategy profiles $x^{(j)} \in \mathrm{X}^*$ $(j=1,2)$ such that

$$\left[\Phi[x^{(1)}] < \Phi[x^{(2)}]\right] \Leftrightarrow \left[\Phi_i[x^{(1)}] < \Phi_i[x^{(2)}] \quad (i \in \mathbb{N})\right], \tag{5.2.12}$$

and *internally stable* otherwise.

Example 5.2.1 Consider the two-player game

$$\langle\{1, 2\}, \{\mathrm{X}_i = [-1, 1]\}_{i=1,2}, \{f_i(x) = -x_i^2 + 2x_1x_2\}_{i=1,2}\rangle. \tag{5.2.13}$$

A strategy profile $x^{\mathrm{e}} = (x_1^{\mathrm{e}}, x_2^{\mathrm{e}}) \in [-1, 1]^2$ is a Nash equilibrium in game (5.2.13) if (see (5.2.11))

$$-x_i^2 + 2x_1x_2^{\mathrm{e}} \leqslant -(x_i^{\mathrm{e}})^2 + 2x_1^{\mathrm{e}}x_2^{\mathrm{e}} \quad \forall\, x_i \in [-1, 1] \;\; (i = 1, 2),$$

which is equivalent to

$$-(x_1 - x_2^{\mathrm{e}})^2 \leqslant -(x_1^{\mathrm{e}} - x_2^{\mathrm{e}})^2, \qquad -(x_1^{\mathrm{e}} - x_2)^2 \leqslant -(x_1^{\mathrm{e}} - x_2^{\mathrm{e}})^2.$$

Therefore, $x_1^{\mathrm{e}} = x_2^{\mathrm{e}} = \alpha \;\forall\alpha = \mathrm{const} \in [-1, 1]$, i.e., in (5.2.13) we have the sets

$$\mathbf{X}^{\mathrm{e}} = \{(\alpha, \alpha) \mid \forall\, \alpha \in [-1, 1]\}$$

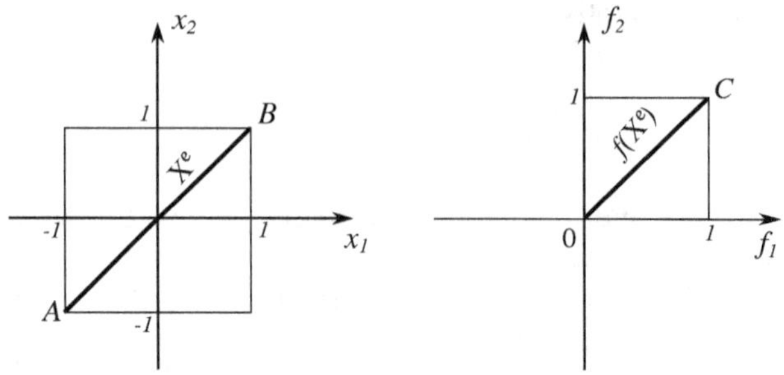

Fig. 5.3 Set $\mathbf{X}^{\mathrm{e}} = AB$. Set $f(\mathbf{X}^{\mathrm{e}}) = \bigcup_{\alpha\in[-1,1]} (\alpha^2, \alpha^2) = OC$

and $f_i(\mathrm{X}^{\mathrm{e}}) = \bigcup_{x^e\in \mathrm{X}^{\mathrm{e}}} f_i(x^e) = \bigcup_{\alpha\in[-1,1]}(\alpha^2, \alpha^2)$, as illustrated in Fig. 5.3. Thus, the set X^{e} is internally unstable, since for game (5.2.13) with $x^{(1)} = (0, 0)$ and $x^{(2)} = (1, 1)$ we obtain $f_i(x^{(1)}) = 0 < f_i(x^{(2)}) = 1$ $(i = 1, 2)$ (see (5.2.12)).

Remark 5.2.3 In the zero-sum setup of game (5.2.10) (i.e., with $\mathbb{N} = \{1, 2\}$ and $f_1 = -f_2 = \overline{f}$), the equality $\overline{f}(x^{(1)}) = \overline{f}(x^{(2)})$ holds for any two saddle points $x^{(k)} \in \mathrm{X}$ $(k = 1, 2)$ (by the equivalence of saddle points). Therefore, the set of saddle points in the zero-sum game is always internally stable. Note that a saddle point is a Nash equilibrium in the zero-sum setup of game (5.2.10).

Remark 5.2.4 In the non-zero-sum setup of game (5.2.10), internal instability (see Example 5.2.1) does not occur if there is a *unique* Nash equilibrium in (5.2.10).

Let us associate with game (5.2.10) an auxiliary N-criteria choice problem of the form

$$\Gamma_{\mathrm{c}} = \langle \mathrm{X}^{\mathrm{e}}, \{\Phi_i[x]\}_{i\in\mathbb{N}}\rangle, \tag{5.2.14}$$

where the set X^{e} of *alternatives* x coincides with the set of Nash equilibria x^{e} of game (5.2.10) and the ith criterion $\Phi_i[x]$ is the payoff function (5.2.8) of player i.

Definition 5.2.2 An alternative $x^{\mathrm{P}} \in \mathbf{X}^{\mathrm{e}}$ is a Pareto-maximal (weakly efficient) alternative in (5.2.14) if for all $x \in \mathrm{X}^{\mathrm{e}}$ the system of inequalities

$$\Phi_i[x] \geqslant \Phi_i[x^{\mathrm{P}}] \quad (i \in \mathbb{N}),$$

with at least one strict inequality, is inconsistent. Denote by X^{P} the set $\{x^{\mathrm{P}}\}$ of all such strategy profiles.

In accordance with Definition 5.2.2, the set $\mathrm{X}^{\mathrm{P}} \subseteq \mathrm{X}^{\mathrm{e}}$ is *internally stable*.

The following **assertion** is obvious. If

$$\sum_{i\in\mathbb{N}} f_i(x) \leqslant \sum_{i\in\mathbb{N}} f_i(x^{\mathrm{P}}) \qquad (5.2.15)$$

for all $x \in \mathrm{X}^{\mathrm{e}}$, then x^{P} is a Pareto-maximal alternative in problem (5.2.14).

Remark 5.2.5 A branch of mathematical programming focused on numerical methods of Nash equilibria design in games (5.2.10) has recently become known as *equilibrium programming*. At Moscow State University, research efforts in this field are being undertaken by the groups of Professors F.P. Vasiliev and A.S. Antipin at the Faculty of Computational Mathematics and Cybernetics. However, the equilibrium calculation methods developed so far yield a Nash equilibrium that is not necessarily Pareto-maximal (in other words, the methods themselves do not guarantee Pareto maximality). At the same time, such a guarantee appears (!) if equilibrium is constructed using the sufficient conditions below—see Theorem 5.2.1.

Formalization of Pareto–Nash Equilibrium

That's a horse of another colour.
—English proverb

Let us return to the noncooperative game (5.2.10)

$$\Gamma = \langle \mathbb{N}, \{\mathrm{X}_i\}_{i\in\mathbb{N}}, \{\Phi_i[x]\}_{i\in\mathbb{N}} \rangle,$$

associating with it the N-criteria choice problem (5.2.14)

$$\langle \mathrm{X}^{\mathrm{e}}, \{\Phi_i[x]\}_{i\in\mathbb{N}} \rangle.$$

Recall that the set of Nash equilibria x^{e} of game (5.2.10) (Definition 5.2.1) is denoted by X^{e}, while the set of Pareto-maximal alternatives x^{P} of problem (5.2.14) (Definition 5.2.2) is denoted by X^{P}.

Definition 5.2.3 A strategy profile $x^* \in \mathbf{X}$ is called a Pareto–Nash equilibrium in game (5.2.10) if x^* is simultaneously

(a) a Nash equilibrium in (5.2.10) (Definition 5.2.1) and
(b) a Pareto-maximal alternative in (5.2.14) (Definition 5.2.2).

Remark 5.2.6 The existence of x^* in game (5.2.10) with $\mathrm{X}^{\mathrm{e}} \neq \varnothing$, compact sets X_i and continuous payoff functions $\Phi_i[x]$ $(i \in \mathbb{N})$ follows directly from the fact that $\mathrm{X}^{\mathrm{e}} \in \mathrm{comp}\mathrm{X}$.

Sufficient Conditions of Pareto–Nash Equilibrium in Game (5.2.10)

Cest tout ce qu'il faut.[13]

Relying on (5.2.11) and (5.2.15), introduce $N + 1$ scalar functions of the form

$$\varphi_i(x, z) = f_i(z\|x_i) - f_i(z) \quad (i \in \mathbb{N}),$$
$$\varphi_{N+1}(x, z) = \sum_{r\in\mathbb{N}} f_r(x) - \sum_{r\in\mathbb{N}} f_r(z), \tag{5.2.16}$$

where $z = (z_1, \dots, z_N)$, $z_i \in \mathbf{X}_i$ $(i \in \mathbb{N})$, $z \in \mathbf{X}$, and $x \in \mathbf{X}$. The Germeier convolution [152, p. 66] of the scalar functions (5.2.16) is given by

$$\varphi(x, z) = \max_{j=1,\dots,N+1} \varphi_j(x, z). \tag{5.2.17}$$

Finally, let us associate with game (5.2.10) and the N-criteria choice problem (5.2.14) *the zero-sum game*

$$\langle \mathrm{X},\ \mathrm{Z} = \mathrm{X},\ \varphi(x, z) \rangle, \tag{5.2.18}$$

in which the first player chooses his strategy $x \in \mathrm{X}$ *to increase* the payoff function, while the opponent (the second player) forms his strategy $z \in \mathrm{X}$, seeking *to decrease* as much as possible the payoff function $\varphi(x, z)$ from (5.2.16) and (5.2.17).

A saddle point $(x^0, z^*) \in \mathrm{X}^2$ in game (5.2.18) is defined by the chain of inequalities

$$\varphi(x, z^*) \leqslant \varphi(x^0, z^*) \leqslant \varphi(x^0, z) \quad \forall\, x, z \in \mathrm{X}. \tag{5.2.19}$$

In this case, the saddle point is formed by the minimax strategy z^*,

$$\left(\min_{z\in\mathrm{X}} \max_{x\in\mathrm{X}} \varphi(x, z) = \max_{x\in\mathrm{X}} \varphi(x, z^*)\right),$$

in combination with the maximin strategy x^0,

$$\left(\max_{x\in\mathrm{X}} \min_{z\in\mathrm{X}} \varphi(x, z) = \min_{z\in\mathrm{X}} \varphi(x^0, z)\right),$$

in game (5.2.18).

The next result provides a sufficient condition for the existence of a Pareto equilibrium in game (5.2.10).

[13] French "All that is needed."

Theorem 5.2.1 *If there exists a saddle point* (x^0, z^*) *in the zero-sum game (5.2.18) (i.e., inequalities* (5.2.19) *hold), then the minimax strategy* z^* *is a Pareto–Nash equilibrium in game (5.2.10).*

Proof Let $z = x^0$ in the right-hand inequality of (5.2.19). Then, using (5.2.16) and (5.2.17), we obtain

$$\varphi(x^0, x^0) = \max_{j=1,\dots,N+1} \varphi_j(x^0, x^0) = 0.$$

In accordance with (5.2.19), it appears that

$$0 \geqslant \varphi(x, z^*) = \max_{j=1,\dots,N+1} \varphi_j(x, z^*) \quad \forall x \in \mathrm{X}.$$

Therefore, the following chain of implications is valid for all $x \in \mathrm{X}$:

$$\left[0 \geqslant \max_{j=1,\dots,N+1} \varphi_j(x, z^*) \geqslant \varphi_j(x, z^*)\right]$$
$$\Longrightarrow \left[\varphi_j(x, z^*) \leqslant 0 \; (j = 1, \dots, N+1)\right]$$
$$\overset{(5.2.16)}{\Longrightarrow} \left\{\left[f_i(z^* \| x_i) - f_i(z^*) \leqslant 0 \quad \forall x_i \in \mathrm{X}_i \quad (i \in \mathbb{N})\right]\right.$$
$$\left.\bigwedge \left[\sum_{i \in \mathbb{N}} f_i(x) - \sum_{i \in \mathbb{N}} f_i(z^*) \leqslant 0 \quad \forall x \in \mathrm{X}^{\mathrm{e}}\right]\right\}$$
$$\Longrightarrow \left\{\left[\max_{x_i \in \mathrm{X}_i} f_i(z^* \| x_i) = f_i(z^*) \; (i \in \mathbb{N})\right]\right.$$
$$\left.\bigwedge \left[\max_{x \in \mathrm{X}^{\mathrm{e}}} \sum_{i \in \mathbb{N}} f_i(x) = \sum_{i \in \mathbb{N}} f_i(z^*)\right]\right\}$$
$$\overset{(5.2.11),(5.2.15)}{\Longrightarrow} \left\{\left[z^* \in \mathrm{X}^{\mathrm{e}}\right] \bigwedge \left[z^* \in \mathrm{X}^P\right]\right\},$$

Due to the inclusion $\mathrm{X}^{\mathrm{e}} \subseteq \mathrm{X}$. ■

Remark 5.2.7 Theorem 5.2.1 suggests the following design method for a Pareto–Nash equilibrium x^* in game (5.2.10).

Step 1. Using the payoff function $\Phi_i[x]$ $(i \in \mathbb{N})$ from (5.2.10) and also the vectors $z = (z_1, \dots, z_N)$, $z_i \in \mathrm{X}_i$, and $x = (x_1, \dots, x_N)$, $x_i \in \mathrm{X}_i$ $(i \in \mathbb{N})$, construct the function $\varphi(x, z)$ by formulas (5.2.16) and (5.2.17).
Step 2. Find the saddle point (x^0, z^*) of the zero-sum game (5.2.18).

Then z^* is the Pareto–Nash equilibrium solution of game (5.2.10).

As far as we know, numerical calculation methods for the saddle point (x^0, z^*) of the Germeier convolution

$$\varphi(x, z) = \max_{j=1,\dots,N+1} \varphi_j(x, z)$$

are lacking; however, they are crucial (see Theorem 5.2.1) for constructing Nash equilibria that are simultaneously Pareto-maximal strategy profiles. Seemingly, this is a new field of equilibrium programming and it can be developed, again in our opinion, using the mathematical tools of Germeier convolution optimization $\max_j \varphi_j(x)$ that were introduced by Professor V. F. Demyanov.

Remark 5.2.8 The next statement follows from results of operations research (see Proposition 5.2.1) and is a basic recipe for proving the existence of a Pareto–Nash equilibrium in mixed strategies in game (5.2.10) (see Sect. 5.2.4). Namely, in game (5.2.10) with the sets $\mathrm{X}_i \in \operatorname{comp} \mathbb{R}^{n_i}$ and the payoff functions $\Phi_i[\cdot] \in C(X)$ $(i \in \mathbb{N})$, the Germeier convolution $\varphi(x, z) = \max_{j=1,\dots,N+1} \varphi_j(x, z)$ (5.2.16), (5.2.17) is continuous on $\mathrm{X} \times \mathrm{X}$.

Formalization of Strongly-Guaranteed Equilibrium in Payoffs and Risks

Cest tout dire![14]

Let us consider the concept of guaranteed equilibrium in game (5.2.1) from the viewpoint of *a risk-neutral player.* Assume each player i exhibits a risk-neutral behavior, i.e., chooses his strategy to increase the payoff (the value of the payoff function $f_i(x, y)$) and simultaneously reduce the risk (the value of the Savage–Niehans risk function $R_i(x, y) = \max_{z \in \mathrm{X}} f_i(z, y) - f_i(x, y)$) under any realization of the uncertainty $y \in \mathrm{Y}$. Hereinafter, we use three N-dimensional vectors $f = (f_1, \dots, f_N)$, $R = (R_1, \dots, R_N)$, and $\Phi = (\Phi_1, \dots, \Phi_N)$ as well as N values $\sigma_i \in (0, 1)$ $(i \in \mathbb{N})$.

Definition 5.2.4 A triplet $(x^{\mathrm{P}}, f^{\mathrm{P}}, R^{\mathrm{P}})$ is called a strongly-guaranteed Nash equilibrium in payoffs and risks in game (5.2.1) if

first, $f^{\mathrm{P}} = f[x^{\mathrm{P}}]$ and $R^{\mathrm{P}} = R[x^{\mathrm{P}}]$;

second, there exist scalar functions $f_i[x] = \min_{y \in \mathrm{Y}} f_i(x, y)$ and $R_i[x] = \max_{y \in \mathrm{Y}} R_i(x, y)$, $R_i(x, y) = \max_{z \in \mathrm{X}} f_i(z, y) - f_i(x, y)$ $(i \in \mathbb{N})$, that are continuous on X;

third, the set X^{e} of all Nash equilibria x^{e} in the game of guarantees

$$\Gamma_3 = \langle \mathbb{N}, \{\mathrm{X}_i\}_{i \in \mathbb{N}}, \{\Phi_i[x] = f_i[x] - \sigma_i R_i[x]\}_{i \in \mathbb{N}} \rangle$$

[14] French "That just goes to show!"

is non-empty at least for one value $\sigma_i \in (0, 1)$, i.e.,

$$\max_{x_i \in \mathrm{X}_i} \Phi_i[x^{\mathrm{e}} \| x_i] = \Phi_i[x^{\mathrm{e}}], \quad (i \in \mathbb{N}),$$

where $X^{\mathrm{e}} = \{x^{\mathrm{e}}\}$ and $[x^{\mathrm{e}}||x_i] = [x_1^{\mathrm{e}}, \ldots, x_{i-1}^{\mathrm{e}}, x_i, x_{i+1}^{\mathrm{e}}, \ldots, x_N^{\mathrm{e}}]$;

fourth, x^{P} is a Pareto-maximal alternative in the N-criteria choice problem of guarantees

$$\langle \mathrm{X}^{\mathrm{e}}, \{\Phi_i[x]\}_{i \in \mathbb{N}} \rangle,$$

i.e., the system of inequalities

$$\Phi_i[x] \geqslant \Phi_i[x^{\mathrm{P}}] \quad (i \in \mathbb{N}) \quad \forall\, x \in \mathrm{X}^{\mathrm{e}},$$

with at least one strict inequality, is inconsistent.

Remark 5.2.9 Let us list a number of advantages of this equilibrium solution.

First, as repeatedly mentioned, economists often divide decision makers (in our game (5.2.1), players) into three groups. The first group includes those who do not like to take risks (*risk-averse players*); the second group, *risk-seeking players*; and the third group, those who consider the payoffs and risks simultaneously (*risk-neutral players*). Definition 5.2.4 treats all players as *risk-neutral* ones, though it would be interesting to analyze the players from different groups (risk-averse, risk-seeking and risk-neutral players). We hope to address these issues in future work.

Second, lower and upper bounds on the payoffs and risks are provided by the inequalities $f_i[x^{\mathrm{P}}] \leqslant f_i(x^{\mathrm{P}}, y)\ \forall y \in \mathrm{Y}$ and $R_i[x^{\mathrm{P}}] \geqslant R_i(x^{\mathrm{P}}, y)\ \forall y \in \mathrm{Y}$, respectively; note that the continuity of the guarantees $f_i[x]$ and $R_i[x]$ follows directly from the inclusions $\mathrm{X}_i \in \mathrm{comp}\ \mathbb{R}^{n_i}$ $(i \in \mathbb{N})$, $\mathrm{Y} \in \mathrm{comp}\ \mathbb{R}^m$, and $f_i[\cdot] \in (\mathrm{X} \times \mathrm{Y})$ (see Proposition 5.2.2).

Third, an increase of the guaranteed payoffs for a separate player (as compared to $f_i[x^{\mathrm{P}}]$) would inevitably cause an increase of the guaranteed risk (again, as compared to $R_i[x^{\mathrm{P}}]$), whereas a reduction of this risk would inevitably cause a reduction of the guaranteed payoff.

Fourth, it is impossible to increase the difference $\Phi_i[x^{\mathrm{P}}]$ for all players simultaneously (this property follows from the Pareto maximality of the strategy profile x^{P}).

Fifth, the best solution has been selected from all guaranteed solutions, as the difference $\Phi_i[x^{\mathrm{P}}]$ takes the largest value (in the sense of vector maximum).

Sixth, under the assumption that the sets X_i $(i \in \mathbb{N})$ and Y are compact and the payoff functions $f_i(x, y)$ are continuous on $\mathrm{X} \times \mathrm{Y}$, the guarantees $f_i[x]$ and $R_i[x]$ exist and are continuous on X (Proposition 5.2.2). Therefore, the existence of solutions formalized by Definition 5.2.4 rests on the existence of Nash equilibria in the game of guarantees. Note that the framework developed in this section of Chap. 5

can be also applied to the concepts of Berge equilibrium, threats and counterthreats, and active equilibrium.

Below we will present a new method of proving the existence of a strongly-guaranteed Nash equilibrium in payoffs and risks. In particular, using the Germeier convolution of the payoff function, we have already established sufficient conditions for the existence of Nash equilibria in pure strategies that are simultaneously Pareto-maximal with respect to all other equilibria (see Theorem 5.2.1).

Concluding this section, we will show the existence of such an equilibrium in mixed strategies under standard assumptions of noncooperative games (compact strategy sets and continuous payoff functions of all players).

5.2.4 Existence of Pareto Equilibrium in Mixed Strategies

Se plaindre que la mariée soit trop belle.[15]

The hope that game (5.2.10) has a Pareto equilibrium in pure strategies (Definition 5.2.3) is delusive. Such an equilibrium may exist only for a special form of the payoff functions, a special structure of the strategy sets, and a special number of players. Therefore, adhering to an approach that stems from Borel [209], von Neumann [261], Nash [257] and their followers, we will establish the existence of a Pareto equilibrium in mixed strategies in game (5.2.10) under standard assumptions of noncooperative games (compact strategy sets and continuous payoff functions).

Thus, suppose that in game (5.2.10) the sets X_i of pure strategies x_i are convex and compact in $\mathbb{R}^{n_i}$ (i.e., convex, closed and bounded; denote this by $\mathrm{X}_i \in \operatorname{cocomp} \mathbb{R}^{n_i}$) while the payoff function $f_i[x]$ of each player i ($i \in \mathbb{N}$) is continuous on the set of all pure strategy profiles $\mathrm{X} = \prod_{i \in \mathbb{N}} \mathrm{X}_i$.

Consider *the mixed extension of game* (5.2.10). For each of the N compact sets X_i ($i \in \mathbb{N}$), construct the Borel σ-algebra $\mathfrak{B}(\mathrm{X}_i)$ and probability measures $\nu_i(\cdot)$ on $\mathfrak{B}(\mathrm{X}_i)$ (i.e., nonnegative countably-additive scalar functions defined on the elements of $\mathfrak{B}(\mathrm{X}_i)$ that are normalized to 1 on X_i). Denote by $\{\nu_i\}$ the set of such measures; a measure $\nu_i(\cdot)$ is called *a mixed strategy of player i* ($i \in \mathbb{N}$) in game (5.2.10). Then, for the same game (5.2.10), construct *mixed strategy profiles*, i.e., the product measures $\nu(dx) = \nu_1(dx_1) \cdots \nu_N(dx_N)$. Denote by $\{\nu\}$ the set of such strategy profiles. Finally, calculate *the expected values*

$$f_i(\nu) = \int_{\mathrm{X}} f_i(x)\nu(dx) \quad (i \in \mathbb{N}). \tag{5.2.20}$$

[15]French "Complaining that the bride is too beautiful." In our book, the advantages of Proposition 5.2.4 have exceeded all expectations.

As a result, we associate with the game Γ_3 (5.2.10) its *mixed extension*

$$\widetilde{\Gamma}_3 = \langle \mathbb{N}, \{\nu_i\}_{i\in\mathbb{N}}, \{f_i(\nu)\}_{i\in\mathbb{N}}\rangle.$$

In the noncooperative game $\widetilde{\Gamma}_3$,

$\nu_i(\cdot) \in \{\nu_i\}$ is a mixed strategy of player i;
$\nu(\cdot) \in \{\nu\}$ is a mixed strategy profile;
$f_i(\nu)$ is the payoff function of player i, defined by (5.2.20).

In what follows, we will use the vectors $z = (z_1, \dots, z_N) \in \mathrm{X}$, where $z_i \in \mathrm{X}_i$ $(i \in \mathbb{N})$, and $x = (x_1, \dots, x_N) \in \mathrm{X}$, as well as the mixed strategy profiles $\nu(\cdot), \mu(\cdot) \in \{\nu\}$ and the expected values

$$\Phi_i(\nu) = \int\limits_{\mathrm{X}} \Phi_i(x)\nu(dx), \qquad \Phi_i(\mu) = \int\limits_{\mathrm{X}} \Phi_i(z)\mu(dz),$$

$$\Phi_i(\mu\|\nu_i) = \int\limits_{\mathrm{X}_1} \cdots \int\limits_{\mathrm{X}_{i-1}} \int\limits_{\mathrm{X}_i} \int\limits_{\mathrm{X}_{i+1}} \cdots \int\limits_{\mathrm{X}_N} \Phi_i(z_1, \dots, z_{i-1}, x_i, z_{i+1},$$
$$\dots, z_N)\mu_N(dz_N)\cdots\mu_{i+1}(dz_{i+1})\nu_i(dx_i)\mu_{i-1}(dz_{i-1})\cdots\mu_1(dz_1). \tag{5.2.21}$$

Once again, take notice that $x_i, z_i \in \mathrm{X}_i$ $(i \in \mathbb{N})$ and $x, z \in \mathrm{X}$.

The following concept of Nash equilibrium in *mixed* strategies $\nu^{\mathrm{e}}(\cdot) \in \{\nu\}$ in game (5.2.10) corresponds to Definition 5.2.1 of a Nash equilibrium in *pure* strategies $x^{\mathrm{e}} \in \mathrm{X}$ in the same game (5.2.10).

Definition 5.2.5 A strategy profile $\nu^{\mathrm{e}}(\cdot) \in \{\nu\}$ is called a Nash equilibrium in the game $\widetilde{\Gamma}_3$ if

$$\Phi_i[\nu^{\mathrm{e}}\|\nu_i] \leqslant \Phi_i[\nu^{\mathrm{e}}] \quad \forall \nu(\cdot) \in \{\nu_i\} \ (i \in \mathbb{N}); \tag{5.2.22}$$

sometimes, the same strategy profile $\nu^{\mathrm{e}}(\cdot) \in \{\nu\}$ will be also called a Nash equilibrium in mixed strategies in game (5.2.10).

By Glicksberg's theorem [30], under the conditions $\mathrm{X}_i \in \operatorname{cocomp} \mathbf{R}^{n_i}$ and $f_i(\cdot) \in C(\mathrm{X})$ $(i \in \mathbb{N})$, there exists a Nash equilibrium in mixed strategies in game (5.2.10). Denote by $\mathfrak{Y}$ the set of such mixed strategy profiles $\{\nu\}$.

With the game $\widetilde{\Gamma}_3$ we associate the N-criteria choice problem

$$\widetilde{\Gamma}_\nu = \langle \mathfrak{Y}, \{\Phi_i[\nu]\}_{i\in\mathbb{N}}\rangle\,. \tag{5.2.23}$$

In (5.2.23), the DM chooses a strategy profile $\nu(\cdot) \in \mathfrak{Y}$ in order to simultaneously maximize all elements of a vector criterion $\Phi(\nu) = (\Phi_1(\nu), \dots, \Phi_N(\nu))$. Here a generally accepted solution is a Pareto-maximal alternative.

Definition 5.2.6 A strategy profile $\nu^{\mathrm{P}}(\cdot) \in \mathfrak{Y}$ is called a Pareto-maximal alternative for the N-criteria choice problem $\widetilde{\Gamma}_{\nu}$ from (5.2.23) if for any $\nu(\cdot) \in \mathfrak{Y}$ the system of inequalities

$$\Phi_i[\nu] \geqslant \Phi_i^{\mathrm{P}}[\nu] \quad (i \in \mathbb{N}),$$

with at least one strict inequality, is inconsistent.

An analog of (5.2.15) states the following. If

$$\sum_{i\in\mathbb{N}} f_i(\nu) \leqslant \sum_{i\in\mathbb{N}} f_i(\nu^{\mathrm{P}}), \tag{5.2.24}$$

for all $\nu(\cdot) \in \mathfrak{Y}$, then the mixed strategy profile $\nu^{\mathrm{P}}(\cdot) \in \mathfrak{Y}$ is a Pareto-maximal alternative in the choice problem $\widetilde{\Gamma}_{\nu}$ (5.2.23).

Definition 5.2.7 A mixed strategy profile $\nu^*(\cdot) \in \{\nu\}$ is called a Pareto–Nash equilibrium in mixed strategies in game (5.2.10) if

1°. $\nu^*(\cdot)$ is a Nash equilibrium in the game $\widetilde{\Gamma}_3$ (Definition 5.2.5);
2°. $\nu^*(\cdot)$ is a Pareto-maximal alternative in the multicriteria choice problem (5.2.23) (Definition 5.2.6).

Now, we will prove the existence of a Nash equilibrium in mixed strategies that is simultaneously Pareto-maximal with respect to all other Nash equilibria.

Proposition 5.2.4 *Consider the noncooperative game* (5.2.10), *assuming that*

1^0. *the set of pure strategies* $\mathbf{X}_i$ *of each player* i *is a nonempty, convex, and compact set in* $\mathbb{R}^{n_i}$ $(i \in \mathbb{N})$;
2^0. *the payoff function* $\Phi_i[x]$ *of player* i $(i \in \mathbb{N})$ *is continuous on the set of all strategy profiles* $\mathbf{X} = \prod_{i\in\mathbb{N}} \mathbf{X}_i$.

Then there exists a Pareto equilibrium in mixed strategies in game (5.2.10).

Proof Using formulas (5.2.16) and (5.2.17), construct the scalar function

$$\varphi(x, z) = \max_{j=1,\dots,N+1} \varphi_j(x, z),$$

where, as before,

$$\varphi_i(x, z) = f_i(z\|x_i) - f_i(z) \quad (i \in \mathbb{N}),$$

$$\varphi_{N+1}(x, z) = \sum_{r\in\mathbb{N}} f_r(x) - \sum_{r\in\mathbb{N}} f_r(z),$$

$z=(z_1,\ldots,z_N)\in\mathbf{X}$, $z_i\in\mathbf{X}_i$ $(i\in\mathbb{N})$, and $x=(x_1,\ldots,x_N)\in\mathbf{X}$, $x_i\in\mathbf{X}_i(i\in\mathbb{N})$. By the construction procedure and Remark 5.2.9, the function $\varphi(x,z)$ is well-defined and continuous on the product of compact sets $\mathbf{X}\times\mathbf{X}$.

Introduce the auxiliary zero-sum game

$$\Gamma_{\mathrm{a}} = \langle\ \{\mathrm{I},\mathrm{II}\}, \mathbf{X}, \mathbf{Z}=\mathbf{X}, \varphi(x,z)\ \rangle.$$

In this game, player I chooses his strategy $x\in\mathbf{X}$ to maximize a continuous payoff function $\varphi(x,z)$ on $\mathbf{X}\times\mathbf{Z}$ ($\mathbf{Z}=\mathbf{X}$) while player II seeks to minimize it by choosing an appropriate strategy $z\in\mathbf{X}$.

Next, we can apply a special case of Glicksberg's theorem [30] to the game Γ_{a}, since the saddle point in the game Γ_{a} coincides with the Nash equilibrium in the noncooperative two-player game

$$\Gamma_2 = \langle\ \{\mathrm{I},\mathrm{II}\}, \{\mathbf{X},\mathbf{Z}=\mathbf{X}\}, \{f_{\mathrm{I}}(x,z)=\varphi(x,z),\ \ f_{\mathrm{II}}(x,z)=-\varphi(x,z)\}\ \rangle.$$

In this game, player I chooses his strategy $x\in\mathbf{X}$ to maximize $f_{\mathrm{I}}(x,z)=\varphi(x,z)$, while player II seeks to maximize $f_{\mathrm{II}}(x,z)=-\varphi(x,z)$. In the game Γ_2, the sets $\mathbf{X}$ and $\mathbf{Z}=\mathbf{X}$ are compact, while the payoff functions $f_{\mathrm{I}}(x,z)$ and $f_{\mathrm{II}}(x,z)$ are continuous on $\mathbf{X}\times\mathbf{Z}$. Therefore, by the aforementioned Glicksberg theorem, there exists a Nash equilibrium (ν^{e},μ^*) in the mixed extension of the game Γ_2, i.e.,

$$\widetilde{\Gamma}_2 = \langle\ \{\mathrm{I},\mathrm{II}\}, \{\nu\}, \{\mu\}, \{f_i(\nu,\mu)=\int\limits_{\mathbf{X}}\int\limits_{\mathbf{X}} f_i(x,z)\nu(dx)\mu(dz)\}_{i=\mathrm{I},\mathrm{II}}\ \rangle.$$

Moreover, (ν^{e},μ^*) obviously represents a saddle point in the mixed extension of the game Γ_{a},

$$\widetilde{\Gamma}_{\mathrm{a}} = \left\langle\ \{\mathrm{I},\mathrm{II}\}, \{\nu\}, \{\mu\}, \varphi(\nu,\mu)=\int\limits_{\mathbf{X}}\int\limits_{\mathbf{X}} \varphi(x,z)\nu(dx)\mu(dz)\ \right\rangle.$$

Consequently, by Glicksberg's theorem, there exists a pair (ν^{e},μ^*) representing a saddle point of $\varphi(v,\mu)$, i.e.,

$$\varphi(\nu,\mu^*)\ \leqslant\ \varphi(\nu^{\mathrm{e}},\mu^*)\ \leqslant\ \varphi(\nu^{\mathrm{e}},\mu),\qquad \forall\ \nu(\cdot),\mu(\cdot)\in\{\nu\}. \tag{5.2.25}$$

Setting $\mu=\nu^{\mathrm{e}}$ in the right-hand inequality in (5.2.25), we obtain $\varphi(\nu^{\mathrm{e}},\nu^{\mathrm{e}})=0$, and hence for all $\nu(\cdot)\in\{\nu\}$ inequalities (5.2.25) yield

$$0\ \geqslant\ \varphi(\nu,\mu^*) = \int\limits_{\mathbf{X}}\int\limits_{\mathbf{X}} \max_{j=1,\ldots,N+1}\varphi_j(x,z)\nu(dx)\mu^*(dz). \tag{5.2.26}$$

As established in [67],

$$\max_{j=1,\dots,N+1} \int_{\mathbf{X}} \int_{\mathbf{X}} \varphi_j(x,z)\nu(dx)\mu(dz)$$

$$\leqslant \int_{\mathbf{X}} \int_{\mathbf{X}} \max_{j=1,\dots,N+1} \varphi_j(x,z)\nu(dx)\mu(dz). \tag{5.2.27}$$

(This is an analog of the property that the maximum of a sum is not greater than the sum of corresponding maxima.) From (5.2.26) and (5.2.27) it follows that

$$\max_{j=1,\dots,N+1} \int_{\mathbf{X}} \int_{\mathbf{X}} \varphi_j(x,z)\nu(dx)\mu^*(dz) \leqslant 0 \quad \forall\, \nu(\cdot) \in \{\nu\},$$

but then for each $j = 1, \dots, N+1$ we surely have

$$\int_{\mathbf{X}} \int_{\mathbf{X}} \varphi_j(x,z)\nu(dx)\mu^*(dz) \leqslant 0 \quad \forall\, \nu(\cdot) \in \{\nu\}. \tag{5.2.28}$$

Recall the normalization conditions of the mixed strategies and mixed strategy profiles, namely,

$$\int_{\mathbf{X}} \nu_i(dx_i) = 1, \int_{\mathbf{X}} \mu_i(dz_i) = 1 \ (i \in \mathbb{N}), \int_{\mathbf{X}} \nu(dx) = 1, \int_{\mathbf{X}} \mu(dz) = 1, \tag{5.2.29}$$

which hold $\forall\, \nu_i(\cdot) \in \{\nu_i\}$ and $\forall\, \mu_i(\cdot) \in \{\mu_i\}$ as well as $\forall\, \nu(\cdot) \in \{\nu\}$ and $\forall\, \mu(\cdot) \in \{\mu\}$. Taking these conditions into account, we will distinguish two cases, $j \in \mathbb{N}$ and $j = N+1$, and further specify inequalities (5.2.28) for each case.

Case 1: $j \in \mathbb{N}$ Using (5.2.16) and (5.2.29) for each $i \in \mathbb{N}$, inequality (5.2.28) is reduced to

$$\int_{\mathbf{X}} \int_{\mathbf{X}} [f_i(z\|x_i) - f_i(z)]\, \nu(dx)\mu^*(dz) = \int_{\mathbf{X}} \int_{\mathbf{X}_i} [f_i(z\|x_i)$$

$$- f_i(z)]\, \nu_i(dx_i)\mu^*(dz) = \int_{\mathbf{X}} \int_{\mathbf{X}_i} f_i(z\|x_i)\nu_i(dx_i)\mu^*(dz)$$

$$- \int_{\mathbf{X}} f_i(z)\mu^*(dz) \int_{\mathbf{X}_i} \nu_i(dx_i) \overset{(5.2.29)}{=} \Bigg[\int_{\mathbf{X}_1} \cdots \int_{\mathbf{X}_{i-1}} \int_{\mathbf{X}_i} \int_{\mathbf{X}_{i+1}} \cdots \int_{\mathbf{X}_N} f_i\,(z_1,$$

$$\dots z_{i-1}, x_i, z_{i+1}, \dots z_N)\mu_N^*(dz_N)\cdots\mu_{i+1}^*(dz_{i+1})\nu_i(dx_i)\mu_{i-1}^*(dz_{i-1})$$
$$\dots\mu_1^*(dz_1)\big] - f_i(\mu^*) = f_i(\mu^*\|\nu_i) - f_i(\mu^*) \leqslant 0 \quad \forall\, \nu_i(\cdot) \in \{\nu_i\}.$$

In combination with (5.2.22), this inequality shows that $\mu^*(\cdot) \in \mathfrak{N}$, i.e., the mixed strategy profile $\mu^*(\cdot)$ is a Nash equilibrium in game (5.2.10) (Definition 5.2.5).

Case 2: $j = N+1$ Now inequality (5.2.28) takes the form

$$\int\limits_{\mathbf{X}}\int\limits_{\mathbf{X}} \varphi_{N+1}(x,z)\nu(dx)\mu^*(dz) = \int\limits_{\mathbf{X}}\int\limits_{\mathbf{X}} \sum_{i\in\mathbb{N}} f_i(x)\nu(dx)\mu^*(dz)$$
$$-\int\limits_{\mathbf{X}}\int\limits_{\mathbf{X}} \sum_{i\in\mathbb{N}} f_i(z)\nu(dx)\mu^*(dz) = \int\limits_{\mathbf{X}} \sum_{i\in\mathbb{N}} f_i(x)\nu(dx)\int\limits_{\mathbf{X}}\mu^*(dz)$$
$$-\int\limits_{\mathbf{X}} \sum_{i\in\mathbb{N}} f_i(z)\mu^*(dz)\int\limits_{\mathbf{X}}\nu(dx) = \sum_{i\in\mathbb{N}}\int\limits_{\mathbf{X}} f_i(x)\nu(dx)$$
$$-\sum_{i\in\mathbb{N}}\int\limits_{\mathbf{X}} f_i(z)\mu^*(dz) \overset{(5.2.24)}{=} \sum_{i\in\mathbb{N}} f_i(\nu) - \sum_{i\in\mathbb{N}} f_i(\mu^*) \leqslant 0 \quad \forall \nu(\cdot) \in \mathfrak{N},$$

since $\mathfrak{N} \subseteq \{\nu\}$. Hence, for $\nu^{\mathrm{P}} = \mu^*$ we directly get (5.2.24), i.e., the strategy profile $\mu^*(\cdot)$ is a Pareto-maximal alternative in the N-criteria choice problem $\widetilde{\Gamma}_{\mathrm{c}}$ (5.2.23) (Definition 5.2.2).

This result, together with the inclusion $\mu^*(\cdot) \in \mathfrak{N}$, completes the proof of Proposition 5.2.4. ■

5.2.5 *De omni re scibili et quibusdam aliis*[16]

Still, this book is devoted to the Golden Rule of ethics, mostly to the altruistic concept of Berge equilibrium. Our intention has been to start Sect. 5.2 with Nash equilibrium, thereby paying tribute to outstanding mathematician J. Nash, who was tragically killed in a car crash on May 23, 2015. However, as easily noticed, all the constructions, and lines of reasoning used in Sect. 5.2.4 can be successfully carried over to the case of Berge equilibrium. We will do this below, concluding Sect. 5.2.

[16]Latin "Of all things that can be known and all kind of other things." The first part of this phrase (de omni re scibili, meaning "of all things that can be known") was the motto of pompous young lad and famous Italian philosopher Pico della Mirandola, who thought this was a fitting description of his encyclopedic knowledge. The second part (et quibusdam aliis, meaning "and even certain other things") was ironically appended by pompous old and famous French philosopher Voltaire, who was somewhat under the impression he was any less full of himself.

To avoid repetitions, we will emphasize the moments in the proof that are dictated by the specifics of Berge equilibrium.

Again consider the N-player game (5.2.1)

$$\langle \mathbb{N}, \{\mathrm{X}_i\}_{i\in\mathbb{N}}, \mathrm{Y}, \{f_i(x, y)\}_{i\in\mathbb{N}}\rangle$$

and, using formulas (5.2.4), define the Savage–Niehans risk functions

$$R_i(x, y) = \max_{z\in\mathrm{X}} f_i(z, y) - f_i(x, y).$$

Next, by formulas (5.2.5) and (5.2.6), construct the strongly-guaranteed payoff $f_i[x]$ of player i and the corresponding guaranteed Savage–Niehans risk $R_i[x]$. As a result, we arrive at the game of guarantees

$$\Gamma^{\mathrm{g}} = \langle \mathbb{N}, \{\mathrm{X}_i\}_{i\in\mathbb{N}}, \{f_i[x], -R_i[x]\}_{i\in\mathbb{N}}\rangle.$$

Then it is natural to pass to the auxiliary game (5.2.10),

$$\langle \mathbb{N}, \{\mathrm{X}_i\}_{i\in\mathbb{N}}, \{\Phi_i[x] = f_i[x] - \sigma_i R_i[x]\}_{i\in\mathbb{N}}\rangle,$$

with a constant $\sigma_i \in (0, 1)$.

Recall that, if in the two-player game ($\mathbb{N} = \{1, 2\}$) the players exchange their payoff functions, then a Nash equilibrium in the new game is a Berge equilibrium in the original game. Therefore, all the properties intrinsic to Nash equilibria remain in force for Berge equilibria. In particular, the set of Berge equilibria is internally unstable. With this instability in mind, let us introduce an analog of Definition 5.2.3 for the auxiliary game (5.2.10). As before, $[x \| z_i] = [x_1, \ldots, x_{i-1}, z_i, x_{i+1}, \ldots, x_N]$.

Definition 5.2.8 A strategy profile $x^{\mathrm{B}} \in \mathrm{X}$ is called a Pareto–Berge equilibrium in game (5.2.10) if $x^{\mathrm{B}} = (x_1^{\mathrm{B}}, \ldots, x_N^{\mathrm{B}})$ is simultaneously

1°. a Berge equilibrium in (5.2.10), i.e.,

$$\max_{x\in\mathrm{X}} \Phi_i[x \| x_i^{\mathrm{B}}] = \Phi_i[x^{\mathrm{B}}] \quad (i \in \mathbb{N}),$$

and

2°. a Pareto-maximal alternative in the N-criteria choice problem

$$\langle \mathrm{X}^{\mathrm{B}}, \{\Phi_i[x]\}_{i\in\mathbb{N}}\rangle,$$

i.e., for any $x \in \mathrm{X}^{\mathrm{B}}$ the system of inequalities

$$\Phi_i[x] \geqslant \Phi_i[x^{\mathrm{B}}] \quad (i \in \mathbb{N}),$$

with at least one strict inequality, is inconsistent.

Denote by X^{B} the set of all $\{x^{\mathrm{B}}\}$.

Sufficient conditions for the existence of a Pareto–Berge equilibrium also involve the Germeier convolution, with the N-dimensional vectors $x = (x_1, \dots, x_N) \in \mathrm{X}$, $z = (z_1, \dots, z_N) \in \mathrm{X}$, $f = (f_1, \dots, f_N)$, $R = (R_1, \dots, R_N)$, and $\Phi = (\Phi_1, \dots, \Phi_N)$, as well as N constants $\sigma_i \in [0, 1]$ $(i \in \mathbb{N})$. Specifically, consider the $N + 1$ scalar functions

$$\begin{aligned}
&\psi_1(x, z) = \Phi_1[z_1, x_2, \dots, x_N] - \Phi_1[z],\\
&\psi_2(x, z) = \Phi_2[x_1, z_2, \dots, x_N] - \Phi_2[z],\\
&\dots\dots\dots\\
&\psi_N(x, z) = \Phi_2[x_1, x_2, \dots, z_N] - \Phi_2[z],\\
&\psi_{N+1}(x, z) = \sum_{j=1}^{N} \Phi_j[x] - \sum_{j=1}^{N} \Phi_j[z],
\end{aligned} \tag{5.2.30}$$

and their Germeier convolution

$$\psi(x, z) = \max_{j=1,\dots,N+1} \psi_j(x, z). \tag{5.2.31}$$

Proposition 5.2.5 *If there exists a saddle point $(x^0, z^{\mathrm{B}}) \in \mathrm{X} \times \mathrm{X}$ in the zero-sum game*

$$\langle \mathrm{X},\ \mathrm{Z} = \mathrm{X},\ \psi(x, z) \rangle,$$

i.e.,

$$\max_{x \in \mathrm{X}} \psi(x, z^B) = \psi(x^0, z^B) = \min_{z \in \mathrm{X}} \psi(x^0, z),$$

then the minimax strategy z^{B} is a Pareto–Berge equilibrium in game (5.2.10).

Like in Proposition 5.2.4, we may establish the existence of a Pareto–Berge equilibrium in mixed strategies.

Proposition 5.2.6 *Consider the noncooperative game* (5.2.10)*, assuming that*

1^0. *the sets $\mathbf{X}_i$ $(i \in \mathbb{N})$ and $\mathbf{Y}$ are nonempty, convex and compact in $\mathbb{R}^{n_i}$ $(i \in \mathbb{N})$;*
2^0. *the payoff functions $f_i(x, y)$ $(i \in \mathbb{N})$ are continuous on the Cartesian product $\mathbf{X} \times \mathbf{Y}$.*

Then there exists a Pareto–Berge equilibrium in mixed strategies in this game.

5.2.6 *A la fin des fins*[17]

Classical scholars believe that the whole essence of mathematical game theory is to provide comprehensive answers to the following three questions:

1. What is an appropriate optimality principle for a given game?
2. Does an optimal solution exist?
3. If yes, how can one find it?

The answer to the first question for the noncooperative N-player game (5.2.1) is the concept of Pareto–Nash equilibrium (Definition 5.2.4) or the concept of Pareto–Berge equilibrium (Definition 5.2.8).

Next, the answer to the second question is given by Propositions 5.2.4 or 5.2.6: if the sets of strategies are convex and compact and the payoff functions of the players are continuous on $\mathbf{X} \times \mathbf{Y}$, then such equilibria exist in mixed strategies.

Finally, the answer to the third question is provided by the following procedure: *first*, construct the guarantees of the outcomes $f_i[x]$ (5.2.5) and risks $R_i[x]$ (5.2.6); *second*, define the functions $\Phi_i[x] = f_i[x] - \sigma_i R_i[x]$ $(i \in \mathbb{N})$; *third*, find the Germeier convolution of the payoff functions $\varphi(x, z)$ using formulas (5.2.16) and (5.2.17) for Nash equilibrium or using formulas (5.2.30) and (5.2.31) for Berge equilibrium; *fourth*, calculate the saddle point (x^0, z^*) of this convolution; then the minimax strategy z^* is the desired Pareto–Berge (or Pareto–Nash) equilibrium.

A Suivre[18] We expect to apply this approach in the game-theoretic problems of Markowitz's portfolio theory.

5.3 Cooperation in a Conflict of N Persons Under Uncertainty

> To negotiate is not to do as one likes.
> — Napoleon Bonaparte

In this section, we introduce a new principle of coalitional equilibration. The integration of individual and collective rationality (from the theory of noncooperative games without side payments) and this principle allows us to formalize the corresponding concept of coalitional equilibrium (CE) for a conflict of N persons under uncertainty. Next, we establish sufficient conditions for the existence of CE, which are reduced to saddle point design for the Germeier convolution of guaranteed payoffs. Following the above-mentioned approach of E. Borel, J. von Neumann and J. Nash, we also prove the existence of CE in

[17] French "At the end after all."

[18] French "To be continued."

the class of mixed strategies under standard assumptions of mathematical game theory (convex and compact sets of uncertainties, convex and compact strategy sets, and continuous payoff functions). At the end of Sect. 5.3, some lines of further research are outlined.

5.3.1 Introduction

The mathematical model of cooperation studied in this section of Chap. 5 is described by a cooperative N-player normal-form game under uncertainty without side payments. As before, assume that the conflicting parties know merely the range of variation (intervals) of uncertain factors, without any probabilistic characteristics. A proper consideration of uncertainties in the models of real conflicts yields more adequate results, which is confirmed by a huge number of publications (the Google Scholar database contains over 1 million research works with keywords "mathematical modeling under uncertainty"). Uncertainty often occurs due to incomplete (inaccurate) knowledge about the strategies implemented by conflicting parties: C. Shannon was used to say, "Information is the resolution of uncertainty."[19] As mentioned earlier, an economic system is often affected by sudden *exogenous* disturbances (e.g., varying numbers and ranges of supplies, product price fluctuations caused by demand-supply dynamics) as well as *endogenous* disturbances (e.g., breakdown and failure of industrial equipment, mistakes in planning and product design); the emergence of new technologies may interfere with ecological systems; disturbances in mechanical systems may be due to weather conditions (temperature, pressure, humidity, etc.). Thus, we naturally face the following question: how to account for the cooperative character of a conflict and as well as for existing uncertainties during strategy choice?

A distinctive feature of each cooperative conflict is a proper consideration of the interests of any admissible coalition—a group of players (conflicting parties) with a coordinated choice of their strategies for achieving the best possible outcomes. Our framework will proceed from several assumptions as follows.

First, if the members of a coalition agree about joint actions by negotiations, then their agreement remains in force during the entire game, i.e., agreements are compulsory.

Second, players cannot distribute any part of their payoffs to other players (i.e., the analysis will be confined to games without side payments also called games with non-transferable payoffs).

Third, the payoff of an empty coalition is zero, i.e., only active players may obtain nonzero payoffs.

[19]Claude Elwood Shannon, (1916–2001), was an American mathematician and electrical engineer who laid the theoretical foundations for digital circuits and information theory.

5.3.2 Game of Guarantees

Consider a mathematical model of a conflict described by a cooperative N-player normal-form game under uncertainty with non-transferable payoffs

$$\Gamma = \langle \mathbb{N} = \{1, \ldots, N\}, \{X_i\}_{i\in\mathbb{N}}, Y^X, \{f_i(x, y)\}_{i\in\mathbb{N}}\rangle.$$

In this game, $\mathbb{N} = \{1, \ldots, N\}$ denotes the set of players; each of the N conflicting parties chooses his *strategy* $x_i \in X_i \subset \mathbb{R}^{n_i}$ $(i \in \mathbb{N})$, thereby forming *a strategy profile* $x = (x_1, \ldots, x_N) \in X = \prod_{i\in\mathbb{N}} X_i \subset \mathbb{R}^n$ $(n = \sum_{i\in\mathbb{N}} n_i)$; regardless of their actions, *an interval uncertainty* $y \in Y \subset \mathbb{R}^m$ is present in Γ; *a payoff function* of player i, $f_i(x, y)$, is defined on all pairs $(x, y) \in X \times Y$, and its value in a specific strategy profile gives the *payoff* of player i. At a conceptual level, each player i in the game Γ seeks to choose a strategy x_i^* $(i \in \mathbb{N})$ that would *maximize* his payoff under any admissible coalition and any realization of the *uncertain factor* $y(x) : X \to Y$, $y(\cdot) \in Y^X$ (in particular, *strategic* uncertainty).

A well-known English proverb states, "Never cackle till your egg is laid." This emphasizes the crucial role of uncertainty; but taking uncertain factors into consideration makes the payoff functions $f_i(x, Y) = \bigcup_{y\in Y} f_i(x, y)$ multivalued. This inevitably complicates further analysis of the cooperative games Γ. Thus, our idea is to assess the performance of each player i in Γ using a lower guarantee $f_i[x]$ of the payoff function $f_i(x, y)$. We suggest that as such guarantees one should take $f_i[x] = \min_{y\in Y} f_i(x, y)$, $\forall y \in Y$. This formula implies $f_i[x] \leqslant f_i(x, y)$ $\forall y \in Y$, and hence the performance of player i in the game Γ can be assigned the lower bound $f_i[x]$ under any strategy profile $x \in X$. In other words, for any uncertainty $y \in Y$ the payoff function $f_i(x, y)$ cannot be smaller than $f_i[x]$. Note that the existence of a continuous scalar function $f_i[x]$ on X follows from the compactness (closedness and boundedness) of the sets X_i $(i \in \mathbb{N})$ and Y and the continuity of $f_i(x, y)$ on $X \times Y$.

5.3.3 Coalitional Equilibrium

Denote by $2^{\mathbb{N}}$ the set of all coalitions (nonempty subsets of the set $\mathbb{N}$), i.e., $2^{\mathbb{N}} = \{K | K \subseteq 2^{\mathbb{N}}\}$. For each coalition $K \in 2^{\mathbb{N}}$, let $-K$ stand for the set $\mathbb{N}\backslash K$, that is, $-K = \mathbb{N}\backslash K$; in particular, $-i = \mathbb{N}\backslash \{i\}$. Then the coalitional structure $\{K, -K\}$ is a partition of the whole player set $\mathbb{N}$. For this coalitional structure, any strategy profile $x = (x_1, \ldots, x_N)$ can be represented as $x = (x_K, x_{-K})$, where $x_K \in X_K = \prod_{j\in K} X_j$ and $x_{-K} \in X_{-K} = \prod_{j\in\mathbb{N}\backslash K} X_j$.

Recall a pair of important properties from the theory of cooperative games without side payments [121]. For a strategy profile $x^* \in X$ in the game of guarantees

$$\Gamma^g = \langle \mathbb{N}, \{X_i\}_{i\in\mathbb{N}}, \{f_i[x]\}_{i\in\mathbb{N}} = \min_{y\in Y} f_i(x, y)\rangle,$$

we say that

(a) the *individual rationality condition* (IRC) holds if

$$f_i[x^*] \geq f_i^0 = \max_{x_i \in X_i} \min_{x_{-i} \in X_{-i}} f_i[x_i, x_{-i}] = \min_{x_{-i} \in X_{-i}} f_i[x_i^0, x_{-i}] \quad (i \in \mathbb{N}),$$

where $x = (x_i, x_{-i})$ and $X_{-i} = \prod_{j \in \mathbb{N} \setminus \{i\}} X_j$; thus, using the maximin strategy x_i^0, player i obtains a payoff $f_i[x_i^0, x_{-i}] \geqslant f_i^0 \; \forall x_{-i} \in X_{-i} \; (i \in \mathbb{N})$;

(b) the *collective rationality condition* (ColRC) holds if the strategy x^* is a Pareto-maximal alternative in the N-criteria choice problem $\Gamma_c^g = \langle X, \{f_i[x]\}_{i \in \mathbb{N}} \rangle$; in other words, for all $x \in X$ the system of inequalities $f_i[x] \geqslant f_i[x^*]$ $(i \in \mathbb{N})$, with at least one strict inequality, is inconsistent. (If for any $x \in X$ we have $\sum_{i \in \mathbb{N}} f_i[x] \leqslant \sum_{i \in \mathbb{N}} f_i[x^*]$, then x^* is a Pareto-maximal alternative in the problem Γ_c^g).

By modifying the concepts of Nash and Berge equilibria [257, 258, 305], we introduce

(c) the *coalitional rationality condition* (CoalRC) for the game of guarantees Γ^g as the inequality

$$f_i[x^*] \geqslant f_i[x_K, x_{-K}^*] \quad \forall x_K \in X_K, \forall K \in 2^{\mathbb{N}} \quad (i \in \mathbb{N}).$$

Definition 5.3.1 A strategy profile $x^* \in X$ is called a coalitional equilibrium *(CE)* if it simultaneously satisfies *IRC*, *ColRC*, and *CoalRC* in the game of guarantees Γ^g.

Remark 5.3.1 In accordance with IRC, a player benefits from building coalitions with other players if the resulting payoff is not smaller than his payoff yielded by the maximin strategy. ColRC gives the largest payoff vector to a player (in the vector sense!). Finally, CoalRC makes his payoff stable against the deviations of separate players or any admissible coalitions from x^*.

5.3.4 *Sufficient Condition*

As dictated by Definition 5.3.1, a CE x^* must satisfy IRC, ColRC, and CoalRC. However, all these conditions follow from $N^2 + 1$ inequalities of the form

$$\begin{aligned} f_i[x^*] &\geqslant f_i[x_j^*, x_{-j}] \quad \forall x_{-j} \in X_{-j} \quad (i, j \in \mathbb{N}), \\ \sum_{i \in \mathbb{N}} f_i[x] &\leqslant \sum_{i \in \mathbb{N}} f_i[x^*] \quad \forall x \in X, \end{aligned} \tag{5.3.1}$$

where $x^* = (x_1^*, \ldots, x_N^*)$.

To formulate sufficient conditions for the existence of CE, we will employ the original approach from [67]. For this purpose, we introduce an N-dimensional vector $z = (z_1, \ldots, z_N) \in X$ and the function $\varphi(x, z)$ as the Germeier convolution [28] of the functions $\varphi_r(x, z)$ $(r = 1, \ldots, N + 1)$:

$$\varphi_j(x, z) = \max_{i \in \mathbb{N}}\{f_i[z_j, x_{-j}] - f_i[z]\} \quad (j \in \mathbb{N}),$$

$$\varphi_{N+1}(x, z) = \sum_{i \in \mathbb{N}} f_i[x] - \sum_{i \in \mathbb{N}} f_i[z], \tag{5.3.2}$$

$$\varphi(x, z) = \max_{r=1,\ldots,N+1} \varphi_r(x, z). \tag{5.3.3}$$

A saddle point $(x^0, z^*) \in X \times Y$ of the scalar function $\varphi(x, z)$ (5.3.2) is given by the chain of inequalities

$$\varphi(x, z^*) \leqslant \varphi(x^0, z^*) \leqslant \varphi(x^0, z) \quad \forall\, x, z \in X. \tag{5.3.4}$$

Theorem 5.3.1 *If $(x^0, z^*) \in X \times X$ is a saddle point of the function $\varphi(x, y)$, then the minimax strategy z^* is the coalitional equilibrium in the game Γ^g.*

Proof Indeed, formula (5.3.2) with $z = x^0$ gives $\varphi(x^0, x^0) = 0$. Then, using transitivity and (5.3.3),

$$[\varphi(x^0, z^*) \leq 0] \Rightarrow [\varphi(x, z^*) \leq 0 \;\; \forall x \in X],$$

and the conclusion follows by (5.3.2). ■

Remark 5.3.2 In accordance with Theorem 5.3.1, CE design is reduced to the calculation of a saddle point (x^0, z^*) for the Germeier convolution $\varphi(x, z)$ (5.3.3). Thus, we have developed a constructive method of CE design in the game Γ, which includes the following steps:

First, define the scalar function $\varphi(x, z)$ using formulas (5.3.2) and (5.3.3);

Second, find a saddle point (x^0, z^*) of the function $\varphi(x, z)$ (see the chain of inequalities (5.3.4));

Third, calculate the values $f_i[x^*]$ $(i \in \mathbb{N})$.

Then the pair $(z^*, f[z^*] = (f_1[z^*], \ldots, f_N[z^*])) \in X \times \mathbb{R}^N$ is a coalitional equilibrium in the game Γ^g: the players should apply their strategies from the profile z^*, thereby obtaining the guaranteed payoffs $f_i[z^*]$.

5.3.5 Existence of Coalitional Equilibrium in Mixed Strategies

One must be very optimistic to look for a coalitional equilibrium in the class of pure strategies, even for the two-player games. Adhering to the approach of Borel [209], von Neumann [261], Nash [257, 258] and their followers, we will establish the existence of mixed strategy CE. Let us begin with a series of auxiliary results laying the foundations of a corresponding existence theorem.

5.3.5.1 Auxiliary Results

Denote by $\operatorname{cocomp} \mathbb{R}^{n_i}$ the set of all convex and compact subsets (convex, closed and bounded sets) of the Euclidean n_i-dimensional space $\mathbb{R}^{n_i}$, and write $f_i(\cdot) \in C(\mathrm{X} \times \mathrm{Y})$ if a scalar function $f_i(x, y)$ is continuous on $\mathrm{X} \times \mathrm{Y}$.

Let us return to the noncooperative game without side payments Γ. Without special mention, we will assume that the elements of the ordered quadruple Γ satisfy the following requirements.

Condition 5.3.1

$$\mathrm{X}_i \in \operatorname{cocomp} \mathbb{R}^{n_i} \ (i \in \mathbb{N}), \quad \mathrm{Y} \in \operatorname{comp} \mathbb{R}^m, \quad f_i(\cdot) \in C(\mathrm{X} \times \mathrm{Y}). \tag{5.3.5}$$

Next, let us pass to the mixed extension of the game Γ^g, which includes mixed strategies, mixed strategy profiles and expected payoffs.

Suppose the game Γ satisfies inequalities (5.3.5); then $f_i(x, y)$ is continuous on the product $\mathrm{X} \times \mathrm{Y}$ of compact sets, where $\mathrm{X} = \prod_{i \in \mathbb{N}} \mathrm{X}_i$. For each compact set $\mathrm{X}_i \subset \mathbb{R}^{n_i}$ $(i \in \mathbb{N})$, construct the Borel σ-algebra $\mathfrak{B}(\mathrm{X}_i)$. Within the framework of mathematical game theory, a *mixed strategy* $\nu_i(\cdot)$ of player i is identified with a *probability measure on the compact set* X_i. A probability measure is a nonnegative scalar function $\nu_i(\cdot)$ defined on the Borel σ-algebra $\mathfrak{B}(\mathrm{X}_i)$ that satisfies the following two conditions:

1. $\nu_i\left(\bigcup_k Q_k^{(i)}\right) = \bigcup_k \nu_i\left(Q_k^{(i)}\right)$ for any sequence $\{Q_k^{(i)}\}_{k=1}^{\infty}$ of pairwise disjoint elements from $\mathfrak{B}(\mathrm{X}_i)$ (*countable additivity*);
2. $\nu_i(\mathrm{X}_i) = 1$ (*normalization*), which implies $\nu_i\left(Q^{(i)}\right) \leq 1$ for all $Q^{(i)} \in \mathfrak{B}(\mathrm{X}_i)$.

Denote by $\{\nu_i\}$ the set of all mixed strategies of player i $(i \in \mathbb{N})$. Construct a mixed strategy profile as the product measure

$$\nu(dx) = \nu_1(dx_1) \cdots \nu_N(dx_N),$$

and let $\{\nu\}$ be the set of all such profiles. In addition, denote by $f_i[\nu] = \int_X f_i[x]\nu(dx)$ the expected payoff of player i. Then *the mixed extension* of the game of guarantees Γ^g has the form

$$\widetilde{\Gamma}^g = \langle \mathbb{N} = \{1, \ldots, N\}, \{\nu_i\}_{i \in \mathbb{N}}, \{f_i[\nu] = \int_X f_i[x]\nu(dx)\}_{i \in \mathbb{N}} \rangle. \tag{5.3.6}$$

Similarly to Definition 5.3.1, introduce

Definition 5.3.2 A mixed strategy profile $\nu^*(\cdot) \in \{\nu\}$ is called a coalitional equilibrium (CE) in the mixed extension (5.3.6) (equivalently, a coalitional equilibrium (CE) in mixed strategies in the game $\widetilde{\Gamma}^{\mathrm{g}}$) if

1. the profile $\nu^*(\cdot)$ is coalitionally rational in game (5.3.6), i.e.,

$$f_i[\nu_1, \ldots, \nu_j^*, \ldots, \nu_N] \leqslant f_i[\nu^*] \;\; \forall \nu_k(\cdot) \in \{\nu_k\} \;\; (k \in \mathbb{N} \setminus \{j\};\; i, j \in \mathbb{N}) \qquad (5.3.7)$$

(denote by $\{\nu^*\}$ the set of all coalitional equilibria in game (5.3.6));

2. $\nu^*(\cdot)$ is a Pareto-maximal alternative in the N-criteria choice problem

$$\widetilde{\Gamma}_{\mathrm{c}}^{\mathrm{g}} = \langle \{\nu^*\}, \{f_i[\nu]\}_{i \in \mathbb{N}} \rangle,$$

i.e., for all $\nu(\cdot) \in \{\nu^*\}$ the system of inequalities

$$f_i[\nu] \geqslant f_i[\nu^*] \quad (i \in \mathbb{N}),$$

with at least one strict inequality, is inconsistent.

An obvious sufficient condition for Pareto maximality is provided in the next remark.

Remark 5.3.3 A mixed strategy profile $\nu^*(\cdot) \in \{\nu^*\}$ is a Pareto-maximal alternative in $\widetilde{\Gamma}_{\mathrm{c}}^{\mathrm{g}} = \langle \{\nu^*\}, \{f_i[\nu]\}_{i \in \mathbb{N}} \rangle$ if

$$\max_{\nu(\cdot) \in \{\nu^*\}} \sum_{i \in \mathbb{N}} f_i[\nu] = \sum_{i \in \mathbb{N}} f_i[\nu^*].$$

Proposition 5.3.1 *In the game Γ^{g} with sets $\mathrm{X}_i \in \operatorname{cocomp} \mathbb{R}^{n_i}$ and $f_i[\cdot] \in C(X)$ $(i \in \mathbb{N})$, the function*

$$\varphi(x, z) = \max_{r=1,\ldots,N+1} \varphi_r(x, z) \qquad (5.3.8)$$

satisfies the inequality

$$\max_{r=1,\ldots,N+1} \int\limits_{\mathrm{X} \times \mathrm{X}} \varphi_r(x, z) \mu(dx) \nu(dz) \leqslant \int\limits_{\mathrm{X} \times \mathrm{X}} \max_{r=1,\ldots,N+1} \varphi_r(x, z) \mu(dx) \nu(dz) \qquad (5.3.9)$$

for any $\mu(\cdot) \in \{\nu\}$ and $\nu(\cdot) \in \{\nu\}$; recall that the scalar functions $\varphi_r(x, z)$ are defined by (5.3.2).

Inequality (5.3.9) was proved in [67].

Remark 5.3.4 In fact, formula (5.3.9) generalizes the well-known property of maximization: the maximum of a sum does not exceed the sum of the individual maxima.

Proposition 5.3.2 *If in the game Γ^{g} conditions (5.3.5) hold, then the function $\varphi(x, z)$ (5.3.8) is continuous on* $\mathrm{X} \times (\mathrm{Z} = \mathrm{X})$.

The proof of a more general result (stating that the maximum of a finite number of continuous functions is continuous) can be found in many textbooks on operations research, e.g., in [136, p. 54, 187].

5.3.5.2 Existence Theorem

The central result of this section—the existence of a coalitional equilibrium in mixed strategies in the game Γ^{g}—is established under conditions (5.3.5).

Theorem 5.3.2 *If in the game Γ^{g} the sets $\mathrm{X}_i \in \operatorname{cocomp} \mathbb{R}^{n_i}$ and $f_i[\cdot] \in C(\mathrm{X})$ $(i \in \mathbb{N})$, then there exists a coalitional equilibrium in mixed strategies in this game.*

Proof Consider the following auxiliary zero-sum two-player game:

$$\Gamma^{\mathrm{a}} = \langle \{1, 2\}, \{\mathrm{X}, \mathrm{Z} = \mathrm{X}\}, \varphi(x, z) \rangle.$$

In the game Γ^{a}, the set X of strategies x chosen by player 1 (which seeks to maximize $\varphi(x, z)$) coincides with the set of strategy profiles of the game Γ^{g}. A solution of the game Γ^{a} is *a saddle point* $(x^0, z^*) \in \mathrm{X} \times \mathrm{X}$; for all $x \in X$ and each $z \in \mathrm{X}$, it satisfies the chain of inequalities

$$\varphi(x, z^*) \leqslant \varphi(x^0, z^*) \leqslant \varphi(x^0, z).$$

Now, associate with the game Γ^{a} its mixed extension

$$\widetilde{\Gamma}^{\mathrm{a}} = \langle \{1, 2\}, \{\mu\}, \{\nu\}, \varphi(\mu, \nu) \rangle,$$

where $\{\nu\}$ and $\{\mu\} = \{\nu\}$ denote the sets of mixed strategies $\nu(\cdot)$ and $\mu(\cdot)$ of players 1 and 2, respectively. The payoff function of player 1 is the expectation

$$\varphi(\mu, \nu) = \int\limits_{\mathrm{X} \times \mathrm{X}} \varphi(x, z) \mu(dx) \nu(dz). \tag{5.3.10}$$

The solution of the game $\widetilde{\Gamma}^{\mathrm{a}}$ is also *a saddle point* (μ^0, ν^*) defined by the two inequalities

$$\varphi(\mu, \nu^*) \leqslant \varphi(\mu^0, \nu^*) \leqslant \varphi(\mu^0, \nu) \tag{5.3.11}$$

for any $\nu(\cdot) \in \{\nu\}$ and $\mu(\cdot) \in \{\nu\}$.

Sometimes, this pair (μ^0, ν^*) is called the *solution of the game* Γ^{a} *in mixed strategies.*

In 1952, I. Gliksberg [30] established the existence of a mixed strategy Nash equilibrium for a noncooperative game of $N \geqslant 2$ players. Applying his theorem to the zero-sum two-player game Γ^{a} as a special case, we obtain the following result. In the game Γ^{a}, let the set $\mathrm{X} \subset \mathbb{R}^n$ be nonempty, convex and compact and let the payoff function $\varphi(x, z)$ of player 1 be continuous on $\mathrm{X} \times \mathrm{X}$ (note that the continuity of $\varphi(x, z)$ is assumed in Proposition 5.3.2). Then the game Γ^{a} has a solution (μ^0, ν^*) defined by (5.3.11), i.e., there exists a saddle point in mixed strategies in this game.

In view of (5.3.10), inequalities (5.3.11) can be written as

$$\begin{aligned}
&\int\limits_{\mathrm{X}\times\mathrm{X}} \max_{r=1,\dots,N+1} \varphi_r(x, z)\mu(dx)\nu^*(dz) \\
&\quad \leqslant \int\limits_{\mathrm{X}\times\mathrm{X}} \max_{r=1,\dots,N+1} \varphi_r(x, z)\mu^0(dx)\nu^*(dz) \\
&\quad \leqslant \int\limits_{\mathrm{X}\times\mathrm{X}} \max_{r=1,\dots,N+1} \varphi_r(x, z)\mu^0(dx)\nu(dz)
\end{aligned}$$

for all $\nu(\cdot) \in \{\nu\}$ and $\mu(\cdot) \in \{\nu\}$. Using the measure $\nu_i(dz_i) = \mu_i^0(dx_i)$ $(i \in \mathbb{N})$ (and hence $\nu(dz) = \mu^0(dx)$) in the expression

$$\varphi(\mu^0, \nu) = \int\limits_{\mathrm{X}\times\mathrm{X}} \max_{r=1,\dots,N+1} \varphi_r(x, z)\mu^0(dx)\nu(dz),$$

we obtain $\varphi(\mu^0, \mu^0) = 0$ on the strength of (5.3.11). Similarly, $\varphi(\nu^*, \nu^*) = 0$, and then it follows from (5.3.11) that

$$\varphi(\mu^0, \nu^*) = 0. \tag{5.3.12}$$

The condition $\varphi(\mu^0, \nu^*) = 0$ and the chain of inequalities (5.3.11) by transitivity give

$$\varphi(\mu, \nu^*) = \int\limits_{\mathrm{X}\times\mathrm{X}} \max_{r=1,\dots,N+1} \varphi_r(x, z)\mu(dx)\nu^*(dz) \leqslant 0 \;\; \forall \mu(\cdot) \in \{\nu\}.$$

In accordance with Proposition 5.3.1, we then have

$$0 \geqslant \int\limits_{\mathrm{X}\times\mathrm{X}} \max_{r=1,\dots,N+1} \varphi_r(x, z)\mu(dx)\nu^*(dz) \geqslant \max_{r=1,\dots,N+1} \int\limits_{\mathrm{X}\times\mathrm{X}} \varphi_r(x, z)\mu(dx)\nu^*(dz).$$

Therefore, for all $r = 1, \dots, N+1$,

$$\int\limits_{X\times X} \varphi_r(x,z)\mu(dx)\nu^*(dz) \leqslant 0 \;\; \forall \mu(\cdot) \in \{\nu\}.$$

Consider the following two cases.

Case I ($r = 1, \dots, N$) Here, by (5.3.2) and the normalization of $\nu(\cdot)$, for $r = 1$ we get

$$\begin{aligned}
0 &\geqslant \int\limits_{X\times X} \varphi_1(x,z)\mu(dx)\nu^*(dz) \\
&= \int\limits_{X\times X} \max_{i\in\mathbb{N}} \{f_i[z_1, x_2, \dots, x_N] - f_i[z]\}\,\mu(dx)\nu^*(dz) \\
&\geqslant \int\limits_{X\times X} f_i[z_1, x_2, \dots, x_N]\mu(dx)\nu^*(dz) - \int\limits_X f_i[z]\nu^*(dz) \int\limits_X \mu(dx) \\
&= f_i[\nu_1^*, \mu_2, \dots, \mu_N] - f_i[\nu^*] \;\; (i \in \mathbb{N}).
\end{aligned}$$

For $r = 2, \dots, N$ and $i \in \mathbb{N}$, the inequalities

$$0 \geqslant f_i[\mu_1, \nu_2^*, \mu_3, \dots, \mu_N] - f_i[\nu^*],$$

$$\dots$$

$$0 \geqslant f_i[\mu_1, \dots, \mu_{N-1}, \nu_N^*] - f_i[\nu^*]$$

are proved in the same way.

By Definition 5.3.2, $\nu^*(\cdot)$ is a coalitionally-rational profile in mixed strategies in the game Γ^g.

Case II ($r = N+1$) Again, using (5.3.2) and the normalization of $\nu(\cdot)$ and $\mu(\cdot)$, we have that

$$\begin{aligned}
0 &\geqslant \int\limits_{X\times X} \left[\sum_{i\in\mathbb{N}} f_i[x] - \sum_{i\in\mathbb{N}} f_i[z]\right] \mu(dx)\nu^*(dz) \\
&= \int\limits_X \sum_{i\in\mathbb{N}} f_i[x]\mu(dx) \int\limits_X \nu^*(dz) - \int\limits_X \mu(dx) \int\limits_X \sum_{i\in\mathbb{N}} f_i[z]\nu^*(dz) \\
&= \sum_{i\in\mathbb{N}} f_i[\mu] - \sum_{i\in\mathbb{N}} f_i[\nu^*].
\end{aligned}$$

In accordance with Remark 5.3.3, the mixed strategy profile $\nu^*(\cdot) \in \{\nu\}$ in the game Γ^{g} is a Pareto-maximal alternative in the N-criteria choice problem

$$\widetilde{\Gamma}_{\mathrm{c}}^{\mathrm{g}} = \langle \{\nu\}, \{f_i[\nu]\}_{i \in \mathbb{N}} \rangle.$$

Thus, we have proved that the mixed strategy profile $\nu^*(\cdot)$ in the game $\widetilde{\Gamma}^{\mathrm{g}}$ is a coalitionally-rational profile in mixed strategies that satisfies Pareto maximality. Hence, by Definition 5.3.2, the pair $(\nu^*, f[\nu^*])$ is a coalitional equilibrium in mixed strategies in the game $\widetilde{\Gamma}^{\mathrm{g}}$. ■

5.3.6 Concluding Remarks

Let us summarize the new results on cooperative games derived in this section of Chap. 5.

First, we have formalized the concept of coalitional equilibrium (CE) considering the interests of any coalition in a cooperative N-player game.

Second, we have developed a constructive method of CE design that is based on the calculation of the minimax strategy for a special Germeier convolution associated with the guaranteed payoffs of the players.

Third, we have proved the existence of CE in mixed strategies under standard assumptions of mathematical game theory (continuous payoff functions, compact strategy sets, and compact uncertainties).

In our view, an important role is also played by qualitative properties that follow directly from the above analysis.

1. The CE $x^* \in X$ is stable against the deviations of any admissible coalitions: by modifying their strategies, the players of a coalition either worsen their guaranteed payoffs, or obtain the same guaranteed payoffs as before;
2. CE is applicable even if the coalitional structures evolve during the cooperative game (in particular, if all coalitions remain invariable);
3. CE can be used to build stable unions (alliances) of players;

these are by far not all advantages of CE!

As a matter of fact, there exists another merit worth mentioning.

The individual and collective rationality conditions have always been in the focus of researchers who study cooperative games. But the individual interests of players fit well the concept of Nash equilibrium with its intrinsic selfish character ("to each his own"), while the collective interests of players fit the concept of Berge equilibrium with its altruism ("help everybody, sometimes ignoring one's own interests"). However, such a behavior goes against the human nature of the players. This negative feature of both concepts is eliminated by coalitional rationality.

5.4 How Can One Combine the Altruism of Berge Equilibrium with the Selfishness of Nash Equilibrium? Hybrid Equilibrium

Love thy neighbour as thyself.[20]

Chacun est artisan de sa bonne fortune.[21]

The answer to this question will be given below. In short, these features can be combined, but in the class of mixed strategies. For a noncooperative N-player normal-form game, we introduce the concept of hybrid equilibrium (HE) by combining the concepts of Nash and Berge equilibria and Pareto maximum. Some properties of hybrid equilibria are explored and their existence in mixed strategies is established under standard assumptions of mathematical game theory (convex and compact strategy sets and continuous payoff functions). Similar results are obtained for noncooperative N-player normal-form games under uncertainty.

5.4.1 Introduction

As repeatedly mentioned throughout the book, in 1949 twenty-one years old Princeton University postgraduate J. F. Nash suggested and proved the existence of a solution [257, 258], which subsequently became known as Nash equilibrium (NE). Nash equilibrium has been widely used in economics, military science, policy and sociology. After 45 years, J. Nash together with R. Selten and J. Harsanyi were awarded the Nobel Prize in Economic Sciences "for their pioneering analysis of equilibria in the theory of non-cooperative games." The point is that NE has stability against arbitrary unilateral deviations of a single player, which explains its success in economic and political applications [199–201].

Almost every issue of modern journals on operations research, systems analysis, or game theory contains papers involving the concept of Nash equilibrium. However, there are spots on the sun: an obvious drawback of NE is its pronounced selfishness, as each player seeks *to increase his own payoff only* [203, 206, 207, 210].

The antipode of NE is the concept of Berge equilibrium (BE): each player makes every effort to maximize the payoffs of the other players, neglecting his individual interests. BE was formalized in 1985 by Zhukovskiy [290] as a possible solution of noncooperative N-player games, after a critical analysis of C. Berge's book *Théorie générale des jeux a n personnes* [202] published in 1957 (which explains the term "Berge equilibrium"). In 1995, Russian mathematician K. Vaisman defended his

[20] *The Old Testament*, Leviticus 19:9–18.

[21] French "Every one is the architect of his own fortune."

Candidate of Sciences Dissertation entitled "Berge equilibrium" [11] at Department of Applied Mathematics and Control Processes (St. Petersburg State University) under the scientific supervision of Zhukovskiy. This dissertation and Vaisman's early papers [12, 13, 280, 281] attracted the attention of researchers, first in Russia and then abroad. As of today, the number of publications related to this equilibrium has exceeded three hundreds, including this book. BE is a good mathematical model for the Golden Rule of ethics ("Behave to others as you would like them to behave to you."). BE is famed for its altruism.

Obviously, these features—selfishness and altruism—are intrinsic (in some proportion) to any individual, including a conflicting party. However, it seems delusive to expect that such a combined solution exists in pure strategies. Therefore, again employing the approach of Borel [209], von Neumann [261], Nash [251] and their followers, we will establish the existence of a combined Nash–Berge equilibrium in mixed strategies. This solution is called a hybrid equilibrium (HE). The main goal of Sect. 5.4 is to prove the existence of HE in mixed strategies. Also note a negative property of NE [57] and BE (see Chap. 2 of this book): the sets of both types of equilibria are internally unstable, i.e., there may exist two (NE or BE) profiles such that the payoff of each player in one of them is strictly greater than in the other. We will remove this undesirable negative feature by adding the Pareto maximality of HE with respect to all other equilibria. Thus, our formalization combines three properties, namely, a HE is

first, a Nash equilibrium [240–250];

second, a Berge equilibrium [228–237];

third, Pareto-maximal with respect to the other equilibria [211, 212, 224].

This section of Chap. 5 proves the following result: if a noncooperative N-player normal-form game has bounded convex and closed strategy sets of players and continuous payoff functions, then there exists a HE in mixed strategies in this game.

In addition, we obtain sufficient conditions for the existence of HE that are reduced to calculation of a saddle point for a special Germeier convolution of payoff functions.

Finally, the derived results are extended to the case of noncooperative N-player normal-form games under strategic uncertainty. A proper consideration of uncertain factors yields more adequate models of real conflicts, which is testified by numerous publications in this field (recall the over 1 million research works with keywords "mathematical modeling under uncertainty" in Google Scholar) [214–222].

5.4.2 *Formalization of Hybrid Equilibrium*

Consider the mathematical model of a conflict as a noncooperative N-player normal-form game described by an ordered triplet

$$\Gamma = \langle \mathbb{N}, \{\mathrm{X}_i\}_{i\in\mathbb{N}}, \{f_i(x)\}_{i\in\mathbb{N}}\rangle.$$

Here $\mathbb{N} = \{1, 2, \dots, N\}$ denotes the set of players ($N > 1$); each of N players chooses his *strategy* $x_i \in \mathrm{X}_i \subseteq \mathbb{R}^{n_i}$, thereby forming a *strategy profile*

$$x = (x_1, \dots, x_N) \in \mathrm{X} = \prod_{i \in \mathbb{N}} \mathrm{X}_i \subseteq \mathbb{R}^n \quad (n = \sum_{i \in \mathbb{N}} n_i)$$

in this game; a *payoff function* $f_i(x)$ is defined on the set X, which gives the *payoff* of player i ($i \in \mathbb{N}$). At a conceptual level, each player i in the game Γ is looking for a strategy x_i that would *maximize* his payoff.

A natural approach is to define a solution of the game Γ using a pair

$$(x^*, f(x^*) = f_1(x^*), \dots, f_N(x^*)) \in \mathrm{X} \times \mathbb{R}^N,$$

where the strategies of a profile $x^* = (x_1^*, \dots, x_N^*) \in \mathrm{X}_1 \times \dots \times \mathrm{X}_N = X$ are determined by an optimality principle while the components of the vector $f(x^*)$ specify the corresponding payoffs of players under these strategies. As noted by N. Vorobiev, the founder of the largest national scientific school on game theory, "...the practice of games shows that all the optimality principles developed so far directly or indirectly reflect the idea of a stable strategy profile that satisfies these principles..." [22, p. 94]. To introduce the concept of hybrid equilibrium, we will adopt three optimality principles, namely, Nash equilibrium, Berge equilibrium (from the theory of noncooperative games) and Pareto maximum (PM, from the theory of multicriteria choice problems). Interestingly, each of these principles has its own *type of stability*: NE is stable against the unilateral deviations of any player i (i.e., the deviations of x_i from x_i^*); BE is stable against the deviations of all players except for one player i with the payoff function $f_i(x)$ (i.e., the deviations of $(x_1, \dots, x_{i-1}, x_{i+1}, \dots, x_N)$ from $(x_1^*, \dots, x_{i-1}^*, x_{i+1}^*, \dots, x_N^*)$); finally PM is stable against the deviations of all players (i.e., the deviation of the whole current profile x from the optimal solution x^*). Using the standard notation $(x \| z_i) = (x_1, \dots, x_{i-1}, z_i, x_{i+1}, \dots, x_N)$ of noncooperative games, we introduce the following notions.

Definition 5.4.2.1 A strategy profile $x^{\mathrm{e}} = (x_1^{\mathrm{e}}, \dots, x_i^{\mathrm{e}}, \dots, x_N^{\mathrm{e}}) \in \mathrm{X}$ is called a Nash equilibrium in the game Γ if

$$\max_{x_i \in \mathrm{X}_i} f_i(x^{\mathrm{e}} \| x_i) = f_i(x^{\mathrm{e}}) \quad (i \in \mathbb{N}). \tag{5.4.2.1}$$

Definition 5.4.2.2 A strategy profile $x^{\mathrm{B}} = (x_1^{\mathrm{B}}, \dots, x_i^{\mathrm{B}}, \dots, x_N^{\mathrm{B}}) \in \mathrm{X}$ is called a Berge equilibrium in the game Γ if

$$\max_{x \in \mathrm{X}} f_i(x \| x_i^{\mathrm{B}}) = f_i(x^{\mathrm{B}}) \quad (i \in \mathbb{N}). \tag{5.4.2.2}$$

Let us associate with the game Γ the N-criteria choice problem

$$\Gamma_c = \langle \mathrm{X},\ f(x) \rangle,$$

where the set of alternatives X coincides with the set of strategy profiles X in the game Γ and the vector criterion has the form $f(x) = (f_1(x), \ldots, f_N(x))$, consisting of the payoff functions $f_i(x)$ of all players $i \in \mathbb{N}$ in the game Γ.

Definition 5.4.2.3 An alternative (here a strategy profile $x \in \mathrm{X}$) is Slater (Pareto)-maximal in the problem Γ_c if, for all $x \in \mathrm{X}$, the system of inequalities $f_i(x) > f_i(x^*)$ $(i \in \mathbb{N})$ $(f_i(x) \geqslant f_i(x^P)$ $(i \in \mathbb{N})$, respectively), with at least one strict inequality, is inconsistent.

Corollary 5.4.2.1 *The following sufficient condition of Pareto maximality is obvious: if*

$$\max_{x\in\mathrm{X}} \sum_{i\in\mathbb{N}} f_i(x) = \sum_{i\in\mathbb{N}} f_i(x^*) \quad \forall x \in \mathrm{X}, \tag{5.4.2.3}$$

then the strategy profile x^* is Pareto-maximal in the problem Γ_c.

Now, we introduce the central concept of Sect. 5.4.

Definition 5.4.2.4 A pair $(x^*, f(x^*)) \in \mathrm{X} \times \mathbb{R}^N$ is called a Pareto hybrid equilibrium (PHE) in the game Γ if the strategy profile x^* is simultaneously a Nash equilibrium and a Berge equilibrium in this game, and also a Pareto-maximal alternative in the multicriteria choice problem Γ_c, i.e., the PHE x^* satisfies the following three conditions:

$$\begin{aligned} &\max_{x_i\in\mathrm{X}_i} f_i(x^*\|x_i) = f_i(x^*) \ \ (i \in \mathbb{N}),\\ &\max_{x\in\mathrm{X}} f_i(x\|x_i^*) = f_i(x^*) \ \ (i \in \mathbb{N}),\\ &x^* \text{ is Pareto-maximal in } \Gamma_c. \end{aligned} \tag{5.4.2.4}$$

Remark 5.4.2.1 By Corollary 5.4.2.1, a strategy profile x^* is a PHE in the game Γ if it simultaneously satisfies the three optimality conditions (5.4.2.1)–(5.4.2.3).

Remark 5.4.2.2 By analogy with Definition 5.4.2.4, we may easily introduce the concept of Slater hybrid equilibrium (SHE), by simply replacing the Pareto maximality of x^* with its Slater maximality in the problem Γ_c.

5.4.3 *Properties of Hybrid Equilibria*

Hereinafter, $\operatorname{cocomp}\mathbb{R}^n$ stands for the set of convex and compact subsets of $\mathbb{R}^n$ and we write $\varphi(\cdot) \in C(\mathrm{X})$ if $\varphi(\cdot)$ is a continuous scalar function defined on X.

In this section, the game Γ is assumed to satisfy the conditions

$$\mathrm{X}_i \in \operatorname{cocomp} \mathbb{R}^{n_i}, \quad f_i(\cdot) \in C(\mathrm{X}) \quad (i \in \mathbb{N}). \tag{5.4.3.1}$$

Property 5.4.3.1 Under conditions (5.4.3.1), any PHE in the game Γ is simultaneously a SHE; the set of all SHE is compact in $\mathrm{X} \times \mathbb{R}^N$ (possibly, empty).

Property 5.4.3.1 directly follows from the fact that a Pareto-maximal alternative in the choice problem Γ_c is also Slater-maximal (in general, the converse is not true), while the set of Slater-maximal alternatives X^S in Γ_c is nonempty and compact in X [152, p. 142].

The sets of Nash and Berge equilibria, X^e and X^B, in the game Γ are also compact in X (perhaps, empty) if assumptions (5.4.3.1) hold. In this case, the intersection of the three compact sets $\mathrm{X}^\mathrm{S} \bigcap \mathrm{X}^\mathrm{e} \bigcap \mathrm{X}^\mathrm{B} = \mathrm{X}^*$ is also a compact set in X (again, it may be empty). The compactness of $f(\mathrm{X}^*) = \{f(x) | x \in \mathrm{X}^*\}$ is an immediate consequence of the continuity of the payoff functions $f_i(x)$ on X $(i \in \mathbb{N})$.

Note that, generally speaking, the set of PHE can be noncompact due to the noncompactness of the set of all Pareto-maximal alternatives X^P in the choice problem Γ_c. Also keep in mind the inclusion $f(\mathrm{X}^\mathrm{P}) \subseteq f(\mathrm{X}^\mathrm{S})$.

Property 5.4.3.2 Under assumptions (5.4.3.1), the PHE x^* satisfies the *individual rationality condition*, i.e.,

$$\begin{aligned} f_i(x^*) &\geqslant \max_{x_i \in \mathrm{X}_i} \min_{x_{\mathbb{N}\setminus\{i\}} \in \mathrm{X}_{\mathbb{N}\setminus\{i\}}} f_i(x_i, x_{\mathbb{N}\setminus\{i\}}) \\ &= \min_{x_{\mathbb{N}\setminus\{i\}} \in \mathrm{X}_{\mathbb{N}\setminus\{i\}}} f_i(x_i^0, x_{\mathbb{N}\setminus\{i\}}) = f_i^0 \quad (i \in \mathbb{N}), \end{aligned} \tag{5.4.3.2}$$

where $x = (x_1, \ldots, x_i, \ldots, x_N) = (x_i, x_{\mathbb{N}\setminus\{i\}})$, $x_{\mathbb{N}\setminus\{i\}} = (x_1, \ldots, x_{i-1}, x_{i+1}, \ldots, x_N)$ and $\mathrm{X}_{\mathbb{N}\setminus\{i\}} = \prod_{j \in \mathbb{N}\setminus\{i\}} \mathrm{X}_j (\mathbb{N} \setminus \{i\} = 1, \ldots, i-1, i+1, \ldots, N)$.

Indeed, each Nash equilibrium x^* in the game Γ has property (5.4.3.2) (individual rationality), i.e., $f_i(x^*) \geqslant f_i^0$ $(i \in \mathbb{N})$, where x_i^0 and f_i^0 are the maximin strategy and the payoff of player i, respectively.

Remark 5.4.3.1 As illustrated by Vaisman's counter-example [56, pp. 68–69], individual rationality generally fails for a Berge equilibrium x^B in the game Γ.

Property 5.4.3.3 A PHE x^* is collectively rational in a cooperative N-player game without side payments. This is a consequence of the Pareto maximality of the alternative x^* in the choice problem Γ_c.

Remark 5.4.3.2 Individual rationality imposes certain requirements to alliances (coalitions) with other players: player i joins a coalition only if his payoff guaranteed by the coalition is not smaller than the maximin value f_i^0, which can be achieved by this player independently using the maximin strategy x_i^0.

Collective rationality drives all players to the largest payoffs (in the vector sense!)—the Pareto maxima.

As x^* is a Nash equilibrium, each player seeks to maximize his payoff.

Berge equilibrium matches an altruistic aspiration of each player to maximize the payoffs of all other players.

Let us note that, the first two requirements (individual and collective rationality) are among the standard criteria of "good" solutions for cooperative N-player games without side payments. At the same time, the properties brought by the Nash and Berge equilibria are new for such games, which (we believe) makes the novel concept of PHE an efficient, "good" solution for the game Γ.

5.4.4 Sufficient Conditions

To formulate sufficient conditions for the existence of PHE in the game Γ, we will ensure Pareto maximality in terms of Definition 5.4.2.3 by satisfying equality (5.4.2.3). The sufficient conditions will be based on the original approach from [67]. Let us introduce an N-dimensional vector $z = (z_1, \dots, z_N) \in \mathrm{X}$ and the Germeier convolution [27, 28] of the form

$$\begin{aligned}\varphi_i(x, z) &= f_i(z \| x_i) - f_i(z) \quad (i \in \mathbb{N}),\\ \varphi_{i+N}(x, z) &= f_i(x \| z_i) - f_i(z) \quad (i \in \mathbb{N}),\\ \varphi_{2N+1}(x, z) &= \sum_{j \in \mathbb{N}} f_j(x) - \sum_{j \in \mathbb{N}} f_j(z),\\ \psi(x, z) &= \max_{r=1,\dots,2N+1} \varphi_r(x, z). \end{aligned} \tag{5.4.4.1}$$

A saddle point $(x^0, z^*) \in \mathrm{X} \times \mathrm{X}$ of the scalar function $\psi(x, z)$ (5.4.4.1) is given by the chain of inequalities

$$\psi(x, z^*) \leqslant \psi(x^0, z^*) \leqslant \psi(x^0, z) \ \ \forall x \in \mathrm{X}, z \in \mathrm{X}. \tag{5.4.4.2}$$

Theorem 5.4.4.1 *If (x^0, z^*) is a saddle point of the function $\varphi(x, y)$ (5.4.4.2) in the zero-sum two-player game*

$$\Gamma_{\mathrm{a}} = \langle \mathrm{X}, \mathrm{Z} = \mathrm{X}, \psi(x, z) \rangle,$$

then the maximin strategy $z^* \in \mathrm{X}$ is a PHE *of the game Γ.*

Proof Indeed, formula (5.4.4.1) with $z = x^0$ gives $\psi(x^0, x^0) = 0$. Then, by transitivity,

$$\psi(x, z^*) \leqslant 0 \ \ \forall x \in \mathrm{X}.$$

Using the fact that $\max_{r=1,\dots,2N+1} \varphi_r(x, z^*) \leqslant 0 \; \forall x \in \mathrm{X}$ and (5.4.4.1), we arrive at a set of $2N + 1$ inequalities of the form

$$f_i(z^* \| x_i) \leqslant f_i(z^*) \;\; \forall x_i \in \mathrm{X}_i \;\; (i \in \mathbb{N}),$$
$$f_i(x \| z_i^*) \leqslant f_i(z^*) \;\; \forall x \in \mathrm{X} \;\; (i \in \mathbb{N}),$$
$$\sum_{j \in \mathbb{N}} f_j(x) \leqslant \sum_{j \in \mathbb{N}} f_j(z^*) \;\; \forall x \in \mathrm{X}.$$

Here the first N inequalities make $z^* \in \mathrm{X}$ a Nash equilibrium in the game Γ (see (5.4.2.1)); the second group of inequalities ensures that z^* is a Berge equilibrium as dictated by (5.4.2.2); finally, the last, $(2N + 1)$th inequality means that z^* is a Pareto-maximal alternative in the choice problem Γ_c. ■

Remark 5.4.4.1 By Theorem 5.4.4.1, the construction of a PHE reduces to the calculation of a saddle point (x^0, z^*) for the Germeier convolution $\psi(x, z)$ (5.4.4.1). Thus, we have developed *a constructive method* of PHE design in the game Γ, which consists of the following steps:

first, define the scalar function $\psi(x, z)$ using formulas (5.4.4.1);

second, find a saddle point (x^0, z^*) of the function $\psi(x, z)$ (see the chain of inequalities (5.4.4.2));

third, calculate the values $f_i(z^*)$ $(i \in \mathbb{N})$.

Then the pair $(z^*, f(z^*) = (f_1(z^*), \dots, f_N(z^*)))$ is a PHE in the game Γ: each player $i \in \mathbb{N}$ should apply his strategy from the profile z^*, thereby obtaining the payoff $f_i(z^*)$.

Remark 5.4.4.2 The whole complexity of constructing a PHE in the game Γ lies in calculation of the saddle point (x^0, z^*) (5.4.4.2) for the Germeier convolution $\psi(x, z) = \max_{r=1,\dots,2N+1} \varphi_r(x, z)$ (5.4.4.1). The reason is that the maximization of a finite number of functions $\varphi_r(x, z)$ $(r = 1, \dots, 2N+1)$ spoils the differentiability and concavity (or convexity) of the functions $\varphi_r(x, z)$, despite the fact that it preserves the continuity of this function on the product $\mathrm{X} \times \mathrm{Z}$ of the compact sets X and Z; see [136, p. 54]. Here we face a situation well described by C. Hermite: "I turn with terror and horror from this lamentable scourge of continuous functions with no derivatives" (see footnote no. 29 in p. 63 of this book). Thus, it is necessary to develop numerical calculation methods for the saddle point (x^0, z^*) of the Germeier convolution $\max_{r=1,\dots,2N+1} \varphi_r(x, z)$. Unfortunately, to this date we were not able to find any literature devoted to this field of research. In particular, the saddle point calculation problem was not solved at the International Conference on Constructive Nonsmooth Analysis and Related Topics (CNSA-2017, St. Petersburg, May 22–27, 2017) dedicated to the Memory of Professor V. Demyanov.

5.4.5 Existence of Pareto Hybrid Equilibrium in Mixed Strategies

One must be a rather optimistic person to look for a game Γ (especially with an explicit form of the payoff function) in which a PHE in pure strategies $x_i^* \in \mathrm{X}_i$ $(i \in \mathbb{N})$ exists (by Definition 5.4.2.4, the desired strategy profile x^* must be simultaneously a Nash equilibrium and a Berge equilibrium in the game Γ and also a Pareto-maximal alternative in the corresponding choice problem). Thus, employing the approach of Borel [208], von Neumann [261], Nash [257] and their followers, we will extend the set X_i of pure strategies x_i to a set of mixed strategies. Then we will establish the existence of appropriately formalized mixed strategy profiles in the game Γ that satisfy the three requirements of hybrid equilibrium.

As before, $\operatorname{cocomp}\mathbb{R}^{n_i}$ stands for the set of all convex and compact (closed and bounded) subsets of the Euclidean n_i-dimensional space $\mathbb{R}^{n_i}$ while $f_i(\cdot) \in C(\mathrm{X})$ means that the scalar function $f_i(x)$ is continuous on X.

Consider again the noncooperative N-player game Γ without side payments. Without special mention, assume that the elements of the ordered triplet Γ satisfy requirements (5.4.3.1), i.e.,

$$\mathrm{X}_i \in \operatorname{cocomp}\mathbb{R}^{n_i}, \quad f_i(\cdot) \in C(\mathrm{X}) \quad (i \in \mathbb{N}).$$

For each compact set $\mathrm{X}_i \subset \mathbb{R}^{n_i}$ $(i \in \mathbb{N})$, consider the Borel σ-algebra $\mathfrak{B}(\mathrm{X}_i)$. Further, consider the Borel σ-algebra $\mathfrak{B}(\mathrm{X})$ for the set $\mathrm{X} = \prod_{i\in\mathbb{N}} \mathrm{X}_i$ of all strategy profiles, such that $\mathfrak{B}(\mathrm{X})$ contains all Cartesian products of elements from the Borel σ-algebras $\mathfrak{B}(\mathrm{X}_i)$ $(i \in \mathbb{N})$.

Within the framework of mathematical game theory, *a mixed strategy* $\nu_i(\cdot)$ of player i is identified with a *probability measure on the compact set* X_i. By definition [122, p. 271], in the notations of [108, p. 284] a probability measure is a *nonnegative* scalar function $\nu_i(\cdot)$ defined on the Borel σ-algebra $\mathfrak{B}(\mathrm{X}_i)$ that satisfies the following two conditions:

1. $\nu_i\left(\bigcup_k Q_k^{(i)}\right) = \bigcup_k \nu_i\left(Q_k^{(i)}\right)$ for any sequence $\{Q_k^{(i)}\}_{k=1}^{\infty}$ of pairwise disjoint elements from $\mathfrak{B}(\mathrm{X}_i)$ (*countable additivity*);
2. $\nu_i(\mathrm{X}_i) = 1$ (*normalization*), which implies $\nu_i\left(Q^{(i)}\right) \le 1$ for all $Q^{(i)} \in \mathfrak{B}(\mathrm{X}_i)$.

Denote by $\{\nu_i\}$ the set of all mixed strategies of player i $(i \in \mathbb{N})$.

The product measures $\nu(dx) = \nu_1(dx_1)\cdots\nu_N(dx_N)$, treated in the sense of the well-known definitions from [122, p. 370] (and in the notations of [108, p. 123]), are probability measures on the strategy profile set X. Let $\{\nu\}$ be the set of such probability measures (strategy profiles). Once again, we emphasize that in the construction of the product measure $\nu(dx)$, the role of the σ-algebra of all subsets of the set $\mathrm{X}_1 \times \cdots \times \mathrm{X}_N = \mathrm{X}$ is played by the *smallest* σ-algebra $\mathcal{B}(\mathrm{X})$ that contains all Cartesian products $Q^{(1)} \times \cdots \times Q^{(N)}$, where $Q^{(i)} \in \mathcal{B}(\mathrm{X}_i)$ $(i \in \mathbb{N})$. The well-known properties of probability measures [41, p. 288], [122, p. 254] imply that the sets of all possible measures $\nu_i(dx_i)$ $(i \in \mathbb{N})$ and $\nu(dx)$ are *weakly closed and weakly compact* (see [122, pp. 212, 254], and [180, pp. 48, 49]). As applied, e.g., to $\{\nu\}$, this means that from any infinite sequence $\{\nu^{(k)}\}$ $(k = 1, 2, \ldots)$ one can extract a subsequence $\{\nu^{(k_j)}\}$ $(j = 1, 2, \ldots)$ which *weakly converges* to a measure $\nu^{(0)}(\cdot) \in \{\nu\}$. In other words, for any continuous scalar function $\psi(x)$ on X,

$$\lim_{j\to\infty} \int_{\mathrm{X}} \psi(x)\nu^{(k_j)}(dx) = \int_{\mathrm{X}} \psi(x)\nu^{(0)}(dx)$$

and $\nu^{(0)}(\cdot) \in \{\nu\}$. Due to the continuity of $\psi(x)$, the integrals $\int_{\mathrm{X}} \psi(x)\nu(dx)$ (the expectations) are well defined; by Fubini's theorem,

$$\int_{\mathrm{X}} \varphi(x)\nu(dx) = \int_{\mathrm{X}_1} \cdots \int_{\mathrm{X}_N} \varphi(x)\nu_N(dx_N)\cdots\nu_1(dx_1),$$

and the order of integration can be interchanged.

Let us associate with the game Γ in pure strategies its *mixed extension*

$$\widetilde{\Gamma} = \langle \mathbb{N}, \{\nu_i\}_{i\in\mathbb{N}}, \{f_i[\nu] = \int_{\mathrm{X}} f[x]\nu(dx)\}_{i\in\mathbb{N}}\rangle, \tag{5.4.5.1}$$

where, like in Γ, $\mathbb{N}$ is the set of players while $\{\nu_i\}$ is the set of mixed strategies $\nu_i(\cdot)$ of player i; in game (5.4.5.1), each conflicting party $i \in \mathbb{N}$ chooses its mixed strategy $\nu_i(\cdot) \in \{\nu_i\}$, thereby forming a mixed strategy profile $\nu(\cdot) \in \{\nu\}$; the payoff function of each player i, i.e., the expectation

$$f_i[\nu] = \int_{\mathrm{X}} f_i[x]\nu(dx),$$

is defined on the set $\{\nu\}$.

For game (5.4.5.1), the notion of a PHE x^* (see Definition 5.4.2.4) has the following analog.

Definition 5.4.5.1 A mixed strategy profile $\nu^*(\cdot) \in \{\nu\}$ is called a hybrid equilibrium *(HE)* in the mixed extension 5.4.5.1 (equivalently, a hybrid equilibrium in mixed strategies in the game Γ) if

1. $\nu^*(\cdot)$ is a Nash equilibrium in the game $\widetilde{\Gamma}$, i.e.,

$$\max_{\nu_i(\cdot)\in\{\nu_i\}} f_i(\nu^* \| \nu_i) = f_i(\nu^*) \quad (i \in \mathbb{N}); \tag{5.4.5.2}$$

2. $\nu^*(\cdot)$ is a Berge equilibrium in game (5.4.5.1), i.e.,

$$\max_{\nu_{\mathbb{N}\setminus\{i\}}(\cdot)\in\{\nu_{\mathbb{N}\setminus\{i\}}\}} f_i(\nu\|\nu_i^*) = f_i(\nu^*) \quad (i \in \mathbb{N}); \tag{5.4.5.3}$$

3. $\nu^*(\cdot)$ is a Pareto-maximal alternative in the N-criteria choice problem

$$\widetilde{\Gamma}_c = \langle\{\nu\}, \{f_i(\nu)\}_{i\in\mathbb{N}}\rangle,$$

i.e., for all $\nu(\cdot) \in \{\nu\}$, the system of inequalities

$$f_i(\nu) \geqslant f_i(\nu^*) \quad (i \in \mathbb{N}),$$

with at least one strict inequality, is inconsistent.

Here and in the sequel,

$$\begin{aligned}\nu_{\mathbb{N}\setminus\{i\}}(dx_{\mathbb{N}\setminus\{i\}}) &= \nu_1(dx_1)\cdots\nu_{i-1}(dx_{i-1})\nu_{i+1}(dx_{i+1})\cdots\nu_N(dx_N),\\ (\nu\|\nu_i^*) &= \nu_1(dx_1)\cdots\nu_{i-1}(dx_{i-1})\nu_i^*(dx_i)\nu_{i+1}(dx_{i+1})\cdots\nu_N(dx_N),\\ \nu^*(dx) &= \nu_1^*(dx_1)\cdots\nu_N^*(dx_N);\end{aligned}$$

in addition, denote by $\{\nu^*\}$ the set of hybrid equilibria $\nu^*(\cdot)$, i.e., the set of strategy profiles that satisfy the three requirements of Definition 5.4.5.1.

Let us state several results used below for proving the existence of HE in mixed strategies. The following sufficient condition of Pareto maximality is obvious.

Proposition 5.4.5.1 *A mixed strategy profile $\nu^*(\cdot) \in \{\nu\}$ is a Pareto-maximal alternative in the choice problem $\Gamma_c = \langle\{\nu\}, \{f_i(\nu)\}_{i\in\mathbb{N}}\rangle$ if*

$$\max_{\nu(\cdot)\in\{\nu\}} \sum_{i\in\mathbb{N}} f_i(\nu) = \sum_{i\in\mathbb{N}} f_i(\nu^*). \tag{5.4.5.4}$$

Proposition 5.4.5.2 *Consider the game Γ under conditions* (5.4.3.1), *i.e., the sets X_i are convex and compact and the payoff functions $f_i(x)$ are continuous on $X = X_1 \times \cdots \times X_N$. Let*

$\{\nu^e\}$ be the set of Nash equilibria $\nu^e(\cdot)$ that satisfy (5.4.5.2) *with $\nu^*(\cdot)$ replaced by $\nu^e(\cdot)$;*

$\{\nu^B\}$ be the set of Berge equilibria $\nu^B(\cdot)$ that satisfy (5.4.5.3) *with $\nu^*(\cdot)$ replaced by $\nu^B(\cdot)$;*

$\{\nu^P\}$ be the set of alternatives $\nu^P(\cdot)$ that satisfy (5.4.5.4) *with $\nu^*(\cdot)$ replaced by $\nu^P(\cdot)$ (i.e., ν^P is a Pareto-maximal alternative in mixed strategies in the N-criteria choice problem $\langle\{\nu\}, \{f_i(\nu)\}_{i\in\mathbb{N}}\rangle$).*

Then the set $\{\nu^*\}$ of hybrid equilibria $\nu^*(\cdot)$ in the mixed extension $\widetilde{\Gamma}$ of the game Γ is a weakly compact subset of the set of mixed strategy profiles $\{\nu\}$ in the game $\Gamma\{\nu^*\}$ (may be empty).

Proof Under conditions (5.4.3.1), we have $\{\nu^e\} \neq \emptyset$ as shown by Gliksberg's theorem [30]. Next, the fact $\{\nu^B\} \neq \emptyset$ has been established in the preceding sections of our book. The non-emptiness of the set of Pareto-maximal alternatives, $\{\nu^P\} \neq \emptyset$, can be proved in analogous manner. The intersection of a finite number of weakly compact sets (in our case, three) is also weakly compact, possibly empty. ■

Corollary 5.4.5.1 *Under conditions* (5.4.3.1), *the set*

$$f(\{\nu^*\}) = \bigcup_{\nu(\cdot)\in\{\nu^*\}} f(\nu), \quad f = (f_1, \dots, f_N),$$

is compact (*bounded and closed*) *in the* N*-dimensional Euclidean criterion space* $\mathbb{R}^N$.

Theorem 5.4.5.1 below establishes the implication (5.4.3.1)$\Rightarrow \{\nu^*\} \neq \emptyset$, which is the central result of Sect. 5.4.

Proposition 5.4.5.3 *Consider game (5.4.5.1) under conditions (5.4.3.1). Then the function* $\varphi_r(x, z)$ *in the formula*

$$\psi(x, z) = \max_{r=1,\dots,2N,2N+1} \varphi_r(x, z) \tag{5.4.5.5}$$

satisfies the inequality

$$\begin{aligned} &\max_{r=1,\dots,2N,2N+1} \int_{X\times X} \varphi_r(x, z)\mu(dx)\nu(dz) \\ &\leqslant \int_{X\times X} \max_{r=1,\dots,2N,2N+1} \varphi_r(x, z)\mu(dx)\nu(dz) \end{aligned} \tag{5.4.5.6}$$

for any $\mu(\cdot) \in \{\nu\}$ *and* $\nu(\cdot) \in \{\nu\}$, *where*

$$\begin{aligned} \varphi_i(x, z) &= f_i(x\|z_i) - f_i(z) \ \ (i \in \mathbb{N}), \\ \varphi_j(x, z) &= f_j(z\|x_i) - f_j(z) \ \ (j \in \{N+1, \dots, 2N\}), \\ \varphi_{2N+1}(x, z) &= \sum_{i\in\mathbb{N}} [f_i(x) - f_i(z)]. \end{aligned} \tag{5.4.5.7}$$

This proposition was proved in [57].

Remark 5.4.5.1 In fact, formula (5.4.5.6) generalizes the well-known property of maximization: the maximum of a sum does not exceed the sum of the maxima.

Let us state an interesting fact from operations research, which plays a crucial role in the proof of Theorem 5.4.5.1. Consider $2N + 1$ scalar functions $\varphi_r(x, z)$

($r = 1, \ldots, 2N, 2N+1$), where $z = (z_1, \ldots, z_N) \in \mathrm{Z} = \mathrm{X}$ and $\varphi_j(x, z)$ ($j = 1, \ldots, 2N+1$) are defined by (5.4.5.7).

Proposition 5.4.5.4 *If $2N+1$ scalar functions $\varphi_j(x, z)$ ($j = 1, \ldots, 2N+1$) are continuous on the product $\mathrm{X} \times (\mathrm{Z} = \mathrm{X})$ of compact sets, then the function*

$$\psi(x, z) = \max_{j=1,\ldots,2N+1} \varphi_j(x, z)$$

is also continuous on $\mathrm{X} \times \mathrm{Z}$.

The proof of a more general result can be found in many textbooks on operations research, e.g., [136, p. 54] and [46].

Finally, let us establish the central result of Sect. 5.4—the existence of a hybrid equilibrium (HE) in mixed strategies under conditions (5.4.3.1).

Theorem 5.4.5.1 *If in the game Γ the sets $\mathrm{X}_i \in \operatorname{cocomp} \mathbb{R}^{n_i}$ and $f_i(\cdot) \in C(\mathrm{X})$ ($i \in \mathbb{N}$), then there exists a hybrid equilibrium in mixed strategies in this game.*

Proof. Consider an auxiliary zero-sum two-player game

$$\Gamma^{\mathrm{a}} = \langle \{1, 2\}, \{\mathrm{X}, \mathrm{Z} = \mathrm{X}\}, \psi(x, z) \rangle.$$

In the game Γ^{a}, the set X of strategies x chosen by player 1 (seeking to maximize $\psi(x, z)$) coincides with the set of strategy profiles of the game Γ; the set Z of strategies z chosen by player 2 (seeking to minimize $\psi(x, z)$) coincides with X. A solution of the game Γ^{a} is *a saddle point* $(x^0, z^{\mathrm{B}}) \in \mathrm{X} \times \mathrm{X}$; for all $x \in \mathrm{X}$ and each $z \in \mathrm{X}$, it satisfies the chain of inequalities

$$\psi(x, z^{\mathrm{B}}) \leqslant \psi(x^0, z^{B}) \leqslant \psi(x^0, z).$$

Now, associate with the game Γ^{a} its mixed extension

$$\widetilde{\Gamma}^{\mathrm{a}} = \langle \{1, 2\}, \{\mu\}, \{\nu\}, \psi(\mu, \nu) \rangle,$$

where $\{\nu\}$ and $\{\mu\} = \{\nu\}$ denote the sets of mixed strategies $\nu(\cdot)$ and $\mu(\cdot)$ of players 1 and 2, respectively. The payoff function of player 1 is the expectation

$$\psi(\mu, \nu) = \int_{X \times X} \psi(x, z)\mu(dx)\nu(dz).$$

The solution of the game $\widetilde{\Gamma}^{\mathrm{a}}$ is also a *saddle point* (μ^0, ν^*) defined by the two inequalities

$$\psi(\mu, \nu^*) \leqslant \psi(\mu^0, \nu^*) \leqslant \psi(\mu^0, \nu), \tag{5.4.5.8}$$

for any $\nu(\cdot) \in \{\nu\}$ and $\mu(\cdot) \in \{\nu\}$.

Sometimes, the pair (μ^0, ν^*) is also called the *solution of the game* Γ^a *in mixed strategies.*

Applying Gliksberg's [30] existence theorem of a mixed strategy Nash equilibrium for a noncooperative game of $N \geqslant 2$ players to the zero-sum two-player game Γ^a, we obtain the following result. In the game Γ^a, suppose the set $\mathrm{X} \subset \mathbb{R}^n$ is nonempty, convex and compact and the payoff function $\psi(x, z)$ of player 1 is continuous on $\mathrm{X} \times \mathrm{X}$ (note that the continuity of $\psi(x, z)$ is assumed in Proposition 5.4.5.4). Then the game Γ^a has a solution (μ^0, ν^*) defined by (5.4.5.8), i.e., there exists a saddle point in mixed strategies in this game.

Using (5.4.5.5), inequalities (5.4.5.8) can be written as

$$\int\limits_{X\times X} \max_{j=1,\dots,2N+1} \varphi_j(x, z)\mu(dx)\nu^*(dz)$$
$$\leqslant \int\limits_{X\times X} \max_{j=1,\dots,2N+1} \varphi_j(x, z)\mu^0(dx)\nu^*(dz)$$
$$\leqslant \int\limits_{X\times X} \max_{j=1,\dots,2N+1} \varphi_j(x, z)\mu^0(dx)\nu(dz) \tag{5.4.5.9}$$

for all $\nu(\cdot) \in \{\nu\}$ and $\mu(\cdot) \in \{\nu\}$. Using the measure $\nu_i(dz_i) = \mu_i^0(dx_i)$ $(i \in \mathbb{N})$ (and hence $\nu(dz) = \mu^0(dx)$) in the expression

$$\psi(\mu^0, \nu) = \int\limits_{\mathrm{X}\times\mathrm{X}} \max_{j=1,\dots,2N+1} \varphi_j(x, z)\mu^0(dx)\nu(dz),$$

we obtain $\psi(\mu^0, \mu^0) = 0$ due to (5.4.5.5). Similarly, $\psi(\nu^*, \nu^*) = 0$, and then it follows from (5.4.5.8) that

$$\psi(\mu^0, \nu^*) = 0.$$

The condition $\psi(\mu^0, \mu^0) = 0$ and the chain of inequalities (5.4.5.8) by transitivity give

$$\psi(\mu, \nu^*) = \int\limits_{\mathrm{X}\times\mathrm{X}} \max_{j=1,\dots,2N+1} \varphi_j(x, z)\mu(dx)\nu^*(dz) \leqslant 0 \quad \forall \mu(\cdot) \in \{\nu\}.$$

By Proposition 5.4.5.3, we then have

$$0 \geqslant \int\limits_{\mathrm{X}\times\mathrm{X}} \max_{j=1,\dots,2N+1} \varphi_j(x, z)\mu(dx)\nu^*(dz)$$
$$\geqslant \max_{j=1,\dots,2N+1} \int\limits_{\mathrm{X}\times\mathrm{X}} \varphi_j(x, z)\mu(dx)\nu^*(dz).$$

Therefore, for all $j = 1, \dots, 2N + 1$,

$$\int\limits_{X \times X} \varphi_j(x, z)\mu(dx)\nu^*(dz) \leqslant 0 \ \forall \mu(\cdot) \in \{\nu\}. \tag{5.4.5.10}$$

Consider three cases as follows.

Case I ($j = N, \dots, 2N$) Here, by (5.4.5.10), (5.4.5.7) and the normalization of $\mu(\cdot)$, we obtain

$$0 \geqslant \int\limits_{X \times X} \varphi_{N+i}(x, z)\mu^0(dx)\nu(dz) = \int\limits_{X \times X} [f_i(z\|x_i) - f_i(z)]\mu^0(dx)\nu(dz)$$
$$= \int\limits_{X \times X} f_i(z\|x_i)\mu^0(dx)\nu(dz) - \int\limits_X f_i(z)\mu^0(dx)\int\limits_X \nu(dz)$$
$$= f_i(\mu^0\|\nu_i) - f_i(\mu^0) \ \forall \nu(\cdot) \in \{\nu\} \ (i \in \mathbb{N}).$$

By (5.4.5.2), $\mu^0(\cdot)$ is a Nash equilibrium in the game $\widetilde{\Gamma}$ (equivalently, a Nash equilibrium in mixed strategies in the game Γ).

Case II ($j = 1, \dots, N$) Again, using (5.4.5.10), (5.4.5.7) and the normalization of $\nu(\cdot)$,

$$0 \geqslant \int\limits_{X \times X} \varphi_i(x, z)\mu(dx)\nu^*(dz) = \int\limits_{X \times X} [f_i(x\|z_i) - f_i(z)]\mu(dx)\nu^*(dz)$$
$$= \int\limits_{X \times X_i} f_i(x\|z_i)\mu(dx)\nu_i^*(dz) - \int\limits_X f_i(z)\mu(dz)\int\limits_X \nu^*(dz)$$
$$= f_i(\mu\|\nu_i^*) - f_i(\nu^*) \ \forall \mu(\cdot) \in \{\nu\} \ (i \in \mathbb{N}).$$

In view of (5.4.5.3), the mixed strategy profile $\nu^*(\cdot)$ is a Berge equilibrium in the game Γ, by Definition 5.4.5.1.

Case III ($j = 2N + 1$) Again, using (5.4.5.10), (5.4.5.7) and the normalization of $\nu(\cdot)$ and $\mu(\cdot)$, we have

$$0 \geqslant \int\limits_{X \times X} \left[\sum_{r \in \mathbb{N}} f_r(x) - \sum_{r \in \mathbb{N}} f_r(z)\right] \mu(dx)\nu^*(dz)$$
$$= \int\limits_X \sum_{r \in \mathbb{N}} f_r(x)\mu(dx) \int_X \nu^*(dz) - \int\limits_X \mu(dx) \int\limits_X \sum_{r \in \mathbb{N}} f_r(z)\nu^*(dz)$$
$$= \sum_{r \in \mathbb{N}} f_r(\mu) - \sum_{r \in \mathbb{N}} f_r(\nu^*) \ \forall \mu(\cdot) \in \{\nu\}.$$

By Proposition 5.4.5.1 and (5.4.5.4), the mixed strategy profile $\nu^*(\cdot) \in \{\nu\}$ of the game Γ (2.3.1) is a Pareto-maximal alternative in the multicriteria choice problem

$$\widetilde{\Gamma}_c = \langle \{\nu\}, \{f_i(\nu)\}_{i \in \mathbb{N}} \rangle.$$

Thus, we have proved that the mixed strategy profile $\nu^*(\cdot)$ in the game Γ is simultaneously a Nash equilibrium and a Berge equilibrium that satisfies Pareto maximality. Hence, by Definition 5.4.5.1, the mixed strategy profile $\nu^*(\cdot)$ is a hybrid equilibrium in the game Γ. ■

5.4.6 Hybrid Equilibrium in Games Under Uncertainty

Let us augment the mathematical model of a conflict

$$\Gamma = \langle \mathbb{N}, \{\mathrm{X}_i\}_{i \in \mathbb{N}}, \{f_i(x)\}_{i \in \mathbb{N}} \rangle$$

by including the influence of uncertain factors $y \in \mathrm{Y}$. Assume that these factors take arbitrary values from given ranges without any probability characteristics (e.g., the distribution of y on Y is absent for some reasons). Once again, we emphasize that a proper consideration of uncertainties gives a more adequate description of the decision-making process in economics, ecology, sociology, management, trade, policy, security, and so on. Uncertain factors occur due to incomplete (inaccurate) knowledge about the realizations of strategies chosen by conflicting parties. "There is no such uncertainty as a sure thing." (R. Burns).[22] For example, an economic system is subject to almost unpredictable *exogenous disturbances* (forces of nature, disruption of supplies, low qualification or incompetence of economic partners, counteractions of rivals, to name a few) as well as *endogenous disturbances* (breakdown and failure of industrial equipment, unplanned additional cost and losses of materials, innovations suggested by employees, etc.). New technologies and also anthropogenic and weather changes may cause uncertainty in ecological systems; in mechanical systems, among the sources of uncertainty are weather conditions. "The only thing that makes life possible is permanent, intolerable uncertainty; not knowing what comes next." (Ursula K. Le Guin).[23] Possible approaches to take the effect of uncertain factors into account were the subject of investigations [73, 74] initiated in 2013, which resulted in the book [68]. In this section of Chap. 5, we will use elementary methods to deal with uncertainty.

[22] Robert Burns, (1759–1796), was a national poet of Scotland, who wrote lyrics and songs in Scots and in English.

[23] Ursula K. Le Guin, original name Ursula Kroeber, (1929–2018), was an American writer best known for tales of science fiction and fantasy.

Consider a noncooperative N-player normal form game under uncertainty

$$\langle \mathbb{N}, \{\mathrm{X}_i\}_{i\in\mathbb{N}}, \mathrm{Y}, \{f_i(x, y)\}_{i\in\mathbb{N}}\rangle. \tag{5.4.6.1}$$

Compared with the game Γ (which shares the first two components of its ordered triplet with game (5.4.6.1), namely, $\mathbb{N} = \{1, 2, \ldots, N\}$ and the set X_i of pure strategies x_i of player $i, i \in \mathbb{N}$), in this game we have an additional set $\mathrm{Y} \subset \mathbb{R}^m$ of uncertain factors y and payoff functions $f_i(x, y)$ that depend on y.

Game (5.4.6.1) runs as follows. Each player $i \in \mathbb{N}$ chooses his individual strategy $x_i \in \mathrm{X}_i \subset \mathbb{R}^{n_i}$ $(i \in \mathbb{N})$, which gives a strategy profile $x = (x_1, \ldots, x_N) \in \in \mathrm{X} = \prod_{j\in\mathbb{N}} \mathrm{X}_j \subset \mathbb{R}^n$ $(n = \sum_{j\in\mathbb{N}} n_j)$ in this game. Regardless of their choice, an arbitrary uncertainty $y \in \mathrm{Y}$ figures in (5.4.6.1). For each player i $(i \in \mathbb{N})$, a payoff function $f_i(x, y)$ is defined on all such pairs $(x, y) \in \mathrm{X} \times \mathrm{Y}$. At a conceptual level, each player i seeks to maximize his *payoff* $f_i(x, y)$ under any unpredictable realization of the uncertainty $y \in \mathrm{Y}$. This last requirement calls for estimating the set

$$f_i(x, Y) = \bigcup_{y\in\mathrm{Y}} f_i(x, y)$$

for each player i $(i \in \mathbb{N})$. In turn, for such a multivalued function $f_i(x, \mathrm{Y})$ $(i \in \mathbb{N})$, it is necessary to choose another function $f_i[x]$ that would act as *a guarantee* for any element $f_i(x, y)$ from the set $f_i(x, \mathrm{Y})$. As defined by the Merriam–Webster dictionary, **guarantee is an assurance for the fulfillment of a condition.** A most obvious guarantee for player i in game (5.4.6.1) is the so-called *strong guarantee* [73], provided by the scalar function

$$f_i[x] = \min_{y\in\mathrm{Y}} f_i(x, y). \tag{5.4.6.2}$$

Indeed, it follows from (5.4.6.2) that, for each strategy profile $x \in \mathrm{X}$,

$$f_i[x] \leqslant f_i(x, y) \;\; \forall y \in \mathrm{Y},$$

i.e., in each strategy profile $x \in \mathrm{X}$ the value $f_i(x, y)$ is not smaller than the guarantee $f_i[x]$ under any realization of the uncertainty $y \in \mathrm{Y}$. Recall an important result from operations research that is repeatedly used in this book.

Proposition 5.4.6.1 ([136, p. 54, 187]) *If a scalar function $F(x, y)$ is continuous on the product $\mathrm{X} \times \mathrm{Y}$ of convex and compact sets X and Y, then the function $f[x] = \min_{y\in\mathrm{Y}} F(x, y)$ is continuous on X.*

Therefore, all the N strong guarantees $f_i[x]$ (5.4.6.2) are continuous on X under the assumptions $\mathrm{X}_i \in \operatorname{comp} \mathbb{R}^{n_i}$ $(i \in \mathbb{N})$, $\mathrm{Y} \in \operatorname{comp} \mathbb{R}^m$ and $f_i(\cdot) \in C(\mathrm{X} \times \mathrm{Y})$.

This approach allows us to associate with game (5.4.6.1) under uncertainty the game of guarantees (without uncertainty)

$$\Gamma^{g} = \langle \mathbb{N}, \{X_i\}_{i\in\mathbb{N}}, \{f_i[x]\}_{i\in\mathbb{N}}\rangle, \tag{5.4.6.3}$$

which coincides with the game Γ from Sect. 5.4.2 provided that $f_i(x)$ is replaced by the strong guarantee $f_i[x] = \min_{y\in \mathrm{Y}} f_i(x, y)$.

In contrast to (5.4.6.1), here the performance of each player i is assessed using the strong guarantee $f_i[x]$ instead of the payoff function $f_i(x, y)$ itself (this seems quite natural for considering arbitrary realizations $y \in \mathrm{Y}$).

Then the following analog of Definition 5.4.2.4 can be suggested for the game under uncertainty (5.4.6.1) with the strong guarantees (5.4.6.2).

Definition 5.4.6.1 A pair $(x^{\mathrm{P}}, f[x^{\mathrm{P}}] = (f_1[x^{\mathrm{P}}], \ldots, f_N[x^{\mathrm{P}}])) \in \mathrm{X} \times \mathbb{R}^N$ is called a strongly-guaranteed Pareto hybrid equilibrium in game (5.4.6.1) if

1. the strong guarantees $f_i[x]$ (5.4.6.2) are continuous on X;
2. the strategy profile x^{P} is simultaneously a Nash equilibrium and a Berge equilibrium in the game of guarantees (5.4.6.3), i.e.,

$$\max_{x_i\in \mathrm{X}_i} f_i[x^{\mathrm{P}}\|x_i] = f_i[x^{\mathrm{P}}] \quad (i \in \mathbb{N}),$$

and

$$\max_{x\in \mathrm{X}} f_i[x\|x_i^{\mathrm{P}}] = f_i[x^{\mathrm{P}}] \quad (i \in \mathbb{N}),$$

respectively;
3. the strategy profile x^{P} is a Pareto-maximal alternative in the N-criteria choice problem $\langle \mathrm{X}, \{f_i[x]\}_{i\in\mathbb{N}}\rangle$.

Similarly to Definition 5.4.5.1, we introduce an analog of Definition 5.4.6.1 with a feature that the players use mixed strategies $\nu_i(\cdot)$ $(i \in \mathbb{N})$ in game (5.4.6.1).

Definition 5.4.6.2 A mixed strategy profile $\nu^{\mathrm{P}}(\cdot) \in \{\nu\}$ is called a strongly-guaranteed Pareto hybrid equilibrium in mixed strategies in game (5.4.6.1) if

1. for each player i $(i \in \mathbb{N})$, there exists the strong guarantee

$$f_i[x] = \min_{y\in \mathrm{Y}} f_i(x, y)$$

that is continuous on X;
2. ν^{P} is simultaneously a Nash equilibrium and a Berge equilibrium in game (5.4.5.1), i.e., equalities (5.4.5.2) and (5.4.5.3) hold with $\nu^*(\cdot)$ replaced by $\nu^{\mathrm{P}}(\cdot)$;
3. ν^{P} in game (5.4.5.1) is a Pareto-maximal alternative in the N-criteria choice problem $\widetilde{\Gamma}_c = \langle\{\nu\}, \{f_i[\nu]\}_{i\in\mathbb{N}}\rangle$.

Finally, the combination of Proposition 5.4.6.1 and Theorem 5.4.4.1 directly leads to the following result on the existence of a strongly-guaranteed Pareto hybrid equilibrium in mixed strategies.

Theorem 5.4.6.1 *Consider game (5.4.6.1) with convex and compact sets* X_i $(i \in \mathbb{N})$, *compact set* Y, *and payoff functions* $f_i(x, y)$ $(i \in \mathbb{N})$ *continuous on* $\mathrm{X}\times\mathrm{Y}$. *Then there exists a strongly-guaranteed Pareto hybrid equilibrium in mixed strategies in this game.*

Remark 5.4.6.1 Our analysis in Sect. 5.4.6 has been confined to the strong guarantees $f_i[x] = \min_{y\in\mathrm{Y}} f_i(x, y)$ $(i \in \mathbb{N})$ as the smallest ones. It is possible to adopt the so-called vector guarantees: the components of an N-dimensional vector $f[x] = (f_1[x], \ldots, f_N[x])$ form *a vector guarantee* for an N-dimensional vector $f(x, y) = (f_1(x, y), \ldots, f_N(x, y))$ if, for all $y \in \mathrm{Y}$ and each $x \in \mathrm{X}$, the N strict inequalities

$$f_i(x, y) < f_i[x] \quad (i \in \mathbb{N})$$

are inconsistent. In other words, the vector guarantee $f[x]$ cannot be reduced simultaneously in all the components by choosing $y \in \mathrm{Y}$. In terms of vector optimization, for each alternative $x \in \mathrm{X}$ the vector $f[x]$ is a Slater minimum (weakly efficient) solution in the N-criteria choice problem $\Gamma(x) = \langle \mathrm{Y}, f(x, y)\rangle$.

In the same fashion, using other concepts of vector optima (minima in the sense of Pareto, Geoffrion, Borwein, cone optimality), we may introduce a whole collection of vector guarantees. These guarantees have the remarkable property that their values, first, are not smaller than the corresponding components of the strong guarantee vector $f[x]$ (5.4.6.2) but, second, can be large. Recall that the goal is to increase the payoffs of players (in particular, by increasing their guarantees!). In this respect, the listed vector guarantees are preferable to their strong counterparts. However, one should keep in mind an important aspect: transition from the game under uncertainty (5.4.6.1) to the game of guarantees Γ^g (with subsequent application of Theorem 5.4.4.1) is possible only if the new payoff functions $f_i[x]$ $(i \in \mathbb{N})$ in the game Γ^g are continuous. This continuity can be ensured in the following way.

Let $\mathrm{X}_i \in \operatorname{comp}\mathbb{R}^{n_i}$, $\mathrm{Y} \in \operatorname{comp}\mathbb{R}^m$ and $f_i(\cdot) \in C(\mathrm{X}\times\mathrm{Y})$ $(i \in \mathbb{N})$ in game (5.4.6.1). In addition, require that for each $x \in \mathrm{X}$ at least one $f_j(x, y)$ $(j \in \mathbb{N})$ is *strictly convex* in y on the set Y. Then the minimum in

$$\min_{y\in Y} f_i(x, y) = f_j[x] \tag{5.4.6.4}$$

is achieved at a unique point $y^*(x)$ for each $x \in \mathrm{X}$, and the m-dimensional vector function $y^*(x)$ itself is continuous on X; see [136, p. 54]. In this case, the superposition of the continuous functions $f_i(x, y)$ and $y^*(x)$ implies the continuity of all scalar functions $f_i[x] = f_i(x, y^*(x))$ $(i \in \mathbb{N})$. We finalize the design of Γ^g with the following fact. Assume for each $x \in X$ the same function $f_j[x]$ is implemented by the minimum in (5.4.6.4). Then for all $x \in \mathrm{X}$ the N-dimensional vector $f[x] = (f_1[x], \ldots, f_N[x])$ is a Slater-minimal alternative in the current N-criteria choice problem $\Gamma(x) = \langle \mathrm{Y}, \{f_i(x, y)\}_{i \in \mathbb{N}} \rangle$. In other words, it is impossible to find $\bar{y} \in \mathrm{Y}$ such that $f_i(x, \bar{y}) < f_i[x]$ $(i \in \mathbb{N})$. A detailed treatment of these issues for Slater, Pareto, Geoffrion, Borwein, and cone optimality will be given in our future publications.

Chapter 6
Conclusion

Game theory is a mathematical framework for strategy analysis and design as well as for optimal decision-making under conflict and behavioral uncertainty. On the one hand, game theory plays a key role for modern economics; on the other, it suggests possible approaches and solutions for complex strategic problems in various fields of human activity.

The logical methods of optimal strategy design in mathematical terms date back to the beginning of the seventeenth century. The problems of production and pricing in oligopolies, i.e., the classical problems of game theory, were studied in the nineteenth century by Cournot [225, 226] and Bertrand [204, 205]. The idea of a game as a mathematical model for a conflict of interests appeared at the beginning of the past century in the works of Lasker, E. Zermelo, and E. Borel [209]. Pioneering results on game theory were published since the 1920s, but a systematic treatment was first presented in 1944 by J. von Neumann and O. Morgenstern in their monograph *Theory of Games and Economic Behavior* [262]. The title and content of this book indicated that game theory was claiming for a revolution in economic sciences with its novel approach. Thus, the year 1944—the first edition of the book—is generally considered as the birth of game theory.

Further development of game theory was associated with the name of American mathematician Nash [257, 258], who formulated the principles of decision dynamics. The cited monograph by von Neumann and Morgenstern became well-known mostly owing to an exploration of zero-sum games, in which a win of one party means a simultaneous loss of the other. However, equal attention in the book was paid to the games with non-opposing interests. Nash analyzed different management strategies in economics and business as well as different behavioral strategies and arrived at an important conclusion. With such strategies, one party is always gaining while the other losing, i.e., they yield victors and vanquished. Nash was wondering: is it possible to find an equilibrium in which nobody wins and also nobody loses? Such strategies would revolutionize negotiations, resolution of conflicts and design of other compromise decisions.

© Springer Nature Switzerland AG 2020

M. E. Salukvadze, V. I. Zhukovskiy, *The Berge Equilibrium: A Game-Theoretic Framework for the Golden Rule of Ethics*, Static & Dynamic Game Theory: Foundations & Applications, https://doi.org/10.1007/978-3-030-25546-6_6

Nash created analysis methods for the games in which *all* parties simultaneously win or lose. An example of such a game is wage negotiations between a labor union and an employer. This situation may result in a lengthy strike (affecting both parties) or a mutually beneficial agreement. Nash modeled a situation (the so-called Nash equilibrium or noncooperative equilibrium, as we know it today) in which both parties use optimal strategies, thereby achieving a stable equilibrium. The players are interested in keeping this equilibrium, since any unilateral deviation would worsen their condition.

Nevertheless, the concept of Nash equilibrium is selfish: it guides each player towards maximization of his/her/its *own* payoff only. The constructive criticism of this selfish approach motivated V. Zhukovskiy and his postgraduate student K. Vaisman to adequately consider the interests of all other parties of a conflict, even at the cost of neglecting the individual interests of each player. In 1994 they formulated the altruistic concept of Berge equilibrium, which is the subject of this book. Here are three English proverbs related to Berge equilibrium.

(a) It is better to give than to take.[1]
(b) Share and share alike.[2]
(c) Live and let live.[3]

Game theory has been evolving through different stages, with different levels of interest from the scientific community. In the 1950s, game-theoretic methods seemed to be very promising, but all excitement gradually faded in the 1960s–1970s, despite the considerable mathematical results established then. However, the 1980s saw an increased utilization of game-theoretic methods in different applications, and today one would hardly find any field of economics and business science (micro- and macro-economics, finance, marketing, management, etc.) which can be studied without a basic background in game theory [109–116, 128, 147, 152, 161–167, 175–188, 190, 193, 196].

In the course of development, game theory has become a general logical-mathematical theory of conflicts. Game-theoretic methods allow us to analyze different conflicts (phenomena and processes), to outline and predict the behavioral scenarios for all conflicting parties, as well as to suggest recommendations on conflict resolution and elimination of dangerous consequences.

During the two or three recent decades, the value of game theory and the interest in game-theoretic research have significantly increased in many fields of economic and social sciences. It is no exaggeration to state that game theory is vital for modern

---

[1]Originates from *The Bible*, Acts 20:35: "In all things I have shown you that by working hard in this way we must help the weak and remember the words of the Lord Jesus, how he himself said, 'It is more blessed to give than to receive.'"

[2]Give equal shares to all. Daniel Defoe appears to be the first to have used this phrase in *The Life and Strange Adventures of Robinson Crusoe* (1719): "He declar'd he had reserv'd nothing from the Men, and went Share and Share alike with them in every Bit they eat."

[3]People should accept the way other people live and behave, especially if they do things in a different way.

economics. In the present time the scientific community is devoting much research effort to extend the scope of game theory. On the one hand, this theory forms a rather abstract branch of mathematics; on the other, a rather efficient analysis tool for economic, political, legal, military, technical and other problems. Applications of game-theoretic methods are found in agriculture, medicine, ecology, sports, anthropology, psychology, pedagogy, sociology, and others.

In modern economics and business science, game theory has a wide variety of applications. Game-theoretic tools and approaches may be fruitful in situations connected with strategic decision-making, competition, cooperation, risks and uncertainty. At the macro-level, game theory is used for decision-making processes in international trade, competition, taxation, protectionist practices and cartelization (e.g., OPEC), including an assessment of contributions for each party and further allocation of profits. At the micro-level, game theory assists, e.g., in advertising cost optimization in a competitive market, efficient production organization or auction design. Using game-theoretic methods, one may choose business partners for joint projects, construct behavioral scenarios for competitors, as well as find mechanisms of interregional interactions and income allocation schemes. Game-theoretic models are widespread in planning and prediction, strategic development design, pricing, negotiations, in particular, coordination of mutual interests and relations of partners, asset owners, employees and employers, and other economic agents. Moreover, game-theoretic methods are used to analyze the behavior of criminal gangs and political struggle.

Game theory provides

– a formal and clear language to analyze different economic phenomena, processes and systems;
– possible tools to check intuitive or rational decisions and solutions in terms of their consistency and applicability to a given problem;
– principles, criteria and methods to find optimal solutions.

A classical and most remarkable example of successful application of game-theoretic methods was the Federal Communications Committee (FCC) spectrum auction held in 1994 [148]. The organizers intended to collect at least \$ 3.5 million but, with the help of game theory experts, the real revenues reached approximately \$ 8 billion [238, 252, 253, 259–265, 270–277, 279, 283–289, 306].

Nowadays, the number of publications (papers, monographs, textbooks) on game theory is into tens of thousands [149, 150]. Despite its long history, game theory has become appreciated by the scientific community only relatively recently. The pioneering research of future Nobel laureates J. Nash, R. Selten, L. Hurwitz, R. Myerson and others took place in the 1950s. Yet the first Nobel Prize in Economic Sciences for the advances in game theory was awarded in 1994, which was the first indication of wide scientific recognition. Since then, during a period of less than 15 years, the Nobel Prize in Economic Sciences was awarded seven times for game-theoretic research; in particular, in 2005 jointly to R. Aumann and T. Schelling "for having enhanced our understanding of conflict and cooperation through game-theory analysis."

As a matter of fact, an explicit polarization can be observed in the monographs and textbooks on game theory. In a considerable part of these publications, the authors give a detailed description of the mathematical framework of game theory, including solution concepts, principles and models, restricting themselves to a few abstract examples. As a result, it may seem that game theory has nothing to do with real economic problems. Such books are characterized by a high level of formal abstraction and a considerable simplification of real situations, which makes the corresponding game-theoretic models unsuitable in practice [193]. This gap between theory and practice often appears in light assumptions and conclusions, which are not accompanied by good interpretations in the context of a given problem or not reduced to specific managerial decisions or behavioral strategies in a given situation.

This state of affairs explains the existing scepticism of practitioners (economists and managers) towards game theory. Another reason of the scepticism that restricts the use of game theory is the relatively high complexity of this theory. The main complexity consists in its logic rather than its mathematical framework.

Another considerable part of the literature is focused on outlining economic situations that can be described by games, without a proper consideration of methods and tools to find solutions. Such an approach conceals the rich capabilities of game theory. As a result, practitioners have a clear idea that this theory is applicable, but do not fully comprehend how. In other words, the practical results included in the monographs and textbooks on the subject are either trivial, or very complicated [147, p. 246].

The authors of this book are far from overestimating the capabilities of game theory, which is often done by some researchers. Game-theoretic models represent a tool that should be properly handled and applied whenever possible. Like any other models, games provide a more or less adequate approximation of real situations and events. This does not mean, however, that the models cannot be efficient in practical problems. Game theory itself is neither a universal description of real life in mathematical terms, nor a universal solution procedure for all problems. For a successful application of game theory, one needs to be facilitated with its logical-mathematical framework and also with the subject under study.

Indeed, game theory and its postulates may seem rather abstract or even unsuitable. But we believe that the major application of game theory is the development of a special "strategic vision" of a current situation, often nonformalizable yet facilitating a qualitative, complete and rigorous analysis.

As a counterweight to the generally accepted selfish Nash equilibrium, this book is devoted to a new solution concept for noncooperative games—the altruistic Berge equilibrium. For over twenty years since its appearance, Berge equilibrium has been facing different troubles. *First*, the sudden death of K. Vaisman at the age of 35, who was the initiator and enthusiast of this concept; *second*, the negative review of Shubik [269] of Berge's book [202] in which the main idea of this equilibrium was described; *third*, the unclear usefulness of Berge equilibrium in real problems (where and how can it be applied?); *fourth*, the easiness of deriving theoretical results on Berge equilibrium in two-player games (for such games, Berge equilibrium design is reduced to Nash equilibrium calculation if the players exchange their payoff functions); and *fifth*, the absence of an authoritative researcher to lead this direction of investigations. These reasons (and probably others not mentioned here) suspended the elaboration of a constructive theory of Berge equilibria at the stage of accumulation of facts, revelation of properties, comparison with Nash equilibrium, and analysis mostly in the context of matrix games.

In our opinion, the forthcoming stage of development will be associated with an heuristic approach to the mathematical theory of Berge equilibrium. No doubt, at this stage it is necessary to answer the following questions of paramount importance:

1^0. How should a Berge equilibrium be constructed?
2^0. Does a Berge equilibrium exist?

Actually, these questions are directly addressed in the present book for the *static setup* of noncooperative N-player games (such games have no dynamics and are time-invariant).

One central result of the book is that the Germeier convolution is involved in answering question 1^0. More specifically, the problem is reduced to a saddle point calculation for a special Germeier convolution (Sect. 2.8.3) of the players' payoff functions, which is efficiently constructed using the original noncooperative game: the minimax strategy at this saddle point is the Berge equilibrium in the original game.

This technique has allowed us to answer question 2^0 about the existence of a Berge equilibrium. Moreover, our existence theorem takes into account the internal instability of the set of Berge equilibria (there may exist two Berge equilibria such that the players' payoffs in one equilibrium are strictly greater than in the other, see Example 2.4.1). To deal with this, the concept of Berge equilibrium has been augmented by Pareto maximality with respect to other Berge equilibria, yielding the so-called Berge–Pareto equilibrium (Definition 2.9.1). Theorem 2.9.1 establishes the existence of a Berge–Pareto equilibrium in mixed strategies for continuous payoff functions and compact strategy sets of all players.

Another central result consists in laying the theoretical foundations of Berge equilibrium design under interval strategic uncertainty, a novel direction of Berge equilibrium-related research. We suggest two decision approaches under such conditions. *First*, the formal definition of a strongly-guaranteed Berge equilibrium, which is reduced to instantaneous minimization of each payoff function and further transition to the game of guarantees. *Second*, the formal definition of Slater-minimal guarantees [82, 83] for each situation, with the same transition to the game of guarantees [33, 109–114]. This approach involves a two-level hierarchical game in which the lower level is formed by strategic uncertainties (under the information discrimination of all players). Both definitions lead to an appropriate modification of the maximin. The latter and former approaches yield existence theorems in mixed strategies, see Theorem 3.5.1 and also the end of Sect. 3.5.3.

Finally, the applications to the Cournot and Bertrand oligopoly models are described in Chap. 4, including the case of import as an uncertain factor.

Note that the material presented in Chap. 4 settles the issue regarding the use of Berge equilibrium in real problems. In addition, as explained in the Preface, the concept of Berge equilibrium completely matches the Golden Rule of ethics: "Behave to others as you would like them to behave to you."

Finally, we should emphasize that the approach adopted in Chap. 3 is not the only possible one. Even for the antagonistic case of noncooperative games, there exist other principles (minimax regret, pessimism–optimism) as well as other criteria (Laplace–Bayes, Hodges–Lehmann, BK-criterion, P-criterion [137]), each having certain advantages and shortcomings. We have not considered Berge equilibrium for differential positional games, although the existence theorem for the separate dynamical system was established earlier in [72] under an appropriately modified formalization of the players' strategies and motions generated by them. (Also see numerous publications of Zhukovskiy's scholars on dynamic programming-based Berge equilibrium design for specific multistage games arising in competitive economics). Other applications-relevant models not covered by this book include differential positional games with time delay, multistage positional setups of the games, and many more. The above-mentioned problems are waiting for thorough study, and the reader will certainly discover many interesting facts getting deeper into them. "On deep paths of mystery unknown creatures leave their spoor."[4]

[4]A fragment from *Ruslan and Lyudmila*, a poem by Aleksandr S. Pushkin, (1799–1837), a Russian poet, novelist, dramatist, and short-story writer. Considered as the greatest poet and founder of modern Russian literature.

Short Biographies

Rich as we are in biography, a well-written life is
almost as rare as a well-spent one; and there are
certainly many more men whose history deserves
to be recorded than persons willing and able to record it.
—Carlyle[1]

This section describes in brief the life and research activities of leading Russian speaking mathematicians and mechanical engineers of the past and present who are little known in the Western countries, but frequently mentioned in the text. The list includes 8 distinguished persons, namely, I. Vekua, N. Vorobiev, Yu. Germeier, N. Krasovskii, L. Pontryagin, B. Pshenichnyi, A. Subbotin, and P. Chebyshev. They are ordered alphabetically. Probably, the short biographies below have some omissions, but the authors hope that this material will be of interest for the reader.

Pafnuty Lvovich Chebyshev (born May, 26, 1821—died December 8, 1894), was a Russian mathematician and mechanical engineer, the founder of St. Petersburg Mathematical School, Assistant Academician (since 1853), Extraordinary Academician (since 1856), and Ordinary Academician (since 1859) of the St. Petersburg Academy of Sciences. Born in the village of Okatovo (Kaluga Governorate, Russian Empire). Received primary education at home. Graduated from Moscow State University (1841). During the period 1847–1882, worked at St. Petersburg State University (since 1850, as Professor). Full Member of the Artillery Division of the Military-Scientific Committee (since 1855), Member of the Scientific Council at the Ministry of Public Education (1856–1873), and Full Member of the Interim Artillery Committee (since 1859).

[1]A quote from *Critical and Miscellaneous Essays* by Thomas Carlyle, (1795–1881), a Scottish historian and essayist.

© Springer Nature Switzerland AG 2020

M. E. Salukvadze, V. I. Zhukovskiy, *The Berge Equilibrium: A Game-Theoretic Framework for the Golden Rule of Ethics*, Static & Dynamic Game Theory: Foundations & Applications, https://doi.org/10.1007/978-3-030-25546-6

His major research was devoted to mathematical analysis, polynomial approximation of functions, number theory, probability theory, theory of machines and mechanisms, theory of surfaces, variational calculus, and some other directions of mathematics and mechanics. A distinctive feature of his research activities was a close connection between theory and practice, as he repeatedly emphasized himself. Studied the integration of algebraic functions: proved the integrability of the differential binomial in the class of elementary functions; elaborated the general theory of orthogonal polynomials; posed the problem of moments and derived quadrature formulas. With the aim of reducing the amount of calculations, in 1873 suggested quadrature formulas with equal coefficients, under the additional requirement that the formulas have good accuracy for any polynomials of degree not greater than $n - 1$, where n denotes the number of nodes. That research was closely related to his work at the Interim Artillery Committee.

Considered quadratic approximations, approximations using trigonometric polynomials and rational functions. Chebyshev's approximation theory has become a major component of the constructive theory of functions. In 1849 and 1850, established important results on the distribution of prime numbers. Believed that the construction of probability theory must have a reliable mathematical foundation. Proved the central limit theorem of probability theory. In 1846, provided a new proof for the Poisson theorem as well as outlined its applicability in practice. In 1867, suggested a simple yet general proof for the law of large numbers. In 1887, demonstrated that the results of his research on the limit values of integrals can be used for proving the Laplace–Poisson theorem (on the probability with which the sum of very many independent random variables lies on a given interval). In the theory of surfaces, in 1878 elaborated theory of nets; in particular, solved the problem of mapping a plane into an arbitrary surface so that the lengths of lines are preserved. Owing to the studies of parabolic interpolation using the least-squares method, in 1867 introduced a new calculus similar to variational calculus.

Chebyshev was the founder of the mathematical theory of mechanisms. In 1854, published his work *Theory of mechanisms known as parallelograms*. Pioneered the structural analysis of plane mechanisms and established a condition for the existence of a mechanism. Constructed a series of mechanisms that execute a given law (including walking and rowing mechanisms), which have been appreciated just recently. Designed a steam engine, an adding machine, and a centrifugal regulator with good mechanical properties. Using his mechanisms for an approximate reproduction of mathematical laws, elaborated a theory of polynomial approximation of functions with smallest deviation from zero. Chebyshev and his followers were at the dawn of major research directions in Russian and Soviet mathematics.

Corresponding Member (since 1860) and Foreign Member (since 1874) of the Paris Academy of Sciences, Member of the London Royal Society (since 1877), Member of the Berlin Academy of Sciences (since 1871), Member of the Bologne Academy of Sciences (since 1873), Member of the Swedish Academy of Sciences (since 1893), as well as Honorary Member of many other academies of sciences, scientific communities and universities. In 1944, the USSR Academy of Sciences established the Chebyshev Medal and Prize for best research in mathematics and the Chebyshev Prize for best research in theory of machines and mechanisms.

Yury Borisovich Germeier (born July 18, 1918, Atkarsk, Saratov oblast—died June 24, 1975, Moscow), was a Russian mathematician. Graduated from the Faculty of Mechanics and Mathematics at Moscow State University (1941). Candidate of Sciences in Physics and Mathematics (1947). Doctor of Sciences in Physics and Mathematics (1963).

His Candidate of Sciences dissertation was entitled "The Derivatives of Riemann and de la Vallée–Poussin, and Their Application to Some Questions in the Theory of Trigonometric Series." His Doctor of Sciences dissertation was devoted to assessing the efficiency of aviation systems in combination with some optimization problems and random process problems.

Since 1968, Professor of the Department of Computational Mathematics at the Faculty of Mechanics and Mathematics (Moscow State University). In 1970–1975, the founder and first head of the Department of Operations Research at the Faculty of Mechanics and Mathematics (Moscow State University). Delivered lectures on mathematical and methodological foundations of operations research and game theory with nonantagonistic interests.

Was awarded the Order of the Red Banner of Labour (1957), the Order of the Badge of Honor (1975), the Medal for Valorous Labour during the Great Patriotic War (1946), the Medal for the 800th Anniversary of Moscow (1948), and the Medal for Valorous Labour (1970). His research interests covered the properties of generalized derivatives and their connection with the summability of trigonometric series. His results became a starting point for further research in this field, particularly for the properties of symmetric second-order derivatives. Studied the efficiency of aircraft launched torpedoes. Conducted fundamental scientific investigations on the design of a universal efficiency assessment method for air-to-air gunnery. Made considerable advances in the assessment of ordinance principles for combat aircrafts, the efficiency and reliability of aviation equipment and systems. Took an important step in reliability theory, passing from the purely probabilistic setups to the maximin ones. Suggested to compensate insufficient information about distribution laws using the principle of guaranteed result. Derived pioneering results on the worst-case distribution laws, which yielded a new interpretation for standard laws of reliability theory. A founder of the national school of operations research. Contributed much to theory of games with nonantagonistic interests. Identified an applications-relevant class of games with hierarchical structure that describes most of economic systems. Introduced and justified punishment strategies. Solved some games with incompletely known interests of cooperating players and games with forbidden situations. Jointly with N.N. Moiseev, laid the foundations of a theory of hierarchical systems. Established the instability of classical solutions in cooperative games and generalized the concept of equilibrium. Published over 150 research works, including two fundamental monographs [28, 29].

Nikolay Nikolaevich Krasovskii (born September 7, 1924—died April 4, 2012), was a Russian mathematician, Academician of the USSR Academy of Sciences (since 1968), Corresponding Member of the USSR Academy of Sciences (since 1964). Born in Sverdlovsk. Graduated from the Ural Polytechnic Institute (1949). In 1949–1959, worked at the Ural Polytechnic Institute (since 1957, as Professor). Since 1959, worked at the Ural State University. In 1970–1977, was the director of the Institute of Mathematics and Mechanics at the Ural Scientific Center of the USSR Academy of Sciences. At present, the Institute is bearing his name.

His major research was devoted to the stability of motion and dynamics of controlled systems as well as to the general qualitative theory of differential equations. Developed the method of Lyapunov functions and solved the existence problem of such functions in basic cases of stability and instability. Created some stability analysis methods for essentially nonlinear systems with large disturbances. Suggested a new functional interpretation of systems with aftereffect and, based on this interpretation, solved stabilization and control problems for such systems. Extended Lyapunov's stability theory to the stochastic systems with Markovian switching. Elaborated the stabilization theory of controlled systems. For linear systems, introduced an analysis method of programmed optimal control in the form of the functional problem of moments. Elaborated a control theory for game-theoretic dynamic problems. Proposed a new setup of differential games with the existence of a saddle point in pairs of appropriately adjusted classes of strategies (in particular, the concept of a feedback mixed strategy), and proved the existence of a saddle point in the class of such strategies. Developed efficient algorithms of optimal strategy design. Conducted research in the field of qualitative theory of ordinary differential equations and equations with delayed argument. Also made some contributions to complex analysis, variational calculus, and approximate calculations. Died on April 4, 2012. Buried in Yekaterinburg.

Lev Semenovich Pontryagin (born September 3, 1908—died May 3, 1988), was a Russian mathematician, Academician of the USSR Academy of Sciences (since 1958), Corresponding Member of the USSR Academy of Sciences (since 1939). Born in Moscow. Lost his sight at the age of 14 in an accident. Graduated from Moscow State University (1929). Was a student of P.S. Aleksandrov. Since 1930, worked at Moscow State University (since 1935, as Professor) and, since 1939, simultaneously at the Steklov Institute of Mathematics (the USSR Academy of Sciences). In 1970, established the Department of Optimal Control at the Faculty of Computational Mathematics and Cybernetics (Moscow State University) and was the head of the department till his death.

His major research was devoted to theory of differential equations, topology, theory of oscillations, control, variational calculus, and algebra. Further developed Alexander's duality law and proved (1932) it, connecting the Betti groups of an arbitrary bounded and closed set in the Euclidean space with the Betti groups of its complement. Solved the calculation problem of Betti groups. In topology and topological algebra, elaborated the theory of characters of commutative topological groups; proved theorems on the structure of rather wide types of topological groups and formulated a new direction in topological algebra; proved the theorem asserting that the only locally bicompact connected fields are the fields of real numbers, complex numbers, and quaternions. Obtained a series of results in homotopy theory (Pontryagin classes). Developed the mathematical theory of optimal processes based on Pontryagin's maximum principle. Also made considerable contributions to the theory of asymptotics of relaxation oscillations, variational calculus, dimension theory, ordinary differential equations, theory of regulation, functional analysis; established and supervised a new direction in theory of differential games of quality.

Honorary Member of the International Academy of Astronautics (since 1966), Vice-President of the International Mathematical Union (1970–1974), Honorary Member of the Hungarian Academy of Sciences (since 1972).

Hero of Socialist Labour (1969).

Laureate of the Lenin Prize (1962), USSR State Prize (1941), International Lobachevsky Prize (1966).

Boris Nikolaevich Pshenichnyi (born April 24, 1937—died October 17, 2000), was a Ukrainian mathematician. In 1959, graduated from Lvov State University with specialization in mathematics. In 1964, defended the Candidate of Sciences dissertation, and in 1969 the Doctor of Sciences dissertation. Professor (since 1974), Corresponding Member of the National Academy of Sciences of Ukraine (since 1985), and Academician of the National Academy of Sciences of Ukraine (since 1992). Till 1996, worked at the Glushkov Institute of Cybernetics (the National Academy of Sciences of Ukraine). Since 1996, was the head of the Department of Numerical Methods of Optimal Control at the Institute of Applied Systems Analysis (the National Academy of Sciences of Ukraine).

Laureate of the State Prize of the Ukrainian SSR (1978), USSR State Prize (1981), Glushkov Prize (1994), and State Prize of Ukraine in Science and Technology (1999).

Headed a scientific school of optimal control (theory and applications), differential games and convex analysis. His results on the necessary conditions of extremum provided a general framework for problems arising in theories of Chebyshev approximations, optimal control, and multivalued mappings. He formulated the maximum principle for differential inclusions and extended it to problems with operator constraints. Developed the linearization method for the numerical resolution of most constrained optimization problems. Studied models of economic dynamics, stability of solutions of differential equation, and numerical methods of nonsmooth optimization. Suggested efficient numerical methods for nonlinear programming and computational mathematics.

B. Pshenichnyi published over 170 research works, including 8 monographs (some translated into German, French and English). For many years, delivered lectures at Kiev National University; prepared over 50 Candidates of Sciences and 10 Doctors of Sciences, who are working in Ukraine and abroad. Was repeatedly invited as a visiting professor to Harvard University, University of Paris (Sorbonne), and Humboldt University of Berlin. Died on October 17, 2000, in Kiev.

Andrei Izmailovich Subbotin (born February, 16, 1945—died October 14, 1997), was a Russian mathematician. Doctor of Sciences in Physics and Mathematics (1973), Corresponding Member of the Russian Academy of Sciences (1991), and Full Member of the Russian Academy of Sciences (1997).

Born in Kirov. After graduation (1967) from the Faculty of Mathematics and Mechanics at Ural State University with specialization in mechanics, worked at the Institute of Mathematics and Mechanics (the Ural Branch of the USSR Academy of Sciences); since 1977, head of the Department of Dynamical Systems at the Institute. Since 1992, Professor at Ural State University. Delivered lectures on theory of differential games and generalized solutions of first-order partial differential equations. Scholar of Academician N.N. Krasovskii.

His major research was devoted to optimal control, positional differential games, and generalized solutions to Hamilton–Jacobi equations. Made fundamental contributions to the concept of positional differential game and proved a basic result of game theory—the alternative theorem for antagonistic games; developed analytical and constructive methods for solving differential games. Together with his followers, elaborated the theory of minimax (generalized) solutions to first-order partial differential equations: proved an existence and uniqueness theorem for the minimax solutions of the Cauchy and Dirichlet boundary-value problems for such equations, and established the well-posedness of the minimax solutions and their equivalence to the viscosity solutions introduced by M.G. Crandall and P.-L. Lions. Introduced analytical and constructive methods for this theory and its applications to dynamic optimization.

Authored over 100 research works, including five monographs. Prepared 12 Candidates of Sciences and three Doctors of Sciences. Was awarded the Prize and Golden Medal of the USSR Academy of Sciences for Young Researchers (1973), the Order of the Red Banner of Labour (1976). Laureate of the Lenin Prize (1976). In 2003, the Presidium of the Ural Branch of the Russian Academy of Sciences established the Subbotin Prize to be awarded annually for outstanding investigations and inventions with a considerable significance for science and practice (the best research work in mathematics).

Ilia Nesterovich Vekua (born April 23, 1907—died December 2, 1977), was a Georgian mathematician, Academician of the USSR Academy of Sciences (since 1958), Corresponding Member of the USSR Academy of Sciences (since 1946), Academician of the Academy of Sciences of the Georgian SSR (since 1946), and the President of the Academy of Sciences of the Georgian SSR (since 1972). Born in the village of Sheshelety, Kutais Governorate, Russian Empire (modern day Ochamchira District, Abkhazia). Graduated from Tbilisi State University (1930).

During the period 1952–1954, worked at Moscow University; in 1953–1958, at the Steklov Institute of Mathematics (the USSR Academy of Sciences); in 1958–1964, at Novosibirsk State University as the rector; in 1965–1972, at Tbilisi State University as the rector.

His major research was devoted to theory of functions, mathematical theory of elasticity, theory of differential equations of mixed type, theory of boundary-value problems for elliptic systems of equations, theory of multidimensional singular integral equations, and fluid mechanics. Simultaneously with L. Bers and D. Hilbert, elaborated the theory of pseudo-analytic functions. Studied the application of the theory of functions of a complex variable and the theory of differential and integral-differential equations to a series of problems arising in physics and mechanics, in particular, in the theory of elasticity. Suggested to treat arbitrary shells of positive curvature using analytic function methods. In 1959, developed the method of infinitely small bendings by demonstrating that some characteristics of a bending of positive curvature in the conjugate-isometric parameterization are generalized analytic functions. Elaborated the theory of singular integral equations and general solution methods for a wide class of elliptic partial differential equations. Created the framework of generalized analytic functions for the solution and analysis of general boundary-value problems.

Hero of Socialist Labour (1969). Laureate of the Stalin Prize (1950), Lenin Prize (1963), USSR State Prize (1984). Was awarded six Orders of Lenin (1959, 1961, 1966, 1969, 1975, 1977), and the Order of the Badge of Honor (1946). Honoured Science Worker of the Georgian SSR (1950).

Nikolay Nikolaevich Vorobiev (born September 18, 1925—died July 14, 1995), was a Russian mathematician, expert in the field of algebra, mathematical logic, and probability theory, as well as the founder of the largest national school of game theory. Born in Leningrad. First entered Leningrad Ship Engineering Institute and then moved to the Faculty of Mechanics and Mathematics at Leningrad State University. Graduated from LSU in 1948. In 1952 defended his Candidate of Sciences dissertation in 1961, his Doctor of Sciences dissertation Professor since 1965. During the period 1951–1965, worked at the Leningrad branch of the Steklov Institute of Mathematics (the USSR Academy of Sciences); in 1965–1975, at the Leningrad branch of the Central Economics and Mathematics Institute (the USSR Academy of Sciences); in 1975–1990, at the Institute of Social and Economic Problems (the USSR Academy of Sciences). Was an active lecturer at Leningrad State University.

After a survey paper published in 1959, developed the framework of cooperative (coalitional) game theory with a randomized behavior of agents and solved a series of associated nonstandard problems of probability theory and combinatorial topology.

Made considerable contributions in many branches of game theory and obtained pioneering results on cooperative games.

Professor N.N. Vorobiev authored the monographs [20, 24], important surveys and several popular brochures. Was an active organizer of conferences on game theory (1968, 1971, 1974). Done much to develop mathematical game theory in Russia, established a large scientific school, which still continues research in the field. Died in Leningrad in 1995.

References

1. Isaacs, R., *Differential Games: A Mathematical Theory with Applications to Warfare and Pursuit, Control and Optimization*, Dover, 1999.
2. Balashev, L.E., *Zolotoe pravilo povedeniya* (The Golden Rule of Behavior), Moscow: Akademiya, 1999.
3. Bardin, A.E., Risks and Regrets of Players in Games, *Tr. III Mezhd. Konf.* (Proc. III Int. Conf.), 2010, Orekhovo-Zuevo: Mosk. Gos. Oblast. Gumanit. Inst., pp. 86–89.
4. Bardin, A.E., Risks and Regrets of Producers in the Market Model with Import under Uncontrollable Factors, *Tr. III Mezhd. Konf.* (Proc. III Int. Conf.), 2010, Orekhovo-Zuevo: Mosk. Gos. Oblast. Gumanit. Inst., pp. 277–281.
5. Barskii, L., *Chelovek! Eto zvuchit gor'ko* (Man! This Sounds Bitterly), Moscow: Izd. Kh.G.S.
6. Blagodatskikh, A.I. and Petrov, N.N., *Sbornik zadach i uprazhnenii po teorii igr* (A Compilation of Problems and Exercises on Game Theory), Moscow–Izhevsk: Inst. Komp. Issled., 2007.
7. Boltyanskii, V.G., *Optimal'noe upravlenie diskretnymi sistemami* (Optimal Control of Discrete Systems), Moscow: Nauka, 1973.
8. *Bol'shaya kniga aforizmov* (A Big Book of Aphorisms), Moscow: Eksmo, 2005.
9. Borisovich, Yu.G., Gel'man, B.D., Myshkis, A.D., and Obukhovskii, V.V., *Vvedenie v teoriyu mnogoznachnykh otobrazhenii* (An Introduction to Theory of Multivalued Maps), Voronezh: Voronezh. Univ., 1986.
10. Boss, V., *Lektsii po matematike. Tom 12. Kontrprimery i paradoksy* (Lectures on Mathematics. Vol. 12. Counter-examples and Paradoxes), Moscow: URSS, Librokom, 2009.
11. Vaisman, K.S., Berge Equilibrium, *Cand. Sci. (Phys.-Math.) Dissertation*, St. Petersburg: St. Petersburg. Gos. Univ., 1995.
12. Vaisman, K.S. and Zhukovskiy, V.I., Specifics of Zero-Sum Games with Vector Payoff Function, in *Mnogokriterial'nye sistemy pri neopredelennosti* (Multicriteria Systems under Uncertainty), Chelyabinsk: Chelyabinsk. Univ., 1988, pp. 22–28.
13. Vaisman, K.S., Berge Equilibrium, in *Lineino-kvadratichnye differentsial'nye igry* (Linear-Quadratic Differential Games), Zhukovskiy, V.I. and Chikrii, A.A., Eds., Kiev: Naukova Dumka, 1994, pp. 119–143.
14. Vasil'ev, F.P., *Metody optimizatsii* (Optimization Methods), Moscow: Faktorial Press, 2002.
15. Vasin, A.A., Krasnoshchekov, P.S., and Morozov, V.V., *Issledovanie operatsii* (Operations Research), Moscow: Akademiya, 2008.
16. Vasil'ev, V.A., A-Equilibrium and Fuzzy A-Core in Pure Exchange Model with Externalities, *Automation and Remote Control*, 2016, vol. 77, no. 11, pp. 2080–2089.

© Springer Nature Switzerland AG 2020

M. E. Salukvadze, V. I. Zhukovskiy, *The Berge Equilibrium: A Game-Theoretic Framework for the Golden Rule of Ethics*, Static & Dynamic Game Theory: Foundations & Applications, https://doi.org/10.1007/978-3-030-25546-6

17. Vatel', I.A. and Ereshko, F.I., *Matematika konflikta i sotrudnichestva* (Mathematics of Conflict and Cooperation), Moscow: Znanie, 1973.
18. Ventsel', E.S., *Issledovanie operatsii* (Operations Research), Moscow: Znanie, 1976.
19. Voevodin, V.V. and Kuznetsov, Yu.A., *Matritsy i vychisleniya* (Matrices and Calculations), Moscow: Nauka, 1984.
20. Vorobiev, N.N., Noncooperative Games, in *Problemy kibernetiki* (Problems of Cybernetics), 1978, vol. 33, pp. 69–90.
21. Vorobiev, N.N., *Issledovanie operatsii*, (Operations Research), Moscow: Nauka, 1972.
22. Vorobiev, N.N., *Osnovy teorii igr. Beskoalitsionnye igry* (Fundamentals of Game Theory. Noncooperative Games), Moscow: Nauka, 1984.
23. Vorobiev, N.N., Modern State of Game Theory, *Uspekhi Mat. Nauk*, 1970, vol. 25, no. 2, pp. 81–140.
24. Vorobiev, N.N., *Teoriya igr dlya ekonomistov-kibernetikov* (Game Theory for Economists-Cyberneticians), Moscow: Nauka, 1985.
25. Gabasov, R. and Kirillova, F.M., *Osnovy dinamicheskogo programmirovaniya* (Fundamentals of Dynamic Programming), Minsk: Belaruss. Gos. Univ., 1975.
26. *Poisk*, 2004, nos. 34–35, p. 7.
27. Gantmacher, F.R., *The Theory of Matrices. Vols. 1 and 2*, New York: Chelsea Publishing, 1959.
28. Germeier, Yu. B., *Vvedenie v teoriyu issledovaniya operatsii* (Introduction to Operations Research), Moscow: Nauka, 1971.
29. Germeier, Yu.B., *Non-antagonistic Games*, Reidel, 1986.
30. Glicksberg, I.L., A Further Generalization of the Kakutani Fixed Point Theorem, with Application to Nash Equilibrium Points, *Proc. Am. Math. Soc.*, 1952, vol. 3, no. 1, pp. 170–174.
31. Gorbatov, A.S. and Zhukovskiy V.I., Berge Equilibrium in Bertrand Oligopoly with Import, *Taurida J. Comp. Sci. Theory and Math.*, 2015, no. 2(27), pp. 55–64.
32. Gorelik, V.A. and Kononenko, A.F., *Teoretiko–igrovye modeli prinyatiya reshenii v ekologo-ekonomicheskikh sistemakh* (Game-Theoretic Models of Decision-Making in Ecological-Economic Systems), Moscow: Radio i Svyaz', 1982.
33. Gorelik, V.A., Gorelov, M.A., and Kononenko, A.F., *Analiz konfliktnykh situatsii v sistemakh upravleniya* (Analysis of Conflict Situations in Control Systems), Moscow: Radio i Svyaz', 1991.
34. Gorobets, B.S., *Pedagogi shutyat tozhe... Tol'ko strozhe* (Teachers Are Joking Too... But in a Stricter Way), Moscow: Librokom, 2011.
35. Hurwicz, L., Programming in Linear Topological Spaces, in *Issledovaniya po lineinomu programmirovaniyu* (Studies in Linear Programming), Arrow, K.J., Hurwicz, L., and Uzawa, H., Moscow: Inostrannaya Literatura, 1962, pp. 65–155.
36. Guseinov, A.A., *Zolotoe pravilo nravstvennosti* (The Golden Rule of Ethics), Moscow: Molodaya Gvardiya, 1988.
37. Guseinov, A.A., *Velikie proroki i mysliteli. Nravstvennye ucheniya ot Moiseya do nashikh dnei* (Great Prophets and Thinkers. Ethical Teachings from Moses to Present), Moscow: Veche, 2009.
38. Guseinov, A.A., *Antichnaya etika* (Antique Ethics), Moscow: Librokom, 2011.
39. Guseinov, A.A., *Filosofiya - mysl' i postupok: stat'i, doklady, lektsii, interv'yu* (Philosophy—Thought and Deed: Papers, Presentations, Lectures, Interviews), St. Petersburg: Gumanit. Univ. Profsoyuzov, 2012.
40. Guseinov, A.A., Philosophy as an Ethical Project, *Voprosy Filosofii*, 2014, no. 4, pp. 16–26.
41. Dunford, N. and Schwartz, J.T., *Linear Operators*, New York: Interscience, 1958. Translated under the title *Lineinye operatory. Obshchaya teoriya*, Moscow: Inostrannaya Literatura, 1962, vol. 1.
42. *Differentsial'nye igry. Ukazatel' literatury za 1968–1976* (Differential Games. Index of Literature for 1968–1976), Ushakov, V.N., Ed., Sverdlovsk: Ural. Nauchn. Tsentr Akad. Nauk USSR, 1978.

43. *Differentsial'nye igry neskol'kikh lits. Ukazatel' literatury za 1968–1983* (Differential Multiperson Games. Index of Literature for 1968–1983), Zhukovskiy, V.I. and Dochev, D.T., Eds., Ruse, Bulgaria: Center for Mathematics, 1985.
44. *Differentsial'nye igry neskol'kikh lits. Ukazatel' literatury za 1984–1988* (Differential Multiperson Games. Index of Literature for 1984–1988), Zhukovskiy, V.I. and Ushakov, V.N., Eds., Sverdlovsk: Ural. Otd. Akad. Nauk USSR, 1990.
45. *Differentsial'nye igry neskol'kikh lits. Ukazatel' literatury za 1989–1994* (Differential Multiperson Games. Index of Literature for 1989–1994), Zhukovskiy, V.I. and Ukhobotov, V.I., Eds., Chelyabinsk: Gos. Univ., 1995.
46. Dmitruk, A.V., *Vypuklyi analiz. Elementarnyi vvodnyi kurs* (Convex Analysis. An Elementary Introductory Course), Moscow: Makspress, 2012.
47. Dubina, I.N., *Osnovy teorii ekonomicheskikh igr* (Theoretical Foundations of Economic Games), Moscow: Knorus, 2010.
48. Egorov, A.I., *Osnovy teorii upravleniya* (Foundations of Control Theory), Moscow: Fizmatlit, 2004.
49. Zhautykov, O.A., Zhukovskiy, V.I., and Zharkynbaev, S.Zh, *Differentsial'nye igry neskol'kikh lits (s zapazdyvaniem vremeni)* (Differential Multiperson Games with Time Delay), Almaty: Nauka, 1988.
50. Zhitomirskii, G.I., Dynamic Problems with Conflict, *Extended Abstract of Cand. Sci. Dissertation (Phys.-Math.)*, Leningrad State Univ., Leningrad, 1989.
51. Zhukovskiy, V.I., *Vvedenie v differentsial'nye igry pri neopredelennosti* (Introduction to Differential Games under Uncertainty), Moscow: Mezhd. Nauchno-Issled. Inst. Probl. Upravlen., 1997.
52. Zhukovskiy, V.I., *Kooperativnye igry pri neopredelennosti i ikh prilozheniya* (Cooperative Games under Uncertainty and Their Applications), Moscow: URSS, 2010, 2nd ed.
53. Zhukovskiy, V.I., *Konflikty i riski* (Conflicts and Risks), Moscow: Ross. Zaochn. Inst. Tekstil. Legk. Promysh., 2007.
54. Zhukovskiy, V.I., *Vvedenie v differentsial'nye igry pri neopredelennosti. Ravnovesie po Neshu* (Introduction to Differential Games under Uncertainty. Nash Equilibrium), Moscow: URSS, 2010.
55. Zhukovskiy, V.I., *Vvedenie v differentsial'nye igry pri neopredelennosti. Ravnovesie ugroz i kontrugroz* (Introduction to Differential Games under Uncertainty. Equilibrium in Threats and Counter-Threats), Moscow: URSS, 2010.
56. Zhukovskiy, V.I., *Vvedenie v differentsial'nye igry pri neopredelennosti. Ravnovesie po Berzhu–Vaismanu* (Introduction to Differential Games under Uncertainty. Berge–Vaisman Equilibrium), Moscow: URSS, 2010.
57. Zhukovskiy, V.I., *Riski pri konfliktnykh situatsiyakh* (Risks in Conflict Situations), Moscow: URSS, LENAND, 2011.
58. Zhukovskiy, V.I. and Bel'skikh, Yu.A., Nash and Berge Equilibria in a Two-Player Linear-Quadratic Game, *Tr. XIII Mezhd. Nauchno-Praktich. Konf. "Otechestvennaya nauka v epokhu izmenenii: postulaty proshlogo i teorii novogo vremeni"* (Proc. XIII Int. Scientific-Practical Conf. "National Science in the Era of Change: Postulates of Past and Theories of New Time"), St. Petersburg, National Association of Scientists, 2015, vol. 13, pp. 41–44.
59. Zhukovskiy, V.I., Bel'skikh, Yu.A., and Samsonov, S.P., Coefficient Criteria for Equilibrium Choice (An Example of a Linear-Quadratic Two-Player Game), *Vestn. South Ural State Univ. Ser. Math. Mech. Phys.*, 2015, vol. 7, no. 4, pp. 20–26.
60. Zhukovskiy, V.I. and Vysokos, M.I., The Golden Rule in the Cournot Duopoly Model, *Taurida J. Comp. Sci. Theory and Math.*, 2015, no. 2(27), pp. 46–54.
61. Zhukovskiy, V.I., Vysokos, M.I., and Gorbatov, A.S., Berge Equilibrium in the Single-Stage Setup of Controlled Cournot Duopoly, *Analiz, Modelirovanie, Upravlenie, Razvitie Sotsial'no-Ekonomicheskikh Sistem: Tr. IX Mezhd. Shkoly-Simpoziuma (AMUR-2015)* (Analysis, Modeling, Management, Development of Socio-economic Systems: Proc. IX Int. School-Symposium (AMMD-2015)), Krymskii Fed. Univ., Simferopol, 2015, pp. 105–109.

62. Zhukovskiy, V.I. and Gorbatov, A.S., Connection between the Golden Rule and Berge Equilibrium, *Control Theory and Theory of Generalized Solutions of Hamilton–Jacobi Equations (CGS'2015): Abstracts of II International Seminar dedicated to the 70th Anniversary of Academician A.I. Subbotin*, Yekaterinburg, Inst. Mat. Mekh. Ural. Otd. Ross. Akad. Nauk, 2015, pp. 58–60.
63. Zhukovskiy, V.I. and Gorbatov, A.S., Zero Risk in Single-Criterion Choice Problems, *Upravlen. Risk.*, 2015, vol. 74, no. 2, pp. 29–36.
64. Zhukovskiy, V.I. and Dochev, D.T., *Vektornaya optimizatsiya dinamicheskikh sistem* (Vector Optimization of Dynamic Systems), Ruse, Bulgaria: Center for Mathematics, 1981.
65. Zhukovskiy, V.I. and Zhiteneva, Yu.M., Bertrand Difference–Differential Game Model in the Framework of Berge Equilibrium, *Analiz, Modelirovanie, Upravlenie, Razvitie Sotsial'no-Ekonomicheskikh Sistem: Tr. IX Mezhd. Shkoly-Simpoziuma (AMUR-2015)* (Analysis, Modeling, Management, Development of Socio-economic Systems: Proc. IX Int. School-Symposium (AMMD-2015)), Krymskii Fed. Univ., Simferopol, 2015, pp. 109–114.
66. Zhukovskiy, V.I. and Zhukovskaya, L.V., *Risk v mnogokriterial'nykh i konfliktnykh sistemakh pri neopredelennosti* (Risk in Multicriteria Choice and Conflict Systems under Uncertainty), Moscow: URSS, 2004.
67. Zhukovskiy, V.I. and Kudryavtsev, K.N., Pareto-Optimal Nash Equilibrium: Sufficient Conditions and Existence in Mixed Strategies, *Automation and Remote Control*, 2016, vol. 77, no. 8, pp. 1500–1510.
68. Zhukovskiy, V.I. and Kudryavtsev, K.N., *Uravnoveshivanie konfliktov i prilozheniya* (Equilibrating Conflicts and Applications), Moscow: URSS, 2012.
69. Zhukovskiy, V.I., Kudryavtsev, K.N., and Smirnova, L.V., *Garantirovannye resheniya konfliktov i prilozheniya* (Guaranteed Solutions of Conflicts and Applications), Moscow: URSS, 2013.
70. Zhukovskiy, V.I., Kudryavtsev, K.N., and Gorbatov, A.S., The Berge Equilibrium in Cournot Oligopoly Model, *Vestn. Udmurt. Univ. Math. Mech. Comp. Sci.*, 2015, vol. 25, no. 2, pp. 147–156.
71. Zhukovskiy, V.I. and Kudryavtsev, K.N., Mathematical Foundations of the Golden Rule. I. Static Case, *Automation and Remote Control*, 2017, vol. 78, no. 10, pp. 1920–1940.
72. Zhukovskiy, V.I., Smirnova, L.V., and Gorbatov, A.S., Mathematical Foundations of the Golden Rule. II. Dynamic Case, *Mat. Teor. Igr Prilozh.*, 2016, vol. 8, no. 1, pp. 27–62.
73. Zhukovskiy, V.I. and Kudryavtsev, K.N., Equilibrating Conflicts under Uncertainty. I. Analogue of a Saddle-Point, *Mat. Teor. Igr Prilozh.*, 2013, vol. 5, no. 1, pp. 27–44.
74. Zhukovskiy, V.I. and Kudryavtsev, K.N., Equilibrating Conflicts under Uncertainty. II. Analogue of a Maximin, *Mat. Teor. Igr Prilozh.*, 2013, vol. 5, no. 2, pp. 3–45.
75. Zhukovskiy, V.I. and Kudryavtsev, K.N., The Existence of Guaranteed Imputation in Counter-Strategies, *Spectral and Evolution problems: Proc. 17th Crimean Autumn Mathematical School-Symposium*, Simferopol, 2007, vol. 17, pp. 28–31.
76. Zhukovskiy, V.I. and Makarkina, T.V., Two Approches to Make Guaranteed Decisions in the Cournot Duopoly Model with Import, *Analiz, Modelirovanie, Upravlenie, Razvitie Sotsial'no-Ekonomicheskikh Sistem: Tr. IX Mezhd. Shkoly-Simpoziuma (AMUR-2015)* (Analysis, Modeling, Management, Development of Socio-economic Systems: Proc. IX Int. School-Symposium (AMMD-2015)), Krymskii Fed. Univ., Simferopol, 2015, pp. 227–232.
77. Zhukovskiy, V.I. and Makarkina, T.V., Consideration of Import in the Cournot Duopoly, *Taurida J. Comp. Sci. Theory and Math.*, 2015, no. 2(27), pp. 65–76.
78. Zhukovskiy, V.I., Makarkina, T.V., Samsonov, S.P., and Gorbatov, A.S., Berge Equilibrium for a Price Competition Model in a Duopoly Market, *Tr. XII Mezhd. Nauchno-Praktich. Konf. "Otechestvennaya nauka v epokhu izmenenii: postulaty proshlogo i teorii novogo vremeni"* (Proc. XII Int. Scientific-Practical Conf. "National Science in the Era of Change: Postulates of Past and Theories of New Time"), Yekaterinburg, National Association of Scientists, 2015, vol. 7, pp. 94–98.

79. Zhukovskiy, V.I. and Molostvov, V.S., *Mnogokriterial'naya optimizatsiya sistem v usloviyakh nepolnoi informatsii* (Multicriteria Optimization of Systems under Incomplete Information), Moscow: Mezhd. Nauchno-Issled. Inst. Probl. Upravlen., 1990.
80. Zhukovskiy, V.I. and Molostvov, V.S., *Mnogokriterial'noe prinyatie reshenii v usloviyakh neopredelennosti* (Multicriteria Decision-Making under Uncertainty), Moscow: Mezhd. Nauchno-Issled. Inst. Probl. Upravlen., 1988.
81. Zhukovskiy, V.I. and Salukvadze, M.E., *Mnogokriterial'nye zadachi upravleniya v usloviyakh neopredelennosti* (Multicriteria Control Problems under Uncertainty), Tbilisi: Metsniereba, 1991.
82. Zhukovskiy, V.I. and Salukvadze, M.E., *Mnogoshagovye pozitsionnye konflikty i ikh prilozheniya* (Multistage Positional Conflicts and Their Applications), Moscow–Tbilisi: Intelekti, 2011.
83. Zhukovskiy, V.I. and Salukvadze, M.E., *Nekotorye igrovye zadachi upravleniya i ikh prilozheniya* (Some Game-Theoretic Problems of Control and Their Applications), Tbilisi: Metsniereba, 1998.
84. Zhukovskiy, V.I. and Salukvadze, M.E., *Optimizatsiya garantii v mnogokriterial'nykh zadachakh upravleniya* (Optimization of Guarantees in Multicriteria Control Problems), Tbilisi: Metsniereba, 1996.
85. Zhukovskiy, V.I. and Salukvadze, M.E., *Riski i iskhody v mnogokriterial'nykh zadachakh upravleniya* (Risks and Outcomes in Multicriteria Control Problems), Tbilisi: Intelekti, 2004.
86. Zhukovskiy, V.I. and Sachkov, S.N., About an Uncommon yet Benevolent Method of Conflict Resolution, *Aktual. Voprosy Razvitiya Innovats. Deyatel'n. Nov. Tysyachelet.*, 2014, no. 10(14), pp. 61–64.
87. Zhukovskiy, V.I. and Sachkov, S.N., An Unusual Way of Resolving Conflicts, *Tr. Konf. "Sovremennye tekhnologii upravleniya-2014"* (Proc. Conf. "Contemporary Technology of Management-2014"), Moscow, 2014, pp. 337–355.
88. Zhukovskiy, V.I., Sachkov, S.N., and Sachkova, E.N., Application of the Golden Rule of Ethics in One Model of Competitive Economy, *Tr. Mezhd. Nauchn.-Praktich. Konf. "Problemy razvitiya sovremennoy nauki i puti ikh resheniya"* (Proc. Int. Scientific-Practical Conf. "Development Problems of Modern Science and Ways to Solve Them"), Yekaterinburg, National Association of Scientists, 2015.
89. Zhukovskiy, V.I. and Smirnova, L.V., Application of the Golden Rule in a Positional Differential Game, *Analiz, Modelirovanie, Upravlenie, Razvitie Sotsial'no-Ekonomicheskikh Sistem: Tr. IX Mezhd. Shkoly-Simpoziuma (AMUR-2015)* (Analysis, Modeling, Management, Development of Socio-economic Systems: Proc. IX Int. School-Symposium (AMMD-2015)), Krymskii Fed. Univ., Simferopol, 2015, pp. 115–120.
90. Zhukovskiy, V.I. and Smirnova, L.V., The Existence of Guaranteed Equilibrium, in *Analiz, modelirovanie, upravlenie, razvitie* (Analysis, Modeling, Management, Development), Vernadskii Gos. Univ., Sevastopol, 2012, pp. 159–163.
91. Zhukovskiy, V.I. and Smirnova, L.V., The Existence of Berge Equilibrium in a Differential Game with Separated Dynamics, *Taurida J. Comp. Sci. Theory and Math.*, 2015, no. 2(27), pp. 77–86.
92. Zhukovskiy, V.I. and Tynyanskii, N.T., *Ravnovesnye upravleniya mnogokriterial'nykh dinamicheskikh sistem* (Equilibrium Control of Multicriteria Dynamic Systems), Moscow: Mosk. Gos. Univ., 1984.
93. Zhukovskiy, V.I. and Chikrii, A.A., *Lineino-kvadratichnye differentsial'nye igry* (Linear-Quadratic Differential Games), Kiev: Naukova Dumka, 1994.
94. Zhukovskiy, V.I., Chikrii, A.A., and Soldatova, N.G., The Berge Equilibrium in the Conflicts under Uncertainty, *XII Vserossiiskoe soveshchanie po problemam upravleniya (VSPU-2014)* (Proc. XII All-Russian Meeting on Control Problems (AMCP-2014)), Moscow: Inst. Probl. Upravlen., 2014, pp. 8290–8302.
95. Zhukovskiy, V.I., Chikrii, A.A., and Soldatova, N.G., Existence of Berge Equilibrium in Conflicts under Uncertainty, *Automation and Remote Control*, 2016, vol. 77, no. 4, pp. 640–655.

96. Zhukovskiy, V.I., Chikrii, A.A., and Soldatova, N.G., Diversification Problem of Deposits, in *Analiz, modelirovanie, upravlenie, razvitie* (Analysis, Modeling, Management, Development), Vernadskii Gos. Univ., Sevastopol, 2015, pp. 134–142.
97. Zhukovskiy, V.I. and Chikrii, A.A., On Discrete Conflict-Controlled Processes Described by Grunvald-Letnikov Fractional Systems, *Journal of Automation and Information Sciences*, 2015, vol. 47, no. 1, pp. 24–34.
98. *Zabavnye anekdoty* (Funny Stories), Saint-Petesrburg: Dilya, 1994.
99. Zadeh, L.A., The Concept of a Linguistic Variable and Its Application to Approximate Reasoning, *Information Sciences*, 1975, vol. 8, no. 4, pp. 301–357.
100. Kapitonenko, V.V., *Finansovaya matematika i ee prilozheniya* (Financial Mathematics and Its Applications), Moscow: Prior, 2000.
101. Karlin, S., *Mathematical Methods and Theory in Games, Programming and Economics*, London-Paris: Pergamon Press, 1959.
102. Kleimenov, A.F., *Neantagonisticheskie pozitsionnye differentsial'nye igry* (Nonantagonistic Positional Differential Games), Yekaterinburg: Nauka, 1993.
103. Kolemaev, V.A., *Matematicheskaya ekonomika* (Mathematical Economics), Moscow: Yuniti, 2002.
104. Kolesnik, G.V., *Teoriya igr* (Game Theory), Moscow: Librokom, 2010.
105. Kolmogorov, A.N. and Fomin, S.V., *Elementy teorii funktsii i funktsional'nogo analiza* (Elements of Theory of Functions and Functional Analysis), Moscow: Nauka, 1976.
106. Korczak, J., *Kak lyubit' detei* (How to Love Children), Moscow: Znanie, 1968.
107. Krasovskii, N.N., *Upravlenie dinamicheskoi sistemoi* (Control of Dynamic System), Moscow: Nauka, 1985.
108. Krasovskii, N.N. and Subbotin, A.I., *Pozitsionnye differentsial'nye igry* (Positional Differential Games), Moscow: Nauka, 1985.
109. Kudryavtsev, K.N., Coordinated Solutions in Multiagent Information Environment, *Extended Abstract of Cand. Sci. Dissertation (Phys.-Math.)*, South-Ural Fed. Univ., Chelyabinsk, 2011.
110. Kudryavtsev, K.N., On the Absence of Maximin Strategies in a Differential Game, *Vestn. Yuzhno-Ural. Gos. Univ. Ser. Mat. Mekh. Fiz.*, 2010, vol. 3, no. 30, pp. 13–20.
111. Kudryavtsev, K.N., One Pricing Problem under Uncertainty, *Tr. Vseross. Konf. "Teoriya upravleniya i matematicheskoe modelirovanie"* (Proc. All-Russia Conf. "Control Theorey and Mathematical Modeling"), Izhevsk, Russia, Udmurt. Univ., 2015, pp. 263–264.
112. Kudryavtsev, K.N., On the Existence of Guaranteed Solutions in Payoffs and Risks for Cooperative Games under Uncertainty, *Sist. Upravlen. Inform. Tekh.*, 2010, no. 1.1(39), pp. 148–152.
113. Kudryavtsev, K.N. and Meshkov, V.M., About One Planning Model of Advertising Budget in Duopoly, *Tr. XV Mezhd. Konf. "Sistemy komp'yuternoy matematiki i ikh prilozheniya"* (Proc. XV Int. Conf. "Systems of Computer Mathematics and Their Applications"), Smolensk: Gos. Univ., 2014, vol. 15, pp. 173–175.
114. Kudryavtsev, K.N. and Stabulit, I.S., Strongly-Guaranteed Equilibrium in One Spatial Competition Problem, *Tr. XVI Mezhd. Konf. "Sistemy komp'yuternoy matematiki i ikh prilozheniya"* (Proc. XVI Int. Conf. "Systems of Computer Mathematics and Their Applications"), Smolensk: Gos. Univ., 2015, vol. 16, pp. 181–183.
115. Kudryavtsev, K.N. and Stabulit, I.S., Pareto-Guaranteed Equilibrium in 2D Hotelling Duopoly, *Vestn. Yuzhno-Ural. Gos. Univ. Ser. Mat. Modelir. Programmir.*, 2012, vol. 14, no. 40, pp. 173–177.
116. Lavrov, P.L., *Sotsial'naya revolyutsiya i zadachi nravstvennosti. Tom 1* (Social Revolution and Ethical Tasks. Vol. 1), Moscow: Mysl', 1965.
117. Labsker, L.G. and Yashchenko, N.A., *Teoriya igr v ekonomike* (Game Theory in Economics), Moscow: Knorus, 2012.
118. Lagunov, V.N., *Vvedenie v differentsial'nye igry* (Introduction to Differential Games), Vilnius: Inst. Mat. Kibern. Akad. Nauk Litovsk. SSR, 1979.
119. *Larets ostroslovov* (Casket of Wisecrackers), Moscow: Izdatel'stvo Politicheskoi Literatury, 1991.

120. Lezhnev, A.V., *Dinamicheskoe programmirovanie v ekonomicheskikh zadachakh* (Dynamic Programming in Economic Problems), Moscow: BINOM, Laboratoriya Znanii, 2006.
121. Luce, R.D. and Raiffa, H., *Games and Decisions: Introduction and Critical Survey*, Dover Books on Mathematics, 1989.
122. Lyusternik, L.A. and Sobolev, V.I., *Elementy funktsional'nogo analiza* (Elements of Functional Analysis), Moscow: Nauka, 1969.
123. Mazalov, V.V., *Mathematical Game Theory and Applications*, Wiley, 2014.
124. Makarov, V.A. and Rubinov, A.M., *Matematicheskaya teoriya ekonomicheskoi dinamiki i ravnovesie* (Mathematical Theory of Economic Dynamics and Equilibrium), Moscow: Nauka, 1973.
125. McConnell, C.R., Brue, S.L., and Flynn, S.M., *Economics: Principles, Problems, and Policies*, McGraw-Hill, 2011.
126. Malafeev, O.A., *Upravlyaemye konfliktnye sistemy* (Controlled Conflict Systems), St.-Petersburg: Gos. Univ., 2005.
127. Malkin, G.E., *Bol'shaya kniga aforizmov dlya ochen' umnykh* (A Big Book of Aphorisms for Really Smart Persons), Moscow: RIPOL Klassik, 2005.
128. Malkin, I.G., *Teoriya ustoychivosti dvizheniya* (Theory of Motion Stability), Moscow: Nauka, 1996.
129. Mamedov, M.B., About Pareto Optimal Nash Equilibrium, *Izv. Akad. Nauk Azerbaidzhana. Ser. Fiz.-Mat. Nauk*, 1983, vol. 4, no. 2, pp. 11–17.
130. Matveev, V.A., *Konechnye beskoalitsionnye igry i ravnovesiya* (Finite Noncoalitional Games and Equilibria), Pskov: Ped. Inst., 2004.
131. Mashchenko, S.O., Concept of Nash Equilibrium and Its Development, *Zh. Obchisloval. Prikladn. Mat.*, 2012, no. 1(107), pp. 40–65.
132. Mishchenko, E.F. and Rozov, N.Kh., *Differentsial'nye uravneniya s malym parametrom i relaksatsionnye kolebaniya* (Small-Parameter Differential Equations and Relaxation Oscillations), Moscow: Nauka, 1975.
133. Morozov, V.V., On Mixed Strategies in Game with Vector Payoff Function, *Tr. III Vsesoyuzn. Konf. po issledovaniyu operatsii* (Proc. III All-Soviet Conf. on Operations Research), Gorky, 1978, pp. 210–211.
134. Morozov, V.V., *Osnovy teorii igr* (Foundations of Game Theory), Moscow: Mosk. Gos. Univ., 2002.
135. Morozov, V.V., Mixed Strategies in a Game with Vector Payoffs, *Vestn. Mosk. Gos. Univ. Vychisl. Mat. Kibern.*, 1978, no. 4, pp. 44–49.
136. Morozov, V.V., Sukharev, A.G., and Fedorov, V.V., *Issledovanie operatsii v zadachakh i uprazhneniyakh* (Operations Research in Problems and Exercises), Moscow: Vysshaya Shkola, 1986.
137. Muschick, E. and Müller, P., *Metody prinyatiya tekhnicheskikh reshenii* (Methods of Technical Decision-Making), Moscow: Mir, 1990.
138. Knight, F.H., *Risk, Uncertainty, and Profit*, Boston: Houghton Mifflin, 1921.
139. Noghin, V.D., Duality in Multiobjective Programming, *Zh. Vychisl. Mat. Matem. Fiz.*, 1977, vol. 17, no. 1, pp. 254–258.
140. Noghin, V.D., *Prinyatie reshenii v mnogokriterial'noi srede* (Decision-Making in Multicriteria Environment), Moscow: Fizmatlit, 2005.
141. Petrov, N.N., *Teoriya igr* (Game Theory), Izhevsk: Udmurt. Univ., 1997.
142. Petrov, N.N., *Matematicheskie igry* (Mathematical Games), Moscow: URSS, 2012.
143. Petrosjan, L.A. and Danilov, N.N., *Kooperativnye differentsial'nye igry i ikh prilozheniya* (Cooperative Differential Games and Their Applications), Tomsk: Gos. Univ., 1985.
144. Petrosjan, L.A. and Zakharov, V.V., *Matematicheskie modeli v ekologii* (Mathematical Models in Ecology), St. Petersburg: Gos. Univ., 1997.
145. Petrosjan, L.A., Zenkevich, N.A., and Semina, E.A., *Teoriya igr* (Game Theory), Moscow: Vysshaya Shkola, 1998.
146. Petrosjan, L.A., Zenkevich, N.A., and Shevkoplyas, E.V., *Teoriya igr* (Game Theory), St. Petersburg: BKhV-Peterburg, 2012.

147. Pecherskii, S.L. and Belyaeva, A.A., *Teoriya igr dlya ekonomistov. Vvodnyy kurs* (Game Theory for Economists. An Introductory Course), St. Petersburg: Evrop. Univ., 2004.
148. Pisarchuk, N.N., *Vvedenie v teoriyu igr* (Introduction to Game Theory), Minsk: Belorus. Gos. Univ., 2011.
149. Podinovskii, V.V., General Zero-Sum Two-Person Games, *Zh. Vychisl. Mat. Matem. Fiz.*, 1981, vol. 21, no. 5, pp. 1140–1153.
150. Podinovskii, V.V., The Principle of Guaranteed Result for Partial Preference Relations, *Zh. Vychisl. Mat. Matem. Fiz.*, 1979, vol. 19, no. 6, pp. 1436–1450.
151. Podinovskii, V.V., Efficient Plans in Multicriteria Choice Problems under Uncertainty, *Model. Prots. Prinyat. Reshen.*, Vladivostok: Dal'nevostochn. Nauchn. Tsentr Akad. Nauk SSSR, 1978, pp. 102–113.
152. Podinovskii, V.V. and Noghin, V.D., *Pareto-optimal'nye resheniya mnogokriterial'nykh zadach* (Pareto Optimal Solutions of Multicriteria Problems), Moscow: Fizmatlit, 2007.
153. Pontryagin, L.S., *Ordinary Differential Equations*, Adiwes International Series in Mathematics, Addison-Wesley, 1962.
154. Pontryagin, L.S., Boltyanskii, V.G., Gamkrelidze, R.V., and Mishchenko, E.F., *The Mathematical Theory of Optimal Processes*, Interscience Publishers, 1962.
155. Ras, A., *Davaite melochit'sya* (Let Us Split Hairs), Moscow: Ripol Klassik, 2006.
156. Rogov, M.A., *Risk–menedzhment* (Risk Management), Moscow: Finansy i Statistika, 2001.
157. Rozen, V.V., *Modeli prinyatiya reshenii v ekonomike* (Models of Decision-Making in Economics), Moscow: Vysshaya Shkola, 2002.
158. Rozen, V.V., Equilibria in Games with Ordered Outcomes, in *Sovremennye napravleniya teorii igr* (Modern Directions of Game Theory), Vilnius: Moklas, 1976, pp. 115–118.
159. Rozen, V.V., Properties of Outcomes in Equilibria, in *Matematicheskie modeli povedeniya* (Mathematical Models of Behavior), Saratov: Gos. Univ., 1975, vol. 2, pp. 45–49.
160. Rosenmüller, J., *The Theory of Games and Markets*, Elsevier, 1981.
161. *Russko–anglo–nemetskii tolkovyi slovar' po biznesu* (Russian–English–German Glossary on Business Science), Kuznetsova, N.N., Novikova, E.V., Plekhanov, S.V., and Chekmezov, N.A., Eds., Moscow: Gorizont, 1992.
162. Rybnikov, K.A., *Matematicheskie modeli konfliktnykh situatsii* (Mathematical Models of Conflict Situations), Moscow: Mosk. Gos. Univ., 1998.
163. Safronov, S.G., Model of Market and Producers of Goods, in *Slozhnye dinamicheskie sistemy* (Complex Dynamic Systems), Pskov: Pedagogich. Inst., 1994, pp. 157–162.
164. *Slovar' inostrannykh slov* (Glossary of Foreign Terms), Moscow: Sovetskaya entsiklopediya, 1964.
165. Smirnova, L.V., Hurwitz Principle in Complex Controlled Systems, *Extended Abstract of Cand. Sci. Dissertation (Phys.-Math.)*, Moscow Pedagogical Stat Univ., Moscow, 1998.
166. Smol'yakov, E.R., *Ravnovesnye modeli pri nesovpadayushchikh interesakh uchastnikov* (Equilibrium Models under Noncoinciding Interests of Parties), Moscow: Nauka, 1986.
167. Smol'yakov, E.R., *Teoriya konfliktnykh ravnovesii* (Theory of Conflict Equilibria), Moscow: URSS, 2004.
168. *Sochineniya Koz'my Prutkova* (Works of Kozma Prutkov), Moscow: Sovetskaya Rossiya, 1981.
169. Subbotin, A.I. and Chentsov, A.G., *Optimizatsiya garantii v zadachakh upravleniya* (Optimization of Guarantees in Control Problems), Moscow: Nauka, 1981.
170. Tynyanskii, N.T. and Zhukovskiy, V.I., Non-Zero-Sum Differential Games (Noncooperative Setup), in *Itogi nauki i tekhniki. Mat. analiz* (Results of Science and Technology. Mathematical Analysis), 1977, vol. 15, pp. 199–266.
171. Ulanov, V.A., On a Differentia Game with Infinite Number of Players, in *Problemy teorii igr v obshchikh sistemakh* (Problems of Game Theory in General Systems), Yakutsk. Gos. Univ., 1988, pp. 32–39.
172. Utkin, E.A., *Risk-menedzhment* (Risk Management), Moscow: Ekmos, 1998.
173. Fedorov, V.V., *Chislennye metody maksimina* (Numerical Methods of Maximin), Moscow: Nauka, 1979.

174. Fel'dbaum, A.A., *Osnovy teorii avtomaticheskikh sistem* (Foundations of Automatic Control Systems), Moscow: Nauka, 1966.
175. Fischer, S., Dornbusch, R., and Schmalensee, R., *Economics*, McGraw-Hill, 1988.
176. Fedin, S.N., Gorobets, B.S., and Zolotov Yu.A., *Uchenye shutyat* (Scientists Are Joking), Moscow: Librokom, 2012
177. von Neumann, J. and Morgenstern, O., *Theory of Games and Economic Behavior*, Princeton: Princeton Univ. Press, 1944.
178. Harsanyi J.C. and Selten R., *A General Theory of Equilibrium Selection in Games*, Cambridge: MIT Press, 1988.
179. Hamming, R.W., *Numerical Methods for Scientists and Engineers*, McGraw-Hill, 1962.
180. Hille, E. and Phillips, R.S., *Functional Analysis and Semi-Groups*, American Mathematical Society Colloquium Publications, vol. 31, Providence: Am. Math. Soc., 1957. Translated under the title *Funktsional'nyi analiz i polugruppy*, Moscow: Inostrannaya Literatura, 1962.
181. Khoroshilova, E.V., *Sovremennye anekdoty i vyskazyvaniya* (Modern Funny Stories and Sayings), Moscow: Maks-Press, 2009.
182. Tsvetkova, E.V. and Arlyukova, I.O., *Riski v ekonomicheskoi deyatel'nosti* (Risks in Economic Activity), St.-Petersburg: Inst. Vneshneekon. Svyav. Ekon. Prav., 2002.
183. Cheremnykh, Yu.N., *Mikroekonomika. Prodvinutyi uroven'* (Microeconomics: Advance Level), Moscow: Info-M, 2008.
184. Cherkasov, V.V., *Delovoi risk v predprinimatel'skoi deyatel'nosti* (Business Risk in Entrepreneurship), Kiev: Libra, 1996.
185. Shakhov, V.V., *Vvedenie v strakhovanie. Ekonomicheskii aspekt* (Introduction to Insurance. Economic Aspect), Moscow: Finansy i Statistika, 1994.
186. Shikin, E.V., *Ot igr k igram* (From Games to Games), Moscow: URSS, 1997.
187. Shokin, Yu.I., *Interval'nyi analiz* (Interval Analysis), Novosibirsk: Sib. Otd. Ross. Akad. Nauk, 1981.
188. Shubik, M., The Present and Past of Game Theory, *Mat. Teor. Igr Prilozh.*, 2012, vol. 4, no. 1, pp. 93–116.
189. *Entsiklopediya kibernetiki. Tom 2.* (Encyclopeadia of Cybernetics. Vol. 2), Kiev: Glavn. Redakts. Ukrainsk. Sovetsk. Entsiklop., 1974.
190. *Entsiklopediya matematiki. Tom 1–5.* (Encyclopeadia of Mathematics. Vols. 1–5), Moscow: Sovetsk. Entsiklop., 1977–1985.
191. Arrow, K.J., Hurwicz, L., and Uzawa, H., *Studies in Linear and Non-linear Programming*, Stanford Mathematical Studies in the Social Sciences, Stanford Univ. Press, 1958.
192. Yanovskaya, E.B., Equilibria in General Noncooperative Games and Their Mixed Extensions, in *Teoretiko-igrovye voprosy prinyatiya resheniy* (Game-Theoretic Issues of Decision-Making), Leningrad: Nauka, 1978, pp. 43–65.
193. A century of mathematics in America 1988–1989, vol. 1–3, ed. AMS (vol. 1. P. 382).
194. Abalo, K.Y. and Kostreva, M.M., Some Existence Theorems of Nash and Berge Equilibria, *Appl. Math. Lett.*, 2004, vol. 17, pp. 569–573.
195. Abalo, K.Y. and Kostreva, M.M., Berge Equilibrium: Some Recent Results from Fixed-Point Theorems, *Appl. Math. and Comp.*, 2005, vol. 169, pp. 624–638.
196. Archibald, R.C., *A Semicentennial History of American Mathematical Society. 1888–1938* (2 vols.), 1938.
197. Aumann, R.J. and Peleg, B., Von Neumann–Morgenstern Solutions to Cooperative Games without Side Payments, *Bull. Amer. Math. Soc.*, 1960, vol. 66, pp. 173–179.
198. Basar, T., A Contrexample in Linear-Quadratic Game: Existence of Nonlinear Nash Solutions, *J. Optimiz. Theory and Appl.*, 1974, vol. 14, no. 4, pp. 425–430.
199. Basar, T. and Olsder, G.J., *Dynamic Noncooperative Game Theory*, London: Academic Press, 1982.
200. Bellman, R., Dynamic Programming and a New Formalism in the Calculus of Variations, *Proc. Nat. Acad. Sci. USA*, 1953, vol. 39, pp. 1077–1082.

201. Bellman, R., On the Application of Dynamic Programming to Variational Problems Arising in Mathematical Economics, *Proc. Symposium in Calculus of Variations and Applications*, New York: McGraw–Hill, 1956, pp. 115–138.
202. Berge, C., *Théorie générale des jeux á n personnes games*, Paris: Gauthier Villars, 1957. (Russian translation: Berge, C., *Obshchaya teoriya igr neskol'kikh lits*, Moscow: Fizmatgiz, 1961).
203. Blackwell, O., An Analog of the Minimax Theorem for Vector-Payoffs, *Pacific J. Math.*, 1956, no. 6, pp. 1–8.
204. Bertrand, J., *Caleul des probabilities*, Paris, 1888.
205. Bertrand, J., Book review of theorie mathematique de la richesse sociale and of recherches sur les principles mathematiques de la theorie des richesses, *Journal de Savants*, 1883, vol. 67, pp. 499–508.
206. Bertrand, J., Review of theory mathematique de la richesse sociale recherches sur les principes mathematique de la theorie des richesses, *Journal des Savants*, 1883, vol. 68, pp. 449–508.
207. Borel, E., La théorie du jeu et les equations intégrales a noyau symétrique, *Compes Rendus de l'Académic des Sciences*, 1921, vol. 173, pp. 1304–1308.
208. Borel, E., Sur les jeux an interviennent l'hasard et l'abilité des joueurs, in *Théorie des probabilite's*, Paris, 1924, pp. 204–224.
209. Borel, E., Sur les systemes de formes lineares a determinant symetrique gauche et la theorie generale du jeu, *Comptes Rendus de l'Academie des Sciences*, 1927, vol. 184, pp. 52–53.
210. Bohnenblust, H.F. and Karlin, S., On a Theorem of Ville, in *Contributions to the Theory of Games*, Princeton: University Press, 1953, pp. 155–160.
211. Born, P., Tijs, S., and van der Aarssen, J., Pareto–Equilibrium in Multi-Objective Games, *Methods of Operations Research*, 1988, vol. 60, pp. 302–312.
212. Borwein, J., Proper Efficient Points for Maximization with Respect to Cones, *SIAM J. Control and Optimiz.*, 1977, vol. 15, no. 1, pp. 57–63.
213. Case, J.H. and Kimeldorf, G., On Nash Equilibrium Points and Games of Imperfect Information, *J. Optimiz. Theory and Appl.*, 1972, vol. 9, no. 5, pp. 302–323.
214. Chan, W.L. and Lau, W.T., Vector Saddle-Point and Distributed Parameter Differential Games, *Comput. and Math. Appl.*, 1989, vol. 18, nos. 1–3, pp. 195–207.
215. Chang, S.S., Lee, G.M., and Lee, B.S., Minimax Inequalities for Vector-Valued Mappings on W-spaces, *J. Math. Anal. and Appl.*, 1996, vol. 198, no. 2, pp. 371–380.
216. Chang, S.S., Yuan, G.X., and Lee, G.M., Saddle-Points and Minimax Theorems for Vector-Valued Multifunctions of H-spaces, *Appl. Math. Lett.*, 1998, vol. 11, no. 3, pp. 101–107.
217. Chen, G.Y. and Li, S.J., Existence of Solutions for a Generalized Vector-Valued Quasivariational Inequality, *J. Optimiz. Theory and Appl.*, 1996, vol. 90, no. 2, pp. 321–334.
218. Chikrii, A.A., Chikrii, G.T., and Zhukovskiy, V.I., Game Problems of Control for Functional-Differential Systems, in *Recent Advances in Information Technology*, Wojcik, W. and Sikora, J., Eds., CRC Press, 2017, pp. 13–49.
219. Chernoff, H., Rational Selection of Decision Function, *Econometrica*, 1954, vol. 22, pp. 422–443.
220. Cochrane, J. and Zeleny, M., *Multiple Criteria Decision Making*, Columbia: University of South Carolina Press, 1973.
221. Cohon, J.L., *Multiobjective Programming and Planning*, New York: Academic Press, 1978.
222. Colman, A.M., Körner, T.W., Musy, O., and Tazdaït, T., Mutual Support in Games: Some Properties of Berge Equilibria, *Journal of Mathematical Psychology*, 2011, vol. 55, no. 2, pp. 166–175.
223. Colman, A.M., Körner, T.W., Musy, O., and Tazdait, T., Mutual Support in Games: Some Properties of Berge Equilibria, *J. Mathematical Psychology*, 2011, vol. 55, pp. 166–175.
224. Corly, H.N., Games with Vector Payoffs, *J. Optimiz. Theory and Appl.*, 1985, vol. 47, pp. 491–498.
225. Cournot, A., *Principes de la theorie des richeses*, Paris, 1863.
226. Cournot, A., *Recherches sur les principes mathématiques de la théorie de richesses*, Paris, 1838.

227. Gaskó, N., Dumitrescu, D., and Lung, R.I., Evolutionary Detection of Berge and Nash Equilibria, in *Nature Inspired Cooperative Strategies for Optimization (NICSO 2011)*, Pelta, D.A. et al. (Eds.), SCI vol. 387, pp. 149–158.
228. Deghdak, M. and Florenzano, M., On the Existence of Berge's Strong Equilibrium, *International Game Theory Review*, 2011, vol. 13, no. 3, pp. 325–340.
229. Deghdak, M., On the Existence of Berge Equilibrium with Pseudocontinuous Payoffs, *ROMAII Journal*, 2014, vol. 10, no. 1, pp. 25–37.
230. Dinh, T.L. and Vargas, C., A Saddle-Point Theorem for Set-Valued Maps, *Nonlinear Anal.-Theor.*, 1992, vol. 18, no. 1, pp. 1–7.
231. Dolezal, J., Some Properties of Nonzero-Sum Multistage Games, in *Lect. Notes Comput. Sci.*, 1975, vol. 27, pp. 451–459.
232. Drâgustin, C., Min-max pour des criteres multiples. Recherche operationelle, *Operations Research*, 1979, vol. 12, no. 2, pp. 169–180.
233. Dumitrescu, D., Lung, R.I., and Gaskó, N., Detecting Strong Berge Pareto Equilibrium in a Non-Cooperative Game Using an Evolutionary Approach, *6th IEEE International Symposium on Applied Computational Intelligence and Informatics*, Timisoara, Romania, 2011.
234. Ferro, F., Minimax Theorem for Vector-Valued Functions, *J. Optimiz. Theory and Appl.*, 1989, vol. 60, pp. 19–31.
235. Ferro, F., Minimax Theorem for Vector-Valued Functions. Part 2, *J. Optimiz. Theory and Appl.*, 1991, vol. 68, no. 1, pp. 35–48.
236. Ferro, F., Minimax Type Theorems for n-Valued Functions, *Annali di Matematica Pura ed Appl.*, 1982, vol. 32, pp. 113–130.
237. Ferro, F., Minimax Theorem for Vector-Valued Functions. Part 2, *Dic. Mat. Univ. Genova*, 1989, vol. 101, pp. 1–17.
238. Geoffrion, A.M., Proper Efficiency and the Theory of Vector Maximization, *J. Math. Anal. and Appl.*, 1968, vol. 22, no. 3, pp. 618–630.
239. Hotelling, H., Stability in Competition, *Economic Journal*, 1929, vol. 39, pp. 41–57.
240. Hurwicz, L., Optimality Criteria for Decision Making under Ignorance, *Cowles Commission Discussion Paper, Statistics*, 1951, no. 370.
241. Jentzsch, G., Some Thoughts on the Theory of Cooperative Games, Advances in Game Theory, *Ann. Math. Studies*, 1964, vol. 52, pp. 407–442.
242. Keskin, K. and Saglam, H.C., On the Existence of Berge Equilibrium: An Order Theoretic Approach, *International Game Theory Review*, 2015, vol. 17, no. 3, pp. 1550007. DOI:10.1142/S0219198915500073
243. Abalo, K.Y. and Kostreva, M.M., Fixed Points, Nash Games and Their Organizations, *Topological Methods in Nonlinear Analysis*, 1996, vol. 8, pp. 205–215.
244. Abalo, K.Y. and Kostreva, M.M., Equi-Well-Posed Games, *Journal of Optimization Theory and Applications*, 1996, vol. 89, no. 1, pp. 89–99.
245. Abalo, K.Y. and Kostreva, M.M., Some Existence Theorems of Nash and Berge Equilibria, *Appl. Math. Lett.*, 2004, vol. 17, pp. 569–573.
246. Abalo, K.Y. and Kostreva, M.M., Intersection Theorems and Their Applications to Berge Equilibria, *Applied Mathematics and Computation*, 2006, vol. 182, pp. 1840–1848.
247. Koopmans, T.C., Analysis of Production as an Efficient Combination of Activities, in *Activity Analysis of Production and Allocation*, New York: Wiley, 1951, pp. 33–97.
248. Larbani, M. and Nessah, R., A Note on the Existence of Berge and Berge–Nash Equilibria, *Mathematical Social Sciences*, 2008, vol. 55, pp. 258–271.
249. Leitmann, G., *Cooperative and Non-Cooperative Many Player Differential Games*, Vienn: Springer Verlag, 1974.
250. Lin, J.G., Maximal Vectors and Multi-Objective Optimization, *J. Optimiz. Theory and Appl.*, 1976, vol. 18, no. 1, pp. 41–68.
251. Li, Z.F. and Wang, S.Y., A Type of Minimax Inequality for Vector-Valued Mappings, *J. Math. Anal. and Appl.*, 1998, vol. 227, no. 1, pp. 68–80.

252. Lung, R.I., Gaskó, N., and Dumitrescu, D., Characterization and Detection of ϵ-Berge-Zhukovskii Equilibria, *PLoS ONE*, 2015, vol. 10, no. 7: e0131983. DOI:10.1371/journal.pone.0131983
253. Van Megen, F., Born, P., and Tijs, S., A Preference Concept for Multicriteria Game, *Mathematical Methods of OR*, 1999, vol. 49, no. 3, pp. 401–412.
254. Milnor, J., Games Against Nature, in *Decision Processes*, New York: Wiley, 1954, pp. 49–60.
255. Musy, O., Pottier, A., and Tazdait, T., A New Theorem to Find Berge Equilibria, *International Game Theory Review*, 2012, vol. 14, no. 1, p. 1250005. DOI:10.1142/S0219198912500053
256. Moore, R.E., *Interval Analysis*, New York: Prentice-Hall, 1966.
257. Nash, J.F., Non-Cooperative Games, *Ann. Math.*, 1951, vol. 54, pp. 286–295.
258. Nash, J.F., Equilibrium Points in N-Person Games, *Proc. Nat. Academ. Sci. USA*, 1950, vol. 36, pp. 48–49.
259. Nessah, R., Larbani, M., and Tazdait, T., A Note on Berge Equilibrium, *Applied Mathematics Letters*, 2007, vol. 20, no. 8, pp. 926–932.
260. Nessah, R., Larbani, M., and Tazdait, T., Coalitional ZP-Equilibrium in Games and Its Existence, *International Game Theory Review*, 2015, vol. 17, no. 3, p. 1550002.
261. Von Neumann, J., Zur Theorie der Gesellschaftspiele, *Math. Ann.*, 1928, vol. 100, pp. 295–320.
262. Von Neumann, J. and Morgenstern, O., *Theory of Games and Economic Behavior*, Princeton Univ. Press, 1944.
263. Pareto, V., *Manuel d'économie politique*, Paris: Geard, 1909.
264. Petrosian, L.A., *Differential Games of Pursuit*, London, Singapore: World Scientific, 1993.
265. Pottier, A. and Nessah, R., Berge-Vaisman and Nash Equilibria: Transformation of Games, *International Game Theory Review*, 2014, vol. 16, no. 4, p. 1450009.
266. Radjef, M., Sur l'existence d'un équilibre de Berge pour un jeu différential n-personnes (On the Existence of a Berge Equilibrium for an n-Person Differential Game), *Cahiers Mathématiques de l'Université d'Oran*, 1998, no. 1, pp. 89–93.
267. Savage, L.Y., *The Foundation of Statistics*, New York: Wiley, 1954.
268. Savage, L.Y., The Theory of Statistical Decision, *J. American Statistic Association*, 1951, no. 46, pp. 55–67.
269. Shubik, M., Review of C. Berge "General theory of n-person games," *Econometrica*, 1961, vol. 29, no. 4, p. 821.
270. Steuer, R., *Multiple Criteria Optimization: Theory, Computation and Application*, New York: John Wiley and Sons, 1986.
271. Smith, D.E. and Ginsburg, Y., *A History of Mathematics in America before 1900 (Three Centuries of Science in America)*, Ayer Co Pub., 1980.
272. Steinhaus, H., A Definition for a Theory of Games and Pursuit, *Nav. Res. Log. Quart.*, 1960, vol. 7, pp. 105–107.
273. Tanaka, T., Existence Theorems for Cone Saddle-Points of Vector-Valued Functions in Infinite- -Dimensional Space, *J. Optimiz. Theory and Appl.*, 1989, vol. 62, no. 1, pp. 127–138.
274. Tanaka, T., Generalized Quasiconvexities Cone Saddle-Points and Minimax Theorem for Vector- -Valued Functions, *J. Optimiz. Theory and Appl.*, 1994, vol. 81, no. 2, pp. 355–377.
275. Tanaka, T., Generalized Semicontinuity and Existence Theorems for Cone Saddle- -Points, *Appl. Math. Opt.*, 1997, vol. 36, no. 3, pp. 313–322.
276. Tanaka, T., Some Minimax Problems of Vector-Valued Function, *J. Optimiz. Theory and Appl.*, 1988, vol. 9, no. 3, pp. 505–524.
277. Tanaka, T., Two Types of Minimax Theorems for Vector-Valued Functions, *J. Optimiz. Theory and Appl.*, 1991, vol. 68, no. 2, pp. 321–334.
278. Tirol, J., *The Theory in Industrial Organization*, Cambridge: MIT Press, 1988.
279. Vaisbord, E.M. and Zhukovskiy, V.I., *Introduction to Multi Player Differential Games and Their Applications*, New York: Gordon and Breach, 1988.
280. Vaisman, K.S., The Berge Equilibrium for Linear–Quadratic Differential Game, *The 3-rd Intern. Workshop on Multiple Criteria Problems under Uncertainty*, Orekhovo-Zuevo, Russia, 1994, p. 96.

281. Vaisman, K.S. and Zhukovskiy, V.I., The Berge Equilibrium under Uncertainty, *The 3-rd Intern. Workshop on Multiple Criteria Problems under Uncertainty*, Orekhovo-Zuevo, Russia, 1994, pp. 97–98.
282. Wald, A., Contribution to the Theory of Statistical Estimation and Testing Hypothesis, *Annuals Math. Statist.*, 1939, vol. 10, pp. 299–326.
283. Wald, A., *Statistical Decision Functions*, New York: Wiley, 1950.
284. Wang, S., An Existence Theorem of a Pareto Equilibrium, *Applied Mathematical Letters*, 1991, vol. 3, pp. 61–63.
285. Won Kyu Kim, On a Generalized Berge Strong Equilibrium, *Commun. Korean Math. Soc.*, 2014, vol. 29, no. 2, pp. 367–377.
286. Zeleny, M., Games with Multiple Payoffs, *International Journal of Game Theory*, 1976, vol. 4, pp. 179–191.
287. Zhao, J., The Equilibrium of a Multiple Objective Game, *International Journal of Game Theory*, 1991, vol. 20, pp. 171–182.
288. Zermelo, E., Über eine Anwendung der Mengenlehre auf die Theorie des Schachspiels, *Proceed. of the Fifth Intern. Congress of Mathematicians (Cambridge, 1912)*, Cambridge University Press, 1913, pp. 501–504.
289. Zhukovskiy, V.I., *Lyapunov Functions in Differential Games*, London and New York: Taylor and Francis, 2003.
290. Zhukovskiy, V.I., Some Problems of Non-Antagonistic Differential Games, in *Mathematical Methods in Operations Research*, Institute of Mathematics with Union of Bulgarian Mathematicians, Rousse, 1985, pp. 103–195.
291. Zhukovskii, V.I. and Gorbatov, A.S., Properties of the Savage Risk Function, *Int. Game Theory Review*, in press.
292. Zhukovskii, V.I., Kudryavtsev, K.N., and Gorbatov, A.S., Berge Equilibrium in Cournot's Model of Oligopoly, *Vestn. Udmurtsk. Univ. Mat. Mekh. Komp. Nauki*, 2015, vol. 25, no. 2, pp. 147–156.
293. Zhukovskiy, V.I., Molostvov, V.S., and Zhukovskaya, L.V., On Existence of Guaranteed Risk, *Intern. Conf. on Appl. Math*, Glushkov Inst. Cybern., Kyiv, 2002, p. 36.
294. Zhukovskiy, V.I., Molostvov, V.S., and Vaisman, K.S., Non-Cooperative Games under Uncertainty, in *Game Theory and Appl.*, 1997, vol. III, pp. 189–222.
295. Zhukovskiy, V.I. and Salukvadze, M.E., *The Vector-Valued Maximin*, New York: Academic Press, 1994.
296. Zhukovskiy, V.I. and Salukvadze, M.E., Sufficient Conditions in Vector-Valued Maximin Problems, *J. Optimiz. Theory and Appl.*, 1996, vol. 90, no. 3, pp. 523–534.
297. Zhukovskiy, V.I., Sachkov, S.N., and Gorbatov, A.S., Mathematical Model of the Golden Rule, *SCIENCE, TECHNOLOGY AND LIFE - 2014: Proceedings of the International Scientific Conference*, Czech Republic, Karlovy Vary, December 27–28, 2014, pp. 17–23.
298. Zhukovskiy, V., Topchishvili, A., and Sachkov, S., Application of Probability Measures to the Existence Problem of Berge–Vaisman Guaranteed Equilibrium, *Model Assisted Statistics and Applications*, 2014, vol. 9, no. 3, pp. 223–239.
299. Zhukovskiy, V.I. and Sachkov, S.N., Bilanciamento conflitti Friendly, *Italian Science Review*, 2014, vol. 9(18), pp. 169–179.
300. Zhukovskiy, V.I., Sachkov, S.N., and Smirnova, L.V., Existence of Berge Equilibrium in Mixed Strategies, *Uchenye Zapiski Tavrich. Natsional. Univ.*, 2014, pp. 261–279.
301. Zhukovskiy, V.I., Sachkov, S.N., and Smirnova, L.V., Berge Equilibrium, *Analiz, Modelirovanie, Upravlenie, Razvitie Sotsial'no-Ekonomicheskikh Sistem: Tr. VIII Mezhd. Shkoly-Simpoziuma (AMUR-2014)* (Analysis, Modeling, Management, Development of Socio-economic Systems: Proc. VIII Int. School-Symposium (AMMD-2014)), Krymskii Fed. Univ., Simferopol, 2014, pp. 124–133.
302. Zhukovskiy, V.I., Salukvadze, M.E., and Vaisman, K.S., *Berge Equilibrium*, Preprint. Tbilisi: Institute of Control Systems, 1994.

303. Zhukovskiy, V.I. and Topchishvili, A.T., Mathematical Model of the Golden Rule in Form of Differential Positional Game of Many Persons, *Int. J. Operations and Quantitative Management*, 2016, vol. 22, no. 3, pp. 203–229.
304. Zhukovskiy, V.I. and Topchishvili, A.T., The Explicit Form of Berge Strongly-Guaranteed Equilibrium in Linear-Quadratic Non-Cooperative Game, *International Journal of Operations and Quantitative Management*, 2015, vol. 21, no. 4, pp. 265–273.
305. Zhukovskiy, V.I. and Larbani, M., Berge Equilibrium in Normal Form Static Games: A Literature Review, *Izv. Inst. Matem. Inform. Udmurt. Gos. Univ.*, 2017, vol. 49, no. 1 (29), pp. 80–110.
306. Zhukovskiy, V.I. and Larbani, M., Alliance in Three Person Games, *Dostizhen. Matem. Mekhan.*, 2017, vol. 22, no. 1 (29), pp. 105–119.

CPSIA information can be obtained
at www.ICGtesting.com
Printed in the USA
LVHW082143230220
647967LV00001B/7

9 783030 255459